S0-ATI-120

Topics in
Current Physics

35

Topics in Current Physics Founded by Helmut K. V. Lotsch

Volume 39 **Nonequilibrium Vibrational Kinetics**
Editor: M. Capitelli

Volume 40 **Microscopic Methods in Metals**
Editor: U. Gonser

Volume 41 **Structure and Dynamics of Surfaces I**
Editors: W. Schommers and P. v. Blanckenhagen

Volume 42 **Metallic Magnetism I**
Editor: H. Capellmann

Volumes 1–38 are listed on the back inside cover

Multiple-Photon Excitation and Dissociation of Polyatomic Molecules

Edited by C. D. Cantrell

With 139 Figures

Springer-Verlag Berlin Heidelberg New York
London Paris Tokyo

Professor Dr. Cyrus D. Cantrell

Center for Applied Optics, University of Texas at Dallas, P.O. Box 830688,
Richardson, TX 75083-0688, USA

ISBN 3-540-13492-1 Springer-Verlag Berlin Heidelberg New York
ISBN 0-387-13492-1 Springer-Verlag New York Berlin Heidelberg

Library of Congress Cataloging-in-Publication Data. Main entry under title: Multiple-photon excitation and dissociation of polyatomic molecules. (Topics in current physics; 35.) Includes index. 1. Multiphoton processes. 2. Molecules. 3. Laser beams. I. Cantrell, C. D. (Cyrus D.), 1940– . QC793.5.P428M85 1986 539.7'217 85-17275

Offset printing and bookbinding: Konrad Triltsch, Graphischer Betrieb, Würzburg.
2153/3150-543210

Preface

In the early 1970s, researchers in Canada, the Soviet Union and the United States discovered that powerful infrared laser pulses are capable of dissociating molecules such as SiF_4 and SF_6. This result, which was so unexpected that for some time the phenomenon of multiple-photon dissociation was not recognized in many circumstances in which we now know that it occurs, was first publicized at a time when the possibility of using lasers for the separation of isotopes had attracted much attention in the scientific community. From the mid-1970s to the early 1980s, hundreds of experimental papers were published describing the multiple-photon absorption of CO_2 laser pulses in nearly every simple molecule with an absorption band in the 9 - 11 μm region. Despite this impressive volume of experimental results, and despite the efforts of numerous theorists, there is no agreement among researchers in the field on many fundamental aspects of the absorption of infrared laser light by polyatomic molecules.

This book is devoted to reviews of the experimental and theoretical research that provides the foundations for our current understanding of molecular multiple-photon excitation, and to reviews of research that is pertinent to the laser separation of isotopes. Although no book could provide complete coverage of the field of molecular multiple-photon excitation and its applications, the reviews published here provide much of the background required for researchers and graduate students in the field to appreciate the current literature. This volume includes a review of experimental measurements of multiple-photon absorption performed with pulsed infrared lasers, with an emphasis on the universality of the general features of the dependence of energy absorption and reaction probability on laser fluence for different molecular species. Other chapters review molecular-beam experiments on multiple-photon excitation, the two-frequency method for obtaining high isotopic selectivity, the observation of sharp multiphoton resonances in the SF_6 molecule, rate-equation theories of molecular multiple-photon excitation, the enhancement of isotopic selectivity through control of the laser-pulse shape, and techniques for the calculation of laser-pulse propagation effects.

I thank all of the authors for their contributions and for their patience during the preparation of this volume.

The support of the United States National Science Foundation under Grants CHE-8017324 and CHE-8215245, the United States Office of Naval Research under Grant N00014-82-K-0628, and the Robert A. Welch Foundation under Grant AT-873, are gratefully acknowledged.

Dallas, April 1986 *C.D. Cantrell*

Contents

1. **Molecular Multiphoton Excitation.** By C.D. Cantrell (With 1 Figure) 1
   1.1 Introductory Comments 1
   References 7

2. **Single-Infrared-Frequency Studies of Multiple-Photon Excitation and Dissociation of Polyatomic Molecules**
   By J.L. Lyman, G.P. Quigley, and O.P. Judd (With 34 Figures) 9
   2.1 Overview 9
   2.2 Data-Reduction Techniques 11
      2.2.1 Absorption of Nonuniform Optical Beams by Nonlinear Media 11
      2.2.2 Data-Analysis Techniques for Reaction Studies 16
   2.3 A Quantitative Comparison of Absorption Data 19
      2.3.1 Interaction Dynamics and Normalization Concepts 20
         a) General Considerations for MPE Processes 20
         b) Molecular Absorption in a Coupled Two-Level Approximation 22
         c) Generalization to the MPE Processes 24
      2.3.2 MPA in Different Polyatomic Molecules 25
         a) Absorption Characteristics of Polyatomic Molecules 26
         b) Calculation of $\langle f \rangle$ for SF_6 29
         c) Discussion 30
   2.4 A Review of MPA Data for SF_6 32
      2.4.1 Low-Fluence Spectral Absorption 33
      2.4.2 Fluence-Dependent Absorption at 300 K 35
      2.4.3 Fluence-Dependent Absorption at 140 K 40
      2.4.4 Effect of Optical Pulse Duration and Shape in MPA Processes 41
      2.4.5 Discussion and Summary 43
   2.5 A Comparison of Multiple-Photon Dissociation Data 44
      2.5.1 Experimental Techniques 45
      2.5.2 Molecular Properties 48
      2.5.3 Comparison of MPD Results 51
      2.5.4 Concluding Remarks 61
   2.6 Collisional Effects in Multiple-Photon Absorption and Dissociation Processes 61

2.6.1 Collisional Effects on Multiple-Photon Absorption 63
2.6.2 The Influence of Collisions on Multiple-Photon Dissociation 68
2.6.3 Collisional Effects in Laser-Induced Isomerization Reactions and Molecular Elimination 69
2.6.4 Secondary Processes 70
2.7 Conclusion 71
2.8 Appendix 72
References 87

3. **Molecular-Beam Studies of Laser-Induced Multiphoton Dissociation**
By A.S. Sudbø, P.A. Schulz, Y.R. Shen, and Y.T. Lee
(With 18 Figures) 95
3.1 Overview 95
3.2 Theory 97
3.3 Experimental Arrangement 106
3.4 Experimental Results 108
3.5 Discussion 113
3.6 Concluding Remarks 120
References 121

4. **Two-Frequency Technique for Multiple-Photon Dissociation and Laser Isotope Separation.** By R.V. Ambartzumian (With 20 Figures) 123
4.1 Background 123
4.2 Basic Concepts of Two-IR-Frequency Dissociation 125
4.3 Selectivity of Dissociation 126
4.3.1 Spectral Measurements 126
4.3.2 Isotope Enrichment Experiments 129
4.4 Investigation of Multiple-Photon Excitation 133
4.4.1 Evaluation of q and ℓ in OsO_4 134
4.5 Interaction of the Nonresonant Pulse with Excited Molecules 138
4.5.1 Absorption Measurements 138
4.5.2 Effects of Variation of ω_2 141
4.6 Concluding Remarks 146
References 146

5. **Excitation Spectrum of SF_6 Irradiated by an Intense IR Laser Field**
By S.S. Alimpiev, N.V. Karlov, E.M. Khokhlov, S.M. Nikiforov, A.M. Prokhorov, B.G. Sartakov, and A.L. Shtarkov (With 7 Figures) 149
5.1 Background 149
5.2 The Experiment 150

5.3 Experimental Results and Discussion 153
5.4 Conclusion 157
References 157

6. **Laser-Induced Decomposition of Polyatomic Molecules: A Comparison of Theory with Experiment**
By M.F. Goodman, J. Stone, and E. Thiele (With 23 Figures) 159
6.1 Brief Historical Review 159
6.2 Models of a Complex Polyatomic Molecule 160
6.2.1 Laser Excitation Mechanisms 160
6.2.2 The Heat Bath Feedback Model 163
6.2.3 Some Microscopic Aspects of the Theory 167
6.3 Comparisons of Theory with Experiment 169
6.3.1 Laser-Induced Decomposition of CF_2HCl 169
a) Coherent Discrete Level Pumping and Interface with Rate Equation 170
b) Rotational Hole Filling by Collisions 172
c) A Comparison of Theory with Experiment for CF_2HCl 173
d) Additional Theoretical Predictions 174
6.3.2 Laser Intensity Versus Fluence Effects 175
6.3.3 High-Pressure Fallof of Reaction Rate 177
6.3.4 Laser-Induced Decomposition of SF_6 179
a) Discrete Level Spectrum and Tuning Curve 179
b) Dissociation Yield and Cross Section 182
6.3.5 Laser Frequency Effects 183
6.4 Effects Specific to Laser Excitation 185
6.4.1 Energy Distribution: Laser Versus Thermal 186
6.4.2 Microstate Formalism 190
a) Intramolecular Relaxation 190
b) Relation to RRK and RRKM Theory 191
c) A Discriminating Reaction Mechanism 193
d) Multiple Reaction Channels and Product Selectivity 196
6.5 Phase Coherence and the Transition to Incoherent Pumping 198
6.5.1 The Generalized Master Equation 200
6.5.2 Fermi Golden Rule Considerations 205
6.5.3 Coherent Quasi-Continuum Pumping - an Open Question 208
6.6 Addendum 209
References 211

7. A Method of Laser Isotope Separation Using Adiabatic Inversion
By G.L. Peterson and C.D. Cantrell (With 6 Figures) 215
7.1 Background 215
7.2 Physical Principles 216
7.3 Multilevel Systems Under the Adiabatic Approximation 218
7.4 Conditions for Effective Adiabatic Laser Isotope Separation 219
References 221

8. Three-Level Superfluorescence. By F.P. Mattar, P.R. Berman, A.W. Matos, Y. Claude, C. Goutier, and C.M. Bowden (With 30 Figures) 223
8.1 Background 224
8.2 Pump Dynamics Effects in Three-Level Superfluorescence 231
8.3 Semiclassical Equations of Motion and Computational Method 233
8.4 Deterministic Effects of Pump Dynamics in the Nonlinear Regime of Superfluorescence 240
8.5 Conclusions Concerning Deterministic Three-Level Superfluorescence .. 262
8.6 Quantum Initiation: Calculational Results and Delay-Time Statistics . 262
8.7 Conclusions Concerning Effects of Resonance Diffraction and Copropagation on Superfluorescence Evolution 266
8.8 The Role of Dispersion 268
8.9 Conclusions Concerning the Mutual Influence of Dispersion and Diffraction on the Superfluorescence Buildup 276
8.10 Final Conclusion 276
References 277
Additional References 281

Subject Index 285

List of Contributors

Alimpiev, Sergei S.

Institute of General Physics, Academy of Sciences of the USSR
Vavilov Street 38, SU-117924 Moscow, USSR

Ambartzumian, Rafael V.

Institute of Spectroscopy, Academy of Sciences of USSR
SU-142092 Moscow Academgorodok, USSR

Berman, Paul R.

Department of Physics, New York University, New York, NY 10003, USA

Bowden, Charles M.

Research Directorate, U.S. Army Missile Laboratory, U.S. Army Missile Command
Redstone Arsenal, AL 35838, USA

Cantrell, Cyrus D.

Center for Applied Optics, University of Texas at Dallas, P.O. Box 830688
Richardson, TX 75083-0688, USA

Claude, Yves

Département d'Informatique, Université de Montreal, P.O. Box 6128
Montreal, P.Q., Canada

Goodman, Myron F.

University of S. California, Dept. of Biological Sciences
Los Angeles, CA 90089-1481, USA

Goutier, Claude

Centre de Calcul, Université de Montreal, P.O. Box 6128
Montreal, P.Q., Canada

Judd, O'Dean P.

Los Alamos National Laboratory, Los Alamos, NM 87545, USA

Karlov, Nikolai V.

Institute of General Physics, Academy of Sciences of the USSR
Vavilov Street 38, SU-117924 Moscow, USSR

Khokhlov, Edward M.

Institute of General Physics, Academy of Sciences of the USSR
Vavilov Street 38, SU-117924 Moscow, USSR

Lee, Yuan T.

University of California, Department of Chemistry, Berkeley, CA 94720, USA

Lyman, John L.

Los Alamos National Laboratory, Los Alamos, NM 87545, USA

Matos, Alain W.

Department of Physics, New York University, New York, NY 10003, USA

Mattar, Farres P.

Department of Physics, New York University, New York, NY 11201, USA, and

George R. Harrison Spectroscopy Lab., Massachussetts Institute of Technology
Cambridge, MA 02139, USA

Nikiforov, S.M.

Institute of General Physics, Academy of Sciences of the USSR
Vavilov Street 38, SU-117924 Moscow, USSR

Peterson, Gary L.

TRW Space and Technology Group, One Space Park, Redondo Beach, CA 90278, USA

Prokhorov, Aleksander M.

Institute of General Physics, Academy of Sciences of the USSR
Vavilov Street 38, SU-117924 Moscow, USSR

Quigley, Gerard P.

Los Alamos National Laboratory, Los Alamos, NM 87545, USA

Sartakov, B.G.

Institute of General Physics, Academy of Sciences of the USSR
Vavilov Street 38, SU-117924 Moscow, USSR

Schulz, Peter A.

Department of Physics, Georgia Tech, Atlanta, GA 30332, USA

Shen, Yuen-Ron

University of California, Department of Physics, Berkeley, CA 94720, USA

Shtarkov, Alexey L.

Institute of General Physics, Academy of Sciences of the USSR
Vavilov Street 38, SU-117924 Moscow, USSR

Stone, James

University of S. California, Dept. of Biological Sciences
Los Angeles, CA 90089-1481, USA

Sudbø, Aasmond S.

Televerkets Forskningsinstitutt, P.O. Box 83, 2007 Kjeller, Norway

Thiele, Everett

University of S. California, Dept. of Biological Sciences
Los Angeles, CA 90089-1481, USA

1. Molecular Multiphoton Excitation

C. D. Cantrell

With 1 Figure

Molecular physics has experienced a renaissance during the past fifteen years, in part because of the availability of new laser tools that make commonplace hitherto inaccessible regimes of molecular dynamics, and also because of the widespread perception that these new regimes of molecular behavior may find economically important applications. The laser separation of isotopes using molecular working media has been, or is actively being, pursued in the United States, Canada, the Soviet Union, France, the Federal Republic of Germany, Italy, the United Kingdom, Japan, Israel, Brazil, Argentina, and the People's Republic of China, to mention only the most prominent nations in this area. Much of what has been learned in the course of research on laser isotope separation may prove to be applicable to the much more general problem of laser chemistry, provided that selective excitation of chemical bonds or moieties proves to be generally possible. The early exuberance of researchers in molecular multiphoton excitation, which was fueled as much by our own hopes as by heavy funding, has given way to a more cautious attitude in the face of difficulties in measuring, calculating, and above all understanding, the fundamental photophysical processes involved in the excitation of polyatomic molecules by intense infrared laser light. A mature scientific field is one that has undergone several such cycles. This field is in its adolescence, a difficult period in which the elevated thoughts of theory are not always in harmony, or even in contact, with the reality represented by the experience of the laboratory, and in which conflicting idealized concepts render a smooth development of the field unlikely. The chapters of this book thus orient themselves naturally along a few major themes: the results of experiments on molecular multiphoton excitation and dissociation; theoretical models; and problems for practical applications, of which propagation effects such as self-focusing and self-defocusing are among the most prominent.

1.1 Introductory Comments

The discovery of infrared molecular multiphoton excitation and dissociation occurred very shortly after the development of TEA CO_2 lasers in 1970 [1.1]. The experiments of *Isenor* and *Richardson* showed clearly that dissociation, and by inference excitation to highly energetic states, occurs in SiF_4 under irradiation with focused 10.6 μm laser light [1.2]. However, others were slow to accept either their conclusion that collisionless dissociation occurs under the conditions of their experiment or their argument that polyatomic molecules must possess a vibrational quasi continuum in which nearly resonant absorption of many laser photons is possible. The striking success of *Ambartzumian* et al. [1.3] and of *Lyman* et al. [1.4] in separating sulfur isotopes using a TEA CO_2 laser focused into a mixture of SF_6 and H_2, together with funding for laser separation of uranium isotopes motivated by optimistic projections of the need for nuclear reactor fuel enriched in ^{235}U, ensured the vigorous experimental exploration of this new field. Among the more striking results of this initial survey was the discovery that both the number of photons absorbed per molecule, $\langle n \rangle$, and the dissociation yield depend only weakly on the laser intensity in comparison with atomic multiphoton processes, where ionization rates proportional to I^N (N lying in the range from 11 to 15) have been reported. In fact, it is the time integral of the intensity (the fluence) that plays the major role in determining $\langle n \rangle$ and the dissociation yield. Intensity-dependent effects, while sometimes dramatic, are not as straightforwardly understood as the intensity-dependent effects that occur in atomic multiphoton processes [1.5,6]. In Chap.2, J.L. Lyman and co-workers show that the existing measurements of $\langle n \rangle$ in polyatomic molecules can be fit to a single universal curve as a function of the laser fluence, for irradiation using a single laser frequency.

The method of crossed molecular beams provides detailed information on the dynamics of chemical reactions through measurements of the energy and angular distributions of the product molecules. For example, the presence of a symmetric angular distribution in the center-of-mass frame of the reaction products indicates the formation of a complex, the lifetime of which exceeds its average rotational period [1.7]. The replacement of one of the molecular beams by a laser beam was an important step permitting the study of certain characteristics of multiphoton molecular excitation that cannot be studied by other means. The conventional scheme by which the steps in the process of multiphoton dissociation are currently classified [1.8], shown in Fig.1.1, divides the process of excitation into two major steps, region I and region II, according to the density of states. Excitation in region I is supposed to be coherent, requiring a dynamical description using the Schrödinger equation or coherent density-matrix equations of motion. Excitation in region II, however, is assumed to be incoherent, permitting a description using rate equations [1.9,10]. The process of dissociation occurs in region III, which includes those states of the molecule with energies above the dissociation threshold. Pre-laser studies of molecular dissociation, which were conducted under conditions of thermal equilibrium, were successfully interpreted using a model due to Rice, Ramsperger, Kassel, and Marcus (RRKM), one feature of which is the assumption that the energy of an activated complex about to undergo dissociation is statistically equiparti-

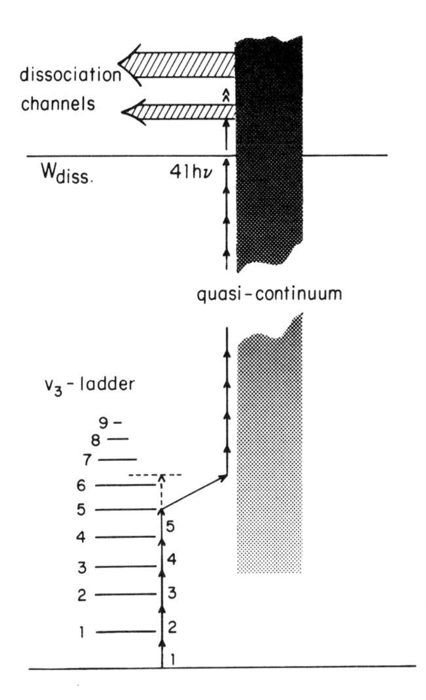

Fig.1.1. Conventional scheme of the excitation and dissociation of SF_6 [1.8]

tioned among all the available degrees of freedom of the complex [1.11]. Molecular-beam experiments have established that the process of infrared-laser-induced molecular photodissociation in SF_6 and other molecules is collisionless, and that the RRKM model provides a quantitative description of the energy and angular distribution of the dissociation fragments; what has not been established to the satisfaction of the community of researchers in this field is the applicability of models that assume the statistical equipartition of energy among all degrees of freedom to the dynamics of excitation in region II, as is suggested in Fig.1.1. Studies of molecular multiphoton excitation using a single-frequency laser beam crossed with a molecular beam are reviewed in Chap.3 by A.S. Sudbø et al.

Selective laser chemistry and laser isotope separation involve three conceptually distinct steps: selective excitation of one species, unimolecular or biomolecular reaction of that species on a time scale and in a manner that preserve the selectivity established in the first step, and collection of the product. It is not required, and may well be impossible, to make a single laser serve in both of the first two steps. Early in the history of molecular isotope separation it was recognized that it would be possible to take the first step with a relatively weak infrared laser the frequency of which is optimal for selective vibrational excitation, and the second step with a more powerful, ultraviolet laser with a less sharply constrained frequency [1.12,13]. Shortly after the successful demonstration of the isotopically selective single-frequency dissociation of SF_6, it was shown experimentally that the selectivity of dissociation can be enhanced by the use of two infrared lasers, one of which provides the selectivity and the other, most of the energy required for dissociation [1.14,15]. Besides its obvious practical advantage in deriving the lion's share of the energy needed for unimolecular reaction from a laser the frequency of which is not required to coincide with a specific, narrow, molecular absorption feature, the two-frequency technique is invaluable for fundamental studies of molecular multiphoton excitation since the intensity and frequency of the first, selective laser and those of the second, dissociating laser may be varied independently. The two-frequency method for infrared multiphoton excitation and laser isotope separation is reviewed in Chap.4 by R.V. Ambartzumian.

The TEA CO_2 lasers used in the first several years of research on infrared multiphoton excitation generally had several characteristics that were undesirable from the point of view of obtaining interpretable results on the fundamental nature of the photophysical processes involved. One of the most important limitations was the impossibility of obtaining multiphoton absorption spectra corresponding to a continuous variation of the laser frequency. Absorption and dissociation data were available only at the discrete CO_2 laser frequencies, which are spaced at intervals of approximately 1.8 cm^{-1} in the 10.6 m laser band. The rather broad and apparently featureless contours of the multiphoton absorption bands observed in this way, especially in the model molecule SF_6, did nothing to discourage speculation that rapid collisionless intramolecular vibrational relaxation occurs at relatively low vibrational levels. In this context, the experiments of Alimpiev and co-workers using a variant of the two-infrared-frequency technique in which the first laser was a high-pressure, continuously tunable CO_2 laser provided a surprising result: the multiphoton absorption spectrum of SF_6, taken as a function of the frequency of the first laser, consists of a series of resonances far too sharp to be resolved by line-selectable lasers [1.16]. At least one of these resonances was definitely assigned as a two-photon resonance to a sublevel of the $v_3 = 2$ vibrational state, as discussed below in Chap.5 by S.S. Alimpiev et al. in the course of their review of the excitation of SF_6 by a continuously tunable and a line-selectable laser. The assignment of the multiphoton resonances observed in SF_6 is also discussed in [1.17, 18]. Subsequent work by *Giardini* and co-workers [1.19] has shown that the existence of closely spaced sharp resonances in multiphoton absorption is a general characteristic of polyatomic molecules, and that many of the resonances observed can be assigned as corresponding to multiphoton transitions to specific excited vibrational states. The work reviewed in Chap.5 and its extension to other molecules have interesting implications for future efforts in laser chemistry and isotope separation. First, at least some of the resonances observed in SF_6 have an extremely narrow width owing to the coalescence of many resonances to different final rotational states [1.18]. Especially for molecules containing heavy elements, it may be possible to employ a multiphoton transition to excite a significant fraction of the population of the selected species to the second vibrational state or higher, and subsequently (with another laser) to dissociation, with higher selectivity than is possible when the first step is a single-photon excitation. Second, it is a corollary that the optimum frequency for the first, selective laser in isotope separation or chemistry may in fact correspond to a multiphoton transition. If this is true, then the linear absorption spectrum of the molecule will not directly reveal the desired frequency, although multiphoton spectroscopy may do so. Further work is needed to explore the consequences of these ideas.

Theories of infrared molecular multiphoton excitation are abundant. However, few of the published formal approaches have been carried sufficiently far to permit, or to warrant, a detailed comparison with experiment. Theories of the excitation of matter by light may very generally be classified as coherent or incoherent, according to whether they do or do not take into account the possibility of a large non-zero expectation value of a physically observable quantity such as the transition dipole operator that couples different energy levels of the system in question. Approaches that make use of the time-dependent Schrödinger equation or the density-matrix equation of motion are therefore coherent, while approaches that use rate equations are incoherent. While coherence is clearly important in region I (Fig.1.1), its importance for a numerically accurate description of excitation in region II is not known, owing to the great difficulty of first-principles computation with the large number of energy levels or quasi levels required for a qualitative representation of region II for a molecule of experimental interest. The most mature of the current rate-equation theories of molecular multiphoton excitation, and the comparison of this theory with experiment, are reviewed in Chap.6 by M.F. Goodman and co-workers, who also give a brief discussion of the possible role of coherence in excitation in region II.

The adiabatic inversion of a spin-1/2 system subjected to both a dc and a radio-frequency magnetic field by sweeping the dc field through its resonant value has long been known in the field of nuclear magnetic resonance. The name "adiabatic rapid passage," which is often applied to this method, derives from the fact that the dc field must be varied on a time scale that is long compared to the reciprocal of the frequency splitting of the levels due to the combined effects of the two fields, but short compared to either the T_1 or T_2 relaxation times of the system [1.20,21]. The possibility of adiabatic inversion in a system of two levels coupled by an optical transition was pointed out by *Treacy* [1.22], who proposed varying the frequency of a laser pulse through the resonant transition frequency. This effect was demonstrated experimentally shortly thereafter [1.23]. Adiabatic inversion on two-photon transitions through varying the amplitude, not the frequency, of a laser pulse was predicted by *Grischkowsky* and his co-workers [1.24], and was observed by *Loy* [1.25]. The possibility of accomplishing adiabatic inversion in more general multilevel systems has been pointed out in [1.26,27]. Given the potential importance of multiphoton transitions in the selective excitation of polyatomic molecules, it is natural to ask whether adiabatic variations of the laser pulse amplitude may be used to enhance the selectivity of excitation. An initial response to this question is essayed by G.L. Peterson and C.D. Cantrell in Chap.7, where the application of adiabatic excitation to isotope separation is illustrated, and the limits of the adiabatic approximation are evaluated quantitatively, for a model multilevel system that schematically represents some of the main features of a ladder of vibration-rotation states in a polyatomic molecule.

Propagation effects, although little considered by the community of researchers in the field of molecular multiphoton excitation and dissociation, may play an important role in the correct interpretation of laboratory experiments on multiphoton absorption and will almost certainly play a major role in practical implementations of laser chemistry and isotope separation. Some of what is currently accepted as experimental fact in the field of molecular multiphoton excitation may be of questionable validity in view of the experimental demonstration of strong self-focusing in SF_6 vapor under the same experimental conditions in which many multiphoton experiments have been performed in absorption cells [1.28]. The unexpected occurrence of self-focusing or self-defocusing in an experiment that is intended to measure the mean number of photons absorbed per molecule, $\langle n \rangle$, can cause a systematic error in the determination of $\langle n \rangle$ if the experiment is carried out in an absorption cell. Normally one determines $\langle n \rangle$ experimentally by measuring the total energy deposited (E_d) and dividing by the product of the energy per photon ($\hbar\omega$), the molecular number density (ρ) and the volume irradiated by the laser beam (V)

$$\langle n \rangle = E_d/(\rho\hbar\omega V) \tag{1.1}$$

In the absence of self-focusing or self-defocusing, V would not depend on the presence or absence of a molecular vapor in the absorption cell, and could therefore be calculated using Gaussian optics, as is usually done. However, in the presence of self-actions of the laser light, V depends significantly on ρ and on the laser power. The $\langle n \rangle$ calculated using the value of V for an empty cell can differ appreciably from the real value of $\langle n \rangle$; worse, the dependence of $\langle n \rangle$ upon the laser fluence or intensity that one deduces from an experiment in which V is not corrected for self-focusing or self-defocusing may be different from the real dependence of the actual $\langle n \rangle$ on the fluence or intensity. Self-actions of the laser light used for chemical or isotopic processes on an industrial scale may, under poorly chosen conditions, prevent the propagation of the laser beam through the working medium. The paucity of experimental data on self-actions of laser light under conditions of molecular multiphoton excitation leads one to turn to theory, where the difficulty of performing calculations on such complex propagation effects is notorious. The numerical techniques that have been evolved for attacking such problems for multilevel systems are illustrated in Chap.8 by F.P. Mattar et al.

References

1.1 A.J. Beaulieu: Appl. Phys. Lett. **16**, 504 (1970)
1.2 N.R. Isenor, M.C. Richardson: Appl. Phys. Lett. **18**, 224 (1971); Opt. Commun. **3**, 360 (1971); Proc. Tenth Intern. Gaseous Electronics Conference, ed. by D. Parsons (Oxford 1971)
1.3 R.V. Ambartzumian, Yu.A. Gorokhov, V.S. Letokhov, G.N. Makarov: Zh. Eksp. Teor. Fiz., Pis'ma Red. **21**, 375 (1975) [English transl.: JETP Lett. **21**, 171 (1975)]
1.4 J.L. Lyman, R.J. Jensen, J. Rink, C.P. Robinson, S.D. Rockwood: Appl. Phys. Lett. **27**, 87 (1975)
1.5 J.L. Lyman, B.J. Feldman, R.A. Fisher: Opt. Commun. **25**, 391 (1978)
1.6 D.S. King, J.C. Stephenson: Chem. Phys. Lett. **66**, 33 (1979)
1.7 W.B. Miller, S.A. Safron, D.R. Herschbach: Discuss. Faraday Soc. **44**, 108 (1967)
1.8 N. Bloembergen, C.D. Cantrell, D.M. Larsen: "Collisionless dissociation of polyatomic molecules by multiphoton infrared absorption", in *Tunable Lasers and Applications*, ed. by A. Mooradian, T. Jaeger, P. Stokseth, Springer Ser. in Opt. Sci., Vol.3 (Springer, Berlin, Heidelberg 1976) pp. 162-176
1.9 J.L. Lyman: J. Chem. Phys. **67**, 1868 (1977)
1.10 E.R. Grant, P.A. Schulz, Aa.S. Sudbø, M.J. Coggiola, Y.T. Lee, Y.R. Shen: "Multiphoton dissociation of polyatomic molecules studied with a molecular beam", in *Laser Spectroscopy III*, ed. by J.L. Hall, J.L. Carsten, Springer Ser. in Opt. Sci., Vol.7 (Springer, Berlin, Heidelberg 1977) pp. 94-101
1.11 P.J. Robinson, K.A. Holbrook: *Unimolecular Reactions* (Wiley, London 1972)
1.12 R.V. Ambartzumian, V.S. Letokhov: Appl. Opt. **11**, 354 (1972)
1.13 V.S. Letokhov: Chem. Phys. Lett. **15**, 221 (1972)
1.14 R.V. Ambartzumian, Yu.A. Gorokhov, V.S. Letokhov, G.N. Makarov, A.A. Puretzky, N.P. Furzikov: Zh. Eksp. Teor. Fiz., Pis'ma Red. **23**, 217 (1976) [English transl.: JETP Lett. **23**, 194 (1976)]
1.15 R.V. Ambartzumian, N.P. Furzikov, Yu.A. Gorokhov, V.S. Letokhov, G.N. Makarov, A.A. Puretzky: Opt. Commun. **18**, 517 (1976)
1.16 S.S. Alimpiev, N.V. Karlov, S.M. Nikiforov, A.M. Prokhorov, B.G. Sartakov, E.M. Khokhlov, A.L. Shtarkov: Opt. Commun. **31**, 309 (1979)
1.17 D.P. Hodgkinson, A.J. Taylor, D.W. Wright, A.G. Robiette: Chem. Phys. Lett. **90**, 230 (1982)
1.18 M. Dilonardo, M. Capitelli, C.D. Cantrell: Appl. Phys. (Berlin) B**29**, 181 (1982);
C.D. Cantrell, W.M. Lee: Chem. Phys. Lett. **93**, 267 (1982)
1.19 E. Borsella, R. Fantoni, A. Giardini-Guidoni, D.R. Adams, C.D. Cantrell: Chem. Phys. Lett. **101**, 86 (1983)
1.20 A. Abragam: *The Principles of Nuclear Magnetism* (Oxford U. Press, Oxford 1961) p. 66
1.21 C.P. Slichter: *Principles of Magnetic Resonance*, Springer Ser. Solid-State Sci., Vol.1 (Springer, Berlin, Heidelberg 1980)
1.22 E.B. Treacy: Phys. Lett. **27**A, 421 (1968)
1.23 E.B. Treacy, A.J. DeMaria: Phys. Lett. **29**A, 369 (1969)
1.24 D. Grischkowsky, M.M.T. Loy, P.F. Liao: Phys. Rev. A**12**, 2514 (1975)
1.25 M.M.T. Loy: Phys. Rev. Lett. **32**, 814 (1974); **41**, 473 (1978)
1.26 M.V. Kuz'min, V.N. Sazonov: Zh. Eksp. Teor. Fiz. **79**, 1759 (1980) [English transl.: Sov. Phys.-JETP **52**, 889 (1980)]; Zh. Eksp. Teor. Fiz. **83**, 40 (1982) [English transl.: Sov. Phys.-JETP **56**, 27 (1982)]
1.27 G.L. Peterson, C.D. Cantrell, R.S. Burkey: Opt. Commun. **43**, 123 (1982)
1.28 A. Nowak, D. Ham: Opt. Lett. **6**, 185 (1981);
P. Bernard, P. Galarneau, S.L. Chin: Opt. Lett. **6**, 139 (1981)

2. Single-Infrared-Frequency Studies of Multiple-Photon Excitation and Dissociation of Polyatomic Molecules

J. L. Lyman, G. P. Quigley, and O. P. Judd

With 34 Figures

In this chapter we review excitation and dissociation experiments performed with pulsed lasers at single infrared frequencies. The review treats five areas of this subject: data-reduction techniques, absorption experiments for a broad range of species, SF_6 absorption experiments, dissociation experiments for a broad range of species, and the effect of gas pressure on both absorption and dissociation. The data and procedures presented in this review allow one to view the absorption and dissociation phenomena resulting from multiple-photon excitation as general phenomena and not just as isolated results from experiments with individual species.

2.1 Overview

In this chapter we consider the multiple-photon excitation (MPE) and dissociation of polyatomic molecules with a single infrared laser frequency. Upon review of a large number of recent publications, we find that laser excitation and dissociation have been studied for a broad enough range of molecular species to make comparisons among these species meaningful. A major objective of this chapter is to make these comparisons in a semiquantitative manner and to identify the molecular and laser radiation properties that most strongly contribute to the absorption of infrared laser radiation by polyatomic molecules, and to the subsequent dissociation (or isomerization) of the molecules. We also seek to identify the role of collisions in both the absorption and reaction processes.

Much of the recent research in MPE is the result of efforts to develop techniques for laser isotope separation. Deuterium [2.1-6], tritium [2.7-9], carbon [2.3,10-13], sulfur [2.14-17], and uranium [2.18-23] have received the most attention. Some of these papers have demonstrated extremely high isotopic selectivities and high reaction efficiencies. Industrial scale processes should follow soon.

A major problem in making these comparisons among molecular species is the diversity in the way different authors have reduced and presented their experimental data. We discuss (Sect.2.2) problems associated with making quantitative absorption and reaction yield measurements, and we suggest techniques that we feel give accurate and usable representations of experimental results. Wherever possible, we use these techniques in the comparisons.

In Sect.2.3 we critically review the available multiple-photon absorption data and suggest an approach based on a two-level oscillator model for comparing experimental results. From this approach we find that, even in the MPE regime, the parameters that most strongly influence the absorption of infrared laser radiation by a polyatomic molecule are the small-signal absorption cross section and the fraction of molecules that interact with the laser at low intensity.

Because of the popularity of SF_6 in MPE experiments, we devote one section (Sect.2.4) to absorption measurements with SF_6. This section gives a critical analysis of the published absorption data, and suggests what we feel are the most reliable measurements of the fluence-dependent absorption cross section (or the number of laser photons absorbed per molecule).

In Sect.2.5 we review multiple-photon dissociation (MPD) experiments. This section contains a discussion of experimental techniques, a list of molecular properties that we feel are important in the laser-induced reaction process, and a table summarizing a large number of MPD experiments. In order to compare MPD experiments, we characterize the reaction for each experiment by the fluence necessary to react 1% of the molecules and the number of laser photons absorbed per molecule at that fluence. We show that the MPD process depends strongly on the density of vibrational states, the bond strength, the small-signal absorption cross section, and other spectroscopic properties. It appears that the amount of absorbed energy influences the reaction probability more strongly than the distribution of that energy within and among the molecules.

We found that sufficient pressure-dependent data existed to give a separate treatment of collisional effects (Sect.2.6). Collisions influence both the absorption process and the composite excitation-dissociation process. We tabulate experimental observations of the effect of collisions on both processes. We discuss the roles of several collisional-energy transfer processes, and we conclude that the relative importance of processes such as rotational-rotational energy transfer and vibrational-translational energy transfer depends to a large extent on the molecular structure.

Our major purpose in this chapter is to review and compare experimental data and not to present or defend any theoretical description of the multiple-photon excitation process. References to theoretical models are, of course, inevitable; we primarily seek, however, to determine what the experiments say (or do not say) about the MPE process. Other chapters in this volume treat the theory of the fundamental processes. A number of papers have augmented the research results treated in this chapter since its earlier versions. These include several reviews [2.24-32] of MPE. *King*'s [2.24] review is particularly good for the details of the absorption process and subsequent dissociation. He also summarized the recently developed methods for directly measuring the rate of dissociation of highly excited molecules (CF_2HCl and CF_2CFCl). This type of measurement [2.24,33-35] is a significant advance in the experimental investigation of MPE. These results have been useful in the development of MPE theory [2.33]. *Letokhov*'s book [2.25] is a general review of MPE. It also treats some related topics. *Danen* and *Jang* [2.26] have completed a review of infrared photochemistry or organic molecules. This review adds significantly to the data on larger molecules in Table 2.10. The two cited reviews [2.27, 28] on the theory of MPE give a current view of the understanding of the absorption and reaction processes.

2.2 Data-Reduction Techniques

2.2.1 Absorption of Nonuniform Optical Beams by Nonlinear Media

Molecular absorption measurements are usually carried out under a variety of different experimental conditions. A minimum set of experimental variables includes the spectral bandwidth, frequency, pulse length, and energy of the optical source; molecular number density; optical path length; optical beam profile; and focusing geometry. In order to compare data from different experiments and to provide physical insight into the absorption process, it is necessary to normalize the data with respect to some of these independent variables. For this purpose it is convenient to characterize the absorption process in terms of the number of photons absorbed per molecule $\eta(\Phi)$, the absorption cross section $\sigma(\Phi)$, and the fluence Φ. The determination of these parameters depends on the beam spatial profile, the amount of absorption along the optical path length, and the beam focusing geometry. Consistency is essential in these calculations.

Errors can result in the calculation of η by simply dividing the net energy absorbed over a given path length by the number of molecules contained in an arbitrary volume defined by the optical beam. Most laser beams possess a nonuniform radial profile. This makes the choice of an appropriate beam area difficult. The determination of η is further complicated if the optical depth is finite in the nonlinear absorber. For example, if the absorption cross section decreases with increasing fluence, the absorption in the wings of the beam may provide a significant, if not the major, contribution to the value of ΔE, the total absorbed energy. The effect is greatly magnified for focused beams where the energy absorbed per molecule is largest in the focal volume, while most of the absorption from the laser beam occurs outside the focal volume.

In a given absorption experiment, the measured parameters are the total input energy and the *average* energy absorbed along a given optical path length of the absorption cell. The problem is then to derive from these data, *local* values of $\eta(\Phi)$, $\sigma(\Phi)$, and Φ that are self-consistent by methods that are reasonably independent of the nature of the absorption process.

Two methods of data analysis will be discussed that provide local values for the absorption parameters in some degree of approximation. The first method assumes that the optical beam spatial profile is Gaussian and that the optical depth of the medium is small. The second method is formulated to apply to an arbitrary optical beam profile; however, the practical implementation of the method is best adopted to Gaussian beams. Focused and unfocused beams can be treated and the method accounts for the finite optical depth of the absorbing medium.

The first method relies on an exact decomposition equation for collimated Gaussian beams that recovers the true intensity dependence of the function [2.36]. Consider a physical parameter $f_r(I)$ that is a nonlinear function of the local optical intensity $I(r)$. Most physical measurements of $f_r(I)$ are averaged over the spatial profile of the beam so that the measured function is related to the real function by

$$f_m(I_0) = I_0 \frac{\int_0^\infty f_r[I(r)] 2\pi r \, dr}{\int_0^\infty I(r) 2\pi r \, dr} \tag{2.1}$$

where I_0 is the central intensity. For a nearly collimated Gaussian beam of radius w_0,

$$I(r) = I_0 \exp(-r^2/w_0^2) \quad . \tag{2.2}$$

By a simple change in variable, (2.1) can be rewritten in the form

$$f_m(I_0) = \int_0^{I_0} f_r(I) \frac{dI}{I} \quad . \tag{2.3}$$

Differentiating both sides of (2.3) with respect to I_0 and changing variables give the result

$$f_r(I) = I \frac{df_m(I)}{dI} = f_m(I) \frac{d \ln f_m(I)}{d \ln I} \quad . \tag{2.4}$$

The real value $f_r(I)$ can therefore be obtained by multiplying the measured function by its logarithmic derivative. As was pointed out in the original reference, this procedure only holds for additive quantities such as total energy and dissociation yield. Other quantities can be made additive, however, by suitable combinations of the variables. For the transmission function $T(I)$ and the absorption coefficient $\alpha(I)$ the additive quantities are $I\,T(I)$ and $I \exp[-\alpha(I)L]$, where L is the optical path length. In terms of the intensity-dependent absorption cross section $\sigma(\Phi)$ and the zero intensity value σ_0, (2.4) can be used to derive the result

$$\tilde{\sigma}_r(I) = \tilde{\sigma}_m(I) + \frac{1}{x} \ln\left(1 + xI \frac{d\tilde{\sigma}_m}{dI}\right) \tag{2.5}$$

where $\tilde{\sigma} = \sigma(I)/\sigma_0$ and x is the optical depth $n\sigma_0\ell$. We note, however, that if $x > 1$, (2.1) must also include an average over the optical path length to account for the different attenuation at the center of the beam and at the low-intensity wings as discussed previously. Consequently, (2.5) is valid only in the limit that $x < 1$.

The second method, which has general applicability, was formulated by *Cotter* [2.37]. The method represents a heuristic approach rather than a rigorous development. Although the full range of applicability has not yet been firmly established, a comparison with the simple limiting cases of a two-level absorber where the functional behavior can be obtained in analytic form indicates that the prescription results in a set of parameters that closely approximates the analytic results. It appears that the validity of this approach is sufficiently good to be useful for this first step in the analysis of nonlinear absorption data.

The essentials of the methods are as follows: the radiation field can be described in terms of a spectral energy density distribution function $f(\mathbf{r},t,\omega)$. It is assumed that a frequency range can be defined in a sufficiently narrow interval that one can integrate $f(\mathbf{r},t,\omega)$ over this interval and obtain an energy density function $\rho(\mathbf{r},t)$. Multiplication of ρ by the velocity of light gives the optical intensity or flux $\phi(\mathbf{r},t)$. The energy fluence $\Phi(r)$ is obtained by integrating $\phi(t)$ over the pulse.

A basic assumption is that $f(\mathbf{r},t,\omega)$ can be factored into a product of separate functions, i.e.,

$$f = \Phi(\mathbf{r}) \cdot F(t) \cdot S(\omega) \tag{2.6}$$

with $F(t)$ and $S(\omega)$ normalized such that

$$\int_t F(t)dt = 1 \quad , \tag{2.7}$$

$$\int_\omega S(\omega)d\omega = 1 \quad . \tag{2.8}$$

This assumption breaks down for large values of the absorption along the optical path length, since the temporal pulse shape $F(t)$ is now a function of the propagation coordinate. For nearly collimated beams the radiation transport along the optical path can be described *formally* in terms of the progagation equation

$$\frac{d\Phi(r,z)}{dz} = -n\sigma(\Phi)\Phi(r,z) \quad , \tag{2.9}$$

where $\sigma(\Phi)$ is a fluence- (or flux-) dependent absorption cross section and n is the molecular density. The physical mechanisms of the absorption process are now implicitly contained in $\sigma(\Phi)$.

The motivation of the proposed prescription is to provide a method for extracting a weighted space average fluence $\bar{\Phi}$ and cross section $\bar{\sigma}(\bar{\Phi})$ from the experimental data that yield a good approximation to the local values of Φ and $\sigma(\Phi)$. The essence of the prescription for determining the various parameters is as follows:

$$\bar{\sigma} = -\frac{1}{nL}\ln\frac{E(L)}{E(0)} = -\frac{1}{nL}\ln T \qquad (2.10)$$

where L is the optical path length, T is the transmittance, and E(L) and E(0) are the total energy measured at the input and output of the absorber, respectively. The fluence $\bar{\Phi}$ corresponding to this value of $\bar{\sigma}$ is obtained from the expression

$$\bar{\Phi} = \frac{1}{L}\int_0 \frac{\int_0^\infty \Phi^2(r,z)2\pi r\,dr}{\int_0^\infty \Phi(r,z)2\pi r\,dr}\,dz \quad . \qquad (2.11)$$

The number of photons absorbed per molecule is given by

$$\eta = \bar{\sigma}\bar{\Phi}/h\nu \quad , \qquad (2.12)$$

where $h\nu$ is the photon energy.

To a first approximation, the spatial average over the radial coordinate of the fluence is simply a weighting of the fluence $\Phi(r,z)$ by the absorbed energy $\eta(\bar{\Phi})$ at that fluence. This is evident if we assume that σ is independent of energy and multiply each integrand of the radial integrals by $n\sigma$. The net effect is to determine an average fluence in the spatial regions of the irradiation volume where the nonlinear optical absorption is large. This prescription therefore applies to focused as well as unfocused irradiation geometries. The definition of $\bar{\Phi}$ also provides a well-defined method for determining the fluence without reference to "the area or volume of the optical beam"; similar comments apply to the determination of η.

In order to implement this method we assume that the radiation field can be apprixmated by a propagating Gaussian beam. The functional form for the fluence can be written as

$$\Phi(r,z) = \frac{E(z)}{\pi w^2}\exp(-r^2/w^2) \quad , \qquad (2.13)$$

where the beam radius governing the geometrical beam expansion or contraction is given by

$$w^2(z) = w_0^2 + \theta^2(z - z_0)^2 \quad . \qquad (2.14)$$

The beam geometry is shown in Fig.2.1; w_0 is the radius of the beam waist at the 1/e intensity point, and θ is the beam divergence angle. The medium absorption is accounted for in E(z). As a first iteration, the functional variation of E(z) is obtained from (2.9) for a fixed value of $\bar{\sigma}(\bar{\Phi})$ and is given by the expression

$$E(z) \cong E(0)\exp(-n\bar{\sigma}z) \quad . \qquad (2.15)$$

Substituting (2.13-15) into (2.11) and performing the indicated integrations yields the result that

$$\bar{\Phi} = \frac{E(0)}{2\pi w_0^2}h(x) \quad , \quad \text{where} \qquad (2.16)$$

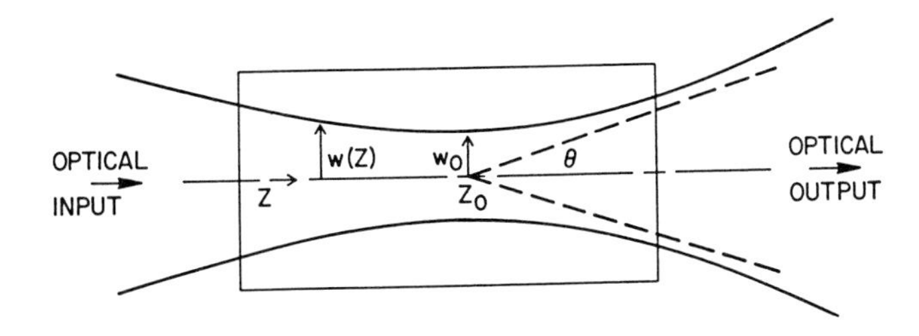

Fig.2.1. Beam geometry and parameter definition for focussed laser radiation

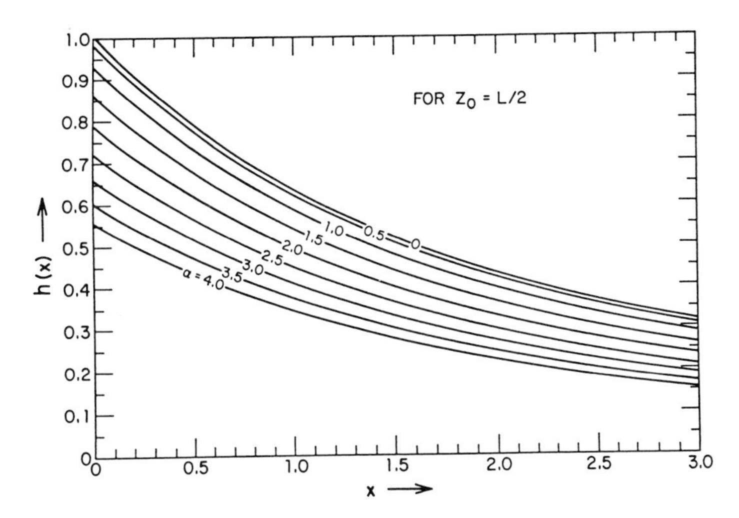

Fig.2.2. Parametric representation of the function h(x) for various values of α (2.17)

$$h(x) = \int_0^1 \frac{e^{-w\xi} d\xi}{1 + \alpha^2(\xi - \xi_0)^2} \quad . \tag{2.17}$$

The parameters involved in the function h(x) are the "effective" optical depth $x = n\bar{\sigma}L$, the normalized position of the beam waist $\xi_0 = z_0/L$, and the normalized beam divergence angle $\alpha = \theta L/w_0$. The function h(x) has been evaluated numerically for several values of the parameters x and α [2.37]. A series of parametric curves for h(x) is shown in Fig.2.2; for this evaluation, $\xi_0 = 0.5$. One can develop this prescription into an iterative scheme to obtain better approximations for $\sigma(\Phi)$ and η. For collimated beams $\alpha = 0$ and h(x) simplifies to the form

$$h = \frac{1 - \exp(-n\bar{\sigma}L)}{n\bar{\sigma}L} \quad . \tag{2.18}$$

In this limit,

$$\bar{\Phi} = \frac{1}{2} \frac{E(0)}{\pi w_0^2} \left(\frac{1 - \exp(-n\bar{\sigma}L)}{(n\bar{\sigma}L)} \right) = -\frac{1}{2} \frac{E(0)}{\pi w_0^2} \frac{1 - T}{\ln T} \quad . \tag{2.19}$$

The various terms in (2.19) can be understood in the following way. From (2.13), the peak on-axis fluence at $z=0$ is $\Phi_0=E(0)/\pi w_0^2$. The correction for the finite optical depth of the medium $n\bar{\sigma}L$ is contained in the bracketed terms. The use of a nonuniform optical beam for absorption measurements in a nonlinear medium necessarily results in average values of the absorption quantities such as $\sigma(\Phi)$ and Φ. The use of the fluence weighting technique indicated in (2.11) for the determination of the local values of $\sigma(\Phi)$ necessarily results in values of $\bar{\Phi}<\Phi_0$. For the particular choice indicated in (2.11), local values of $\sigma(\Phi)$ can be determined only for $\Phi\leqq\Phi_0/2$. This effect accounts for the factor of two in (2.19).

From (2.19) we note that the value of $\bar{\Phi}$ is not simply $E/\pi w_0^2$, but is corrected for the optical depth in the medium. Equations (2.10,12,19) constitute a self-consistent set. Errors have occurred in the literature in the determination of absorption parameters because σ is calculated from (2.10), Φ is taken to be $E/\pi w_0^2$, and η is computed from (2.12). Such a procedure can lead to inconsistencies in the relations of σ, Φ, and η as well as in the functional behavior of these parameters.

2.2.2 Data-Analysis Techniques for Reaction Studies

The large number of molecular species investigated and the wide diversity of experimental techniques and data-reduction methods complicate the comparison of laser-induced reaction data from different laboratories. The reaction (MPD) experiments require the same careful analysis that the absorption experiments do. The set of variables involved in the experiment includes the laser properties: pulse energy, pulse length, frequency, linewidth, mode quality, coherence, spatial profile, and focusing geometry. To this set we add the gas sample properties: absorber pressure, diluent pressure, temperature, sample cell size, cell wall condition, reaction diagnostic technique, and time scale of the diagnostic. In addition, the wide range of molecular properties from the many species that have been studied further complicates comparison and interpretation. Furthermore, the methods used to reduce and report the experimental observations are by no means uniform. The experimental observations range from the time dependence of some signal, where the dynamics of the reaction is the information sought, to the total amount of reaction after irradiation of a sample where the desired information is the reaction probability. In some cases the experimenters report only qualitative observations, and in other cases they attempt to give a microscopic interpretation.

We will now review the data-analysis techniques that have been used for MPD experiments. Much of this section follows the discussion in [2.38,39]. These techniques complement the methods for absorption experiments discussed previously.

Because many of the available data that one can use to estimate reaction probability come from stable-species analysis of multiply irradiated static samples, we approach our discussion from this point of view. We consider an experiment where one uses a pulsed CO_2 laser to irradiate a sample of gas i times in a cell of volume V_0. We assume that experimental details such as beam geometry within the cell (focused or unfocused), pulse energy and temporal profile, gas pressure and temperature, etc., are known. After (or perhaps during) the irradiation, one obtains a measurement of the fraction of the reactant that remains, f. If we treat the number of pulses, i, as a continuous (time) variable, the rate of reaction of species A is

$$\frac{d[A]}{di} = \frac{-V}{V_0}[A] \tag{2.20}$$

where V is the equivalent volume of A reacting per pulse.

If V is independent of i, integration gives

$$V = -\frac{V_0 \ln(f)}{i} \quad . \tag{2.21}$$

The isotopic selectivity is just the ratio of the Vs for two isotopic forms of the molecule [2.38].

The reaction probability for a given fluence, $P(\Phi)$, is related to V by

$$V = \int_0 g(\Phi)P(\Phi)d\Phi \quad , \tag{2.22}$$

where $g(\Phi)$ is the fluence distribution function or the volume per unit fluence at fluence Φ. If $G(\Phi)$ is the volume within the cell with fluence greater than Φ, then

$$g(\Phi) = -dG(\Phi)/d\Phi \quad . \tag{2.23}$$

Let us consider several irradiation geometries. The simplest would be if the experiment were so arranged that the laser irradiated a volume G_0 with uniform fluence Φ_0. In this case

$$\begin{aligned} G(\Phi) &= G_0 \quad , \quad \leq \Phi_0 \\ &= 0 \quad , \quad \Phi \geqq \Phi_0 \end{aligned} \tag{2.24}$$

which from (2.23) gives

$$g(\Phi) = \delta(\Phi_0) \quad . \tag{2.25}$$

Integration and rearrangement of (2.22) give

$$P(\Phi_0) = V/G_0 \quad . \tag{2.26}$$

Clearly, the closer one comes to this ideal situation, the better the reaction probability will be determined. In a real experiment, however, the beam fluence will have some radial dependence, and the absorbing gas will attenuate the beam. Consider a beam with a Gaussian profile that passes through an absorbing gas with mean absorption cross section $\bar{\sigma}$, given by (2.10). The fluence at a distance z from the entrance window and a radial distance r from the center line is

$$\Phi(r,z) = \Phi_0 \exp[-(r/w_0)^2]\exp(-n\bar{\sigma}z) \tag{2.27}$$

where Φ_0 is the fluence at the beam center and entrance window. The function $g(\Phi)$ is just

$$g(\Phi) = -\frac{\partial}{\partial\Phi}\int_{\Phi} dV \quad , \tag{2.28}$$

where the integral is evaluated over the volume with fluence greater than Φ [from (2.23)]. Combining (2.27,28) gives

$$\begin{aligned} g(\Phi) &= G_0/\Phi \quad , & 0 < \Phi < \Phi_0\, e^{-x} \\ &= G_0/[x\Phi \ln(\Phi/\Phi_0)] \quad , & \Phi_0 e^{-x} < \Phi < \Phi_0 \\ &= 0 \quad , & \Phi > \Phi_0 \end{aligned} \tag{2.29}$$

where

$$G_0 = \pi w_0^2 L \quad \text{and}$$
$$x = n\bar{\sigma}L \quad . \tag{2.30}$$

Because the fluence is relatively uniform within the cell, we assume for purposes of application that the local reaction probability depends on the mth power of the fluence or

$$P(\Phi) = C\Phi^m \quad . \tag{2.31}$$

With $g(\Phi)$ and $P(\Phi)$ given by (2.29,31), we integrate (2.22) to get

$$V = G_0 C\Phi_0^m[1 - \exp(-mx)]/m^2x \quad . \tag{2.32}$$

The reaction probability at fluence Φ_0 is therefore

$$P(\Phi_0) = V/G' \tag{2.33}$$

where

$$\begin{aligned} G' &= G_0[1 - \exp(-mx)]/m^2x \quad , \\ &\cong G_0/m \quad , \end{aligned} \tag{2.34}$$

where the approximation is valid for x near zero.

Deconvolution is a more serious problem for focused laser radiation. For a soft focus or for an experiment where the reaction yield is very small, one may assume that the focal region is uniform or Gaussian and use the methods described above. However, this will not give much more than a rough estimate of the reaction probability.

For experiments where a significant fraction of the reaction occurs outside the focal volume, we follow [2.38]. Consider a cylindrically-symmetric laser beam of uniform fluence focused to a point at the center of a cell of length 2L. The fluence within the cell will be zero except within the double cone illuminated by the laser where

$$\Phi(z,r) = \Phi_W (L/z)^2 \tag{2.35}$$

where Φ_W is the fluence at the entrance window, z is the distance along the center line from the focal point, and r is the radial distance from the center line. With this expression and (2.28) we obtain

$$g(\Phi) = (3/2)G_0\Phi_W^{3/2}/\Phi^{5/2} \quad , \qquad \Phi \geqq \Phi_W \tag{2.36}$$
$$= 0 \quad , \qquad \Phi < \Phi_W$$

where G_0 is the irradiated volume within the cell. If we assume a reaction probability that increases with the mth power of the fluence at low fluence, but approaches 1.0 at high fluence with the functional form

$$P(\Phi) = 1 - \exp[-(c\Phi)^m] \quad , \tag{2.37}$$

we can calculate V from (2.22).

To good approximation, when m is greater than about 2, the quantity is

$$V = G_0(c\Phi_W)^{3/2} \quad . \tag{2.38}$$

This 3/2 functional dependence is purely a geometrical effect [2.38,40].

For the purposes of this review (Sect.2.5.4), where we wish to extract the fluence necessary to give a 1% reaction probability from a large body of experimental data obtained with focused laser pulses, we use (2.38) to obtain c for a given experiment. This quantity, with (2.37) and an estimate for m, allows one to obtain $\Phi_{1\%}$, the fluence necessary to give 1% reaction probability. From previous direct measurements of the fluence dependence of reaction probability, we estimate m to be 3 for species where $\Phi_{1\%}$ is small (less than 10 J cm^{-2}), and 4 for species where $\Phi_{1\%}$ is larger. This procedure, of course, gives a less precise value of $\Phi_{1\%}$ than a direct measurement with unfocused pulses. It does, however, allow one to obtain an estimate of $\Phi_{1\%}$ from many experiments. A more precise method of obtaining $P(\Phi)$ from data of this type is given in [2.38], which also contains examples for other irradiation geometries.

2.3 A Quantitative Comparison of Absorption Data

In this section we make a quantitative comparison of absorption data from MPE experiments with many molecular species. This comparison allows us to determine the molecular and laser properties that most strongly influence the absorption process. By the procedure outlined in this section we determine which features of the absorption data require further theoretical investigation and a more refined experimental technique. We follow, for the most part, the initial report of this study given in [2.41].

At first thought, a quantitative comparison appears difficult since (1) the various molecules differ in symmetry, density of states, and the matrix elements for the transition; (2) the experiments are performed under different conditions of laser beam quality, pulse duration, spectral bandwidth, etc.; and (3) the conditions of the gas such as pressure and material constituents could be different. It is possible, however, to eliminate the major differences in the dependence of these independent variables in the multiphoton absorption (MPA) process through the use of an appropriate normalization of the absorption and fluence variables. Use of these normalized variables eliminates many nonessential aspects of the interaction and provides a universal plot of multiple-photon absorption as a function of fluence for a number of polyatomic molecules that have been discussed in the literature.

2.3.1 Interaction Dynamics and Normalization Concepts

a) General Considerations for MPE Processes

The experimental parameters that are usually reported in multiple-photon absorption experiments are the absorption cross section $\sigma(\Phi)$ or the number of photons absorbed per molecule $\eta(\Phi)$ as a function of the optical fluence Φ or intensity ϕ. In order to effect a quantitative comparison of absorption data for different molecules, it is necessary to normalize these variables in such a way as to eliminate any additional independent variables of the experiments such as the intrinsic parameters of the molecule, the properties of the optical radiation field, and the thermodynamic conditions of the gas. The problem is then essentially one of specifying a quantitative measure of the strength of the initial interaction between the molecule and the optical field and the related molecular response to this interaction. In this section we develop the rationale for the normalization procedure and provide an interpretation for the quantities used in the normalization.

As a guide in this development, we consider some fundamental absorption properties of SF_6. A plot of the fluence-dependent absorption cross section for the Q branch of the ν_3 vibrational mode of SF_6 is shown in Fig.2.3 over several decades of the fluence [2.42,43]. Two points are evident: (1) the cross section is a smooth function of the fluence; (2) the departure from the limit of linear absorption (small fluence limit) occurs at a very low fluence (5 μJ cm^{-2}). We denote this value by Φ_t. The fact that the functional dependence of the cross section is nearly $\Phi^{-1/3}$ from the low-fluence limit to almost dissociation of the molecule implies that the threshold for MPE of the molecule may occur at Φ_t and proceed in a continuous fashion with increasing fluence up to the dissociation limit. In the absence of a MPE process in the molecule, the excitation of the ground-state vibrational transition of the ν_3 fundamental mode could be viewed as an "effective" two-level system. For this situation, the cross section $\sigma(\Phi)$ would be a constant value for increasing fluence until a significant fraction of the ground-state population is

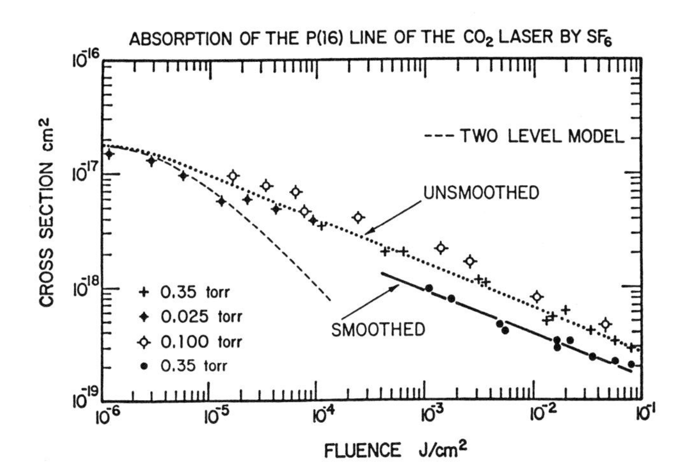

Fig.2.3. Fluence-dependent absorption cross section at 947.7 cm^{-1} for SF_6. The *dashed curve* is the prediction of the coupled two-level model with $\Phi_s = 5\ \mu J\ cm^{-2}$. The *smoothed* and *unsmoothed curves* were obtained with and without a low-pressure gain cell in the CO_2 laser resonator, respectively [2.42,43]

transferred to the upper level of the transition. At this fluence value, Φ_s, saturation of the transition occurs and $\sigma(\Phi) \propto \Phi^{-1}$ for increasing values of Φ. This behavior is indicated by the dotted curve in Fig.2.3. A view of MPE in a first approximation begins with the interaction of the optical field with an isolated two-level system at low fluence. As the fluence increases, population is transferred from the lower level to the upper level of the initial absorbing state(s) where (1) photon absorption out of the excited state occurs in the pumped mode or (2) the population transfers rapidly to other vibrational levels by intramolecular V-V' processes and excited-state photon absorption occurs from these levels. The onset of MPE would then be associated with the population of the upper level of the initial absorbing state(s), which occurs at a fluence $\Phi_s \cong (2\sigma)^{-1}$ in a nondegenerate two-level system. In a first approximation, $\phi_t \cong \Phi_s$. Although the presence of rotational levels and strong coupling to other vibrational levels complicates the concept of an isolated two-level interaction, these effects do not qualitatively change the above description and they can be included in a straightforward manner in the formulation of a generalized two-level model.

A first step in any comparison study of MPE in different molecules would be to normalize the absorption parameters in a manner similar to that for a generalized two-level system. This normalization is strictly valid for $\Phi \leq \Phi_s$. For $\phi > \phi_s$, the normalization removes the zero-order dependence of the independent variables and allows a comparison of the essential differences of the MPE process in different molecules.

b) Molecular Absorption in a Coupled Two-Level Approximation

The coupled two-level absorber model described in [2.41] contains the following features. A finite bandwidth radiation field interacts with one or several rotational states of a vibrational level of a molecule to promote transitions to an upper vibrational state. We consider this interaction in a rate equation approximation. This interaction results from the direct spectral overlap of the radiation field and the absorbing transition of the molecule. Other rotational states that are not coupled directly to the radiation field constitute a set of reservoir states that may be indirectly coupled to the interacting rotational levels through collisions. Also included in the reservoir states are other vibrational levels that are coupled to the interacting states either by collisional or collisionless intermodal V-V' transfer processes.

An approximate solution to the differential equations that describe optical and collisional transitions among the four levels of the model (two absorber levels and two reservoir levels) is given in [2.41]. The results of this analysis give

$$\frac{\eta(\Phi)}{\langle f\rangle} = 1 - \exp\left(-\frac{\sigma_0\Phi}{\langle f\rangle}\right) \tag{2.39}$$

for the relationship between the average number of photons absorbed per molecule, $\eta(\Phi)$, and the three variables Φ (fluence), σ_0 (small-signal cross section), and $\langle f\rangle$ (the effective fraction of the population for a given vibrational transition that is coupled to the radiation field). Note that $\eta(\Phi)$ is not a spatial average, but a local value that one can relate to the experimental values of η through (2.12). In the strong fluence limit the quantity $\langle f\rangle$ is

$$\langle f\rangle = df_i\left[1 - \exp\left(\frac{-f_r}{1-f_r}\frac{\tau_p}{\tau}\right)\right] \tag{2.40}$$

where f_i is the fraction of molecules in the absorbing (usually ground) vibrational level and f_r is the fraction of the f_i molecules in the initial distribution that interact directly with the radiation field. The quantity d is

$$d = \frac{\beta}{1+\beta} \quad , \tag{2.41}$$

where β is the ratio of the degeneracies of the upper and lower vibrational levels; τ_p is the optical pulse length; and τ is the equilibration time of the absorber level and the reservoir level (usually taken as the rotational relaxation time). If the pressure is sufficiently high, the value of $\langle f\rangle$ exceeds df_if_r due to collisional coupling of the absorber and reservoir levels.

Table 2.1 gives model predictions of $\eta(\Phi)$ for four limiting cases. The approximate expression

$$\begin{aligned}\langle f\rangle = {} & df_i\left\{1 - \exp\left[-\frac{f_r}{1-f_r}\frac{\tau_p}{\tau}\left(1+\frac{f_rf_i}{\sigma_0\Phi}\frac{\tau_p}{\tau}\right)\right]\right\} \\ & \times\left[1-\exp\left(-\frac{\tau_p}{\tau}\right)\right] + df_rf_i\exp\left(-\frac{\tau_p}{\tau}\right)\end{aligned} \tag{2.42}$$

Table 2.1. Functional dependence for $\eta(\Phi)$ and $\langle f\rangle$ for a coupled two-level model

	$\frac{\tau_p}{\tau}$	$\frac{\sigma_0\Phi}{f_r f_i}\frac{\tau}{\tau_p}$	$\eta(\Phi)$	$\langle f\rangle$
Weak coupling (collisionless)	< 1	> 1	$\langle f\rangle\bar{G}\,\frac{\sigma_0\Phi}{\langle f\rangle}$	$df_r f_i$
Strong coupling	> 1			
(1) Low flux		< 1	$\langle f\rangle\bar{G}\,\frac{\sigma_0\Phi}{\langle f\rangle}$	df_i
(2) High flux		> 1	$\langle f\rangle$	$df_i\bar{G}\left(\frac{f_r\tau_p}{\tau}\right)$

$d = \frac{\beta}{1+\beta}$; $\bar{G}(x) = 1 - \exp(-x)$

is a smoothly varying function that is correct in all four of the limits in Table 2.1.

Cast in this general form, the optical interaction in a coupled two-level system can apparently be described in terms of two parameters, a fluence parameter $\sigma_0\Phi/\langle f\rangle$ and the effective fraction of states that interact with the radiation field $\langle f\rangle$. All effects due to collisions (τ), intramolecular V-V' coupling (τ), laser pulse duration (τ_p), gas temperature (f_i, f_r, τ), pressure (τ), and laser spectral bandwidth (f_r) can be regarded in a first approximation as affecting the effective fraction of absorber molecules $\langle f\rangle$ that can be coupled to the radiation field. Equations (2.39,42), therefore, constitute a general form for the optical absorption in a coupled two-level system in terms of a set of normalized dependent and independent variables.

The simplest application of the model is to optical absorption from a set of rotational transitions occurring between a pair of vibrational levels that are isolated from other vibrational levels of the molecule. In this situation, f_i is the fractional population of the lower vibrational level and f_r is the fraction of the rotational population that is coupled directly to the radiation field. For an optical interaction with well-defined rotational lines such as occur at low values of J in a P(J) or R(J) branch transition, f_r is less than or equal to the Boltzmann population factor λ_J for the particular rotational level being accessed. For monochromatic radiation and a homogeneously broadened transition, $f_r = \lambda_J$. In many situations, the bandwidth of the optical source Δ is finite. Under these conditions, the value of f_r is obtained by a convolution of the spectral intensity distribution of the source with the absorption features of the molecule.

For a Q-branch transition, the absorption results from a number of overlapped rotational transitions. In addition, each rotational line may be split into a number of components due to the Coriolis interaction in the molecule. The spectral distribution of the Q-branch absorption in SF_6, for example, results from this type of splittings. In this situation the spectroscopic absorption cross section $\sigma_0(\nu)$ is directly proportional to the total fraction of the population with rotational levels (or split components) being accessed at frequency ν. If the absorption band strength is uniformly distributed in frequency, one can define a spectral population distribution function $f(\nu)$. The value of $f(\nu)$ is normalized such that $\int_{-\infty}^{\infty} f(\nu)d\nu = 1$. In terms of $\sigma_0(\nu)$,

$$f(\nu) \cong \frac{\sigma_0(\nu)}{\int \sigma_0(\nu)d\nu} \tag{2.43}$$

where $\int \sigma_0(\nu)d\nu$ is related to the integrated band strength. For ground-state transitions in tetrahedral $(T_d)XY_4$ and octahedral $(O_h)XY_6$ molecules, this relation is rigorously correct [2.44]. If the bandwidth of the optical source Δ is much less than the width of $f(\nu)$, the value of f_r is given directly by $f_r = f(\nu)\Delta$.

For other transitions involving overlapping vibrational bands and coupling to other vibrational levels, the situation is more complex. This aspect is discussed in greater detail in [2.41].

c) Generalization to the MPE Processes

In the previous section we summarized general expressions for absorption in a coupled two-level system. These results should correctly account for molecular absorption at fluence levels for which $\Phi \lesssim \Phi_s$, where $\Phi_s \cong \langle f \rangle/\sigma_0$. For $\Phi > \Phi_s$, MPE processes become important and the two-level approximation breaks down; (2.39) predicts that $\eta(\Phi)$ approaches a constant value at high fluence whereas it is observed experimentally that $\eta(\Phi)$ continues to increase in large polyatomic molecules. The model may be modified in a conceptual sense, however, to establish a plausible basis for extension of these results into the MPE regime.

As $\Phi/\Phi_s \to 1$, a significant fraction of the population in the initial absorbing vibrational state is transferred to the upper vibrational level of the transition by the optical field. In the MPE process we assume that once $\Phi \gtrsim \Phi_s$, successive photon absorption and depletion of the population of the upper level occur on a time scale that is short compared to the transfer time between the lower level of the active state and the reservoir state. This process may involve a direct absorption out of the upper level of the active state to other vibrational levels and a direct photon absorption out of these final states. The net result is that on the average the upper state population of the transition is always small compared to the lower state population. The dynamics of the interaction is then in a first approximation controlled by the lower vibrational level of the absorbing transition. The results obtained above (2.39,42) may then be applied directly to this revised model by the formal procedure of taking the limit $\beta \to \infty$.

In view of the above comments, we consider in a first approximation that the dominant functional behavior of $\eta(\Phi)$ is determined by the dynamics of the lower level of the molecular transition that interacts with the radiation field. We therefore consider as an ansatz, a generalization of (2.39) which may be written in the normalized functional form

$$\frac{\eta(\Phi)}{\langle f \rangle} = G\,\frac{\sigma_0\Phi}{\langle f \rangle} \quad , \tag{2.44}$$

which is valid for $(\sigma_0\Phi/\langle f \rangle) > 1$ and approaches the two-level result $\bar{G}(x)$ as summarized in Table 2.1 for $(\sigma_0\Phi/\langle f \rangle) < 1$. The value of $\langle f \rangle$ is determined from (2.42). Experimental absorption data for SF_6, for example, indicate that $G(x) \rightarrow x^{2/3}$ for $x \gg 1$ (Fig.2.3). In the high-fluence regime, the value of $\langle f \rangle$ as defined in this analysis is independent of fluence. However, the fraction of states that interact with the radiation field may increase with increasing fluence because of power broadening. In typical applications, this effect becomes important for $(\sigma_0\Phi/\langle f \rangle) > 1$. Consequently, it may also be necessary to include this contribution to the fraction of molecules that interact with the radiation field. This effect was not included in the determination of $\langle f \rangle$.

2.3.2 MPA in Different Polyatomic Molecules

In this section we compare MPA in different polyatomic molecules using the normalization procedure summarized above. The required absorption parameters are the fluence-dependent absorption cross section $\sigma(\Phi)$ or the number of photons absorbed per molecule $\eta(\Phi)$. The quality of such a study is compromised by several factors having to do with the quality of existing data for $\sigma(\Phi)$ or $\eta(\Phi)$. There are only a few molecules that have been studied over a sufficient fluence range for a study of this type. In order to effect a valid quantitative comparison of different molecules, it is necessary to compare absorption data that extend from the linear absorption range (Beer's law limit) to near dissociation of the molecule. Differences in experimental procedure, laser-pulse characteristics, and data-reduction methods also tend to make comparisons less reliable. Consequently, the results of the study should be regarded as uncertain by at least a factor of two. The particular choice of molecules for this study results primarily from the collection of investigations reported in the literature for which the data are of sufficient quality and taken over a sufficient range to make the results useful. In some cases, it was necessary to infer some of the experimental parameters in the analysis.

For purposes of comparison, we will normalize the absorption data in terms of $\eta(\Phi)$ by use of (2.44). At the present time, $\langle f \rangle$ can be calculated directly for only a few molecules. An empirical procedure will be developed to determine the relative value of $\langle f \rangle$ for the different molecules. A value of $\langle f \rangle$ will then be calculated directly for SF_6 to establish a quantitative benchmark for the study. A general evaluation of the derived values of $\langle f \rangle$ will be considered in a concluding section.

a) Absorption Characteristics of Polyatomic Molecules

The above analysis indicates that we should see a considerable simplification of η-versus-Φ data for many species if we were to replot the data in the normalized form of (2.44). Table 2.2 summarized the species we used in the comparisons that follow. The table also lists experimental parameters and references to the primary data. As indicated in [2.41], a normalization of this type is only valid at frequencies near the center frequency ν_0 of the molecular absorption feature.

A summary of MPA measurements of $\eta(\Phi)$ for several molecules is shown in Fig.2.4. For purposes of comparison, measurements near the peak of the spectroscopic absorption feature have been selected if available. The lines are representative of the range of the primary data points. This particular form of the data summary indicates the general nature of the MPE process and the fluence range over which these processes are observed. The extent of the range for MPE is only apparent, however, because the fluence variable alone does not constitute a good measure of the interaction of the radiation field with the molecule.

A considerable simplification results if these absorption data are replotted in terms of $\sigma_0\Phi$. The results are shown in Fig.2.5. With the exception of the triatomic molecules, the fluence behavior of the different molecules can be classified into two types: (1) one in which a deviation from linear absorption begins to occur; and (2) one in which $\eta \propto \Phi^\gamma$ where $\gamma \cong 2/3$. The general functional behavior of each molecule is consistent with (2.44) depending on whether $\sigma_0\Phi/\langle f\rangle$ is greater than or less than unity. The factor that has not been taken into account for the different molecules in Fig.2.5 is $\langle f\rangle$. If $\langle f\rangle$ could be calculated for each molecule and the absorption data could be replotted in the normalized form of (2.44), the curves in Fig.2.5 would be expected to merge into a single curve. In practice, $\langle f\rangle$ is difficult to calculate from first principles. Since $\langle f\rangle$ is a divisor in both the dependent and independent variables in (2.44), we can obtain the ratio of $\langle f\rangle$ for two molecules by simply shifting each plot of $\eta(\Phi)$ versus $\sigma_0\Phi$ in Fig.2.5 along the linear absorption line until the curves merge. This procedure is tantamount to multiplication or division of the ordinate and abscissa of a given logarithmic graph by a constant. For convenience, we take the curve for S_2F_{10} as a reference and shift the remaining curves accordingly. The results are shown in Fig.2.6; the relative values of $\langle f\rangle$ with

$$\langle f\rangle_{S_2F_{10}} = 1$$

are summarized in Table 2.2. As expected on physical grounds, all of the data can be fitted to a single curve for $(\sigma_0\Phi/\langle f\rangle) \lesssim 1$. It is surprising, however, that nearly the same functional dependence is obtained for all molecules at high fluence. With the exception of SF_6 and the triatomics, all molecules dissociate at some point on the curve. For the large polyatomic molecules, CH_3COCF_3, S_2F_{10}, and SF_5NF_2, significant dissociation occurs at $(\sigma_0\Phi/\langle f<) \lesssim 1$, i.e., in the range of linear absorp-

Table 2.2. Sumary of species, experimental parameters, and <f> used in comparison study

Molecule	Frequency [cm^{-1}]	σ_0 [$10^{-18} cm^2$]	Pressure [Torr]	τ_p [ns]	<f>	Reference	Symbol
S_2F_{10}	924-948	17.5-0.4	0.025-0.1	150[a]	1	[2.42]	---
SF_6	944-948		0.025-0.35	150[a]	5×10^{-3}	Sect.2.4	—
SF_6	947.7	83	0.3-1.0	150[a]	10^{-4}	Sect.2.4	✡
	944.2	4.6			2×10^{-5}		$\bar{\Delta}$
	952.2	2.5			2×10^{-5}		△
UF_6	620.3-631.4	9.6-4.5	0.2-1.0	50	0.5	[2.45]	▲
SF_5NF_2	908-958	2.8-1.3		150[a]	10	[2.46]	□
SBA[b]	1033.5	0.3		150[a]	2	[2.47]	■
C_2H_3Cl(V.C.)[c]	933.0	0.07	1.0	45	2.5×10^{-3}	[2.48]	⊖
	942.4	0.31		180	10^{-2}		ϕ
	942.4	0.31		45	10^{-3}		⊙
	933.0	0.07		180	10^{-3}		○
CF_3I[d]	1076.0	7.13	0.5	60[a]	0.2	[2.49,50]	●
CF_3COCF_3	970.6	1.34	0.1	---	30	[2.51,52]	▽
C_2H_4[e]	952.9,938.7	0.03	0.7	85[a]	2.5×10^{-3}	[2.53]	▼
D_2O[e]	1079.8	0.022	0.7	120	0.66	[2.53]	+
OCS[e]	1045.0	0.0034	1.0	120	0.3	[2.53]	−

[a] Pulse includes a nominal 1-μs tail.
[b] SBA ↔ *sec*-butyl acetate.
[c] The exact placement of these data is subject to interpretation.
[d] The curve has been oriented slightly from the original reference.
[e] The absolute values of either $\eta(\Phi)$ or Φ at low values of Φ appear to be inconsistent (by a factor of 10) with measured spectroscopic values of σ_0. The values of σ_0 in the table are those derived from the data presented in the reference.

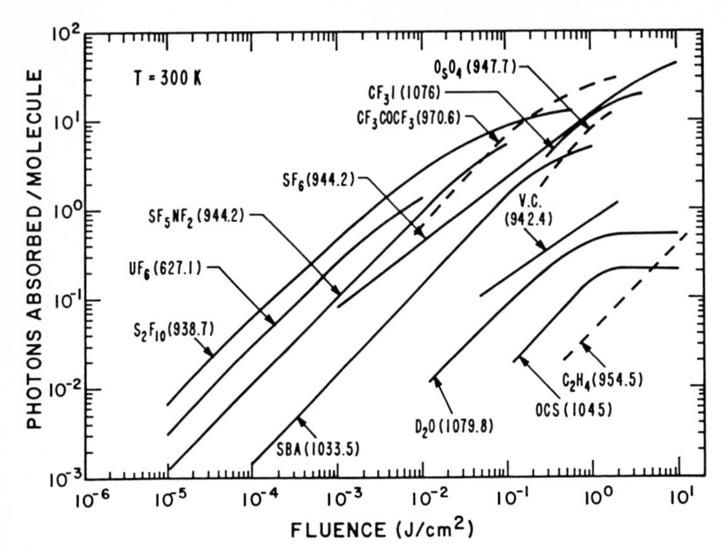

Fig.2.4. Summary of fluence-dependent absorption data for a number of polyatomic molecules. For most molecules, the frequencies correspond to irradiation near the peak of the spectroscopic absorption feature. The units of ν are cm^{-1}; SBA = *sec*-butyl acetate; V.C. = vinyl chloride; see notes in Table 2.2 concerning D_2O, OCS, and C_2H_4

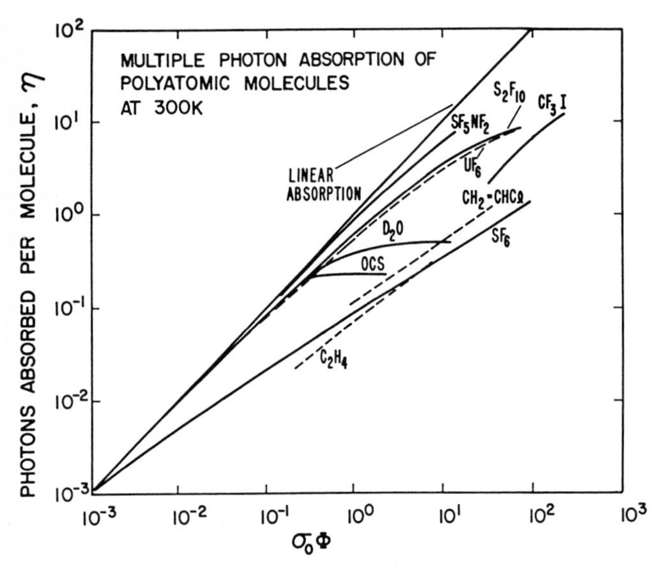

Fig.2.5. Summary of fluence-dependent absorption data for a number of polyatomic molecules plotted as a function of the normalized fluence $\sigma_0\Phi$

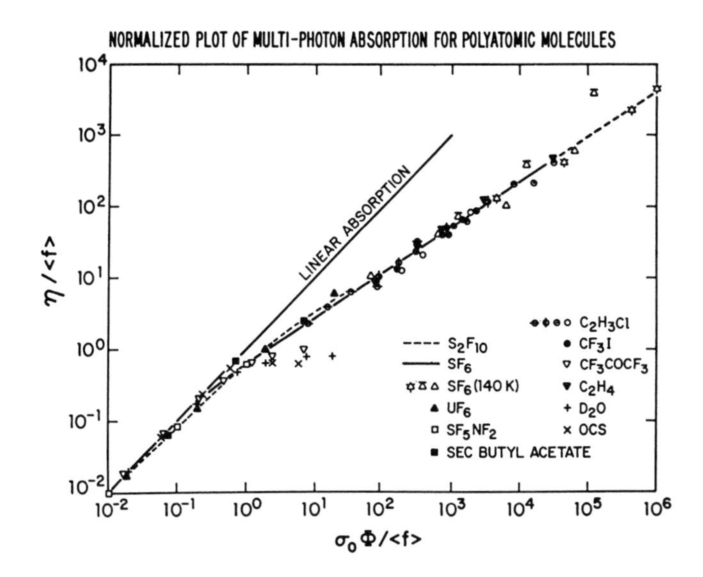

Fig.2.6. Plot of the multiphoton absorption for a number of polyatomic molecules in terms of the normalized variables $\eta(\Phi)/\langle f\rangle$ and $\sigma_0\Phi/\langle f\rangle$. For $(\sigma_0\Phi/\langle f\rangle) > 1$, the slope of the absorption curve is nominally 2/3

tion (Sect.2.5). For values of $(\sigma_0\Phi/\langle f\rangle) > 1$, it appears that most molecules, when excited at a frequency near the peak of the absorbing transition, absorb energy with a fluence dependence of $\Phi^{2/3}$.

b) Calculation of $\langle f\rangle$ for SF_6

It is of interest to establish an absolute calibration for $\langle f\rangle$. We review here several methods used in [2.41] to calculate $\langle f\rangle$ for SF_6 irradiated by the P(16) line of the CO_2 laser at 947.7 cm^{-1}. At this frequency, the interaction is primarily with the Q-branch of the ground-state ν_3 vibrational mode [2.54]. The absorption by hot bands is less than 15% of that resulting from the ground-state transition [2.55,56].

The spectroscopic structure of the ν_3-mode ground-state Q-branch in SF_6 is now understood in detail as a result of several experimental and theoretical studies [2.54]. The population at a given frequency, ν, of the transition can therefore be calculated directly to derive a population distribution function $f(\nu)$ as discussed in Sect.2.3.1 (2.43). The value of f_r is then given by $f_r = f(\nu)\Delta$, where Δ is the approximate bandwidth (FWHM) of the optical source. At low pressure, $\langle f\rangle = f_i f_r$ where f_i is the fraction of population in the vibrational ground state.

This analysis of the spectroscopic structure of the SF_6 Q-branch gives the values of $\langle f\rangle$ summarized in Table 2.3. Figure 2.3 (and [2.43]) shows measurements of σ versus Φ for single-mode (smoothed) and multimode (unsmoothed) laser radiation at 947.7 cm^{-1} for SF_6 at 300 K. The 140 K data are given in [2.57]. One can obtain

Table 2.3. Estimates of <f> for SF_6 at 947.7 cm^{-1}

Temp. [K]	Longitudinal mode	<f> calculated from $f(\nu)\Delta$	<f> estimated from Φ_s
300	Multimode	3×10^{-3}	5×10^{-3}
300	Single mode	3×10^{-4}	5×10^{-4}
140	Single mode	5×10^{-4}	5×10^{-4}

an independent estimate of <f> from these data by estimating the saturation fluence Φ_s (the point where the variation of absorbed energy with fluence deviates from a linear dependence). Table 2.3 also summarizes these estimates. Considering a factor of two uncertainty in the various quantities, the values of <f> computed from $f(\nu)\Delta$ are in good agreement with those estimated from Φ_s. From the values in the table it appears that the scales for $\sigma_0\Phi/\langle f\rangle$ and $\eta(\Phi)/\langle f\rangle$ in Fig.2.6 are nearly absolute.

c) Discussion

Several recent papers [2.11,58,59] report some particularly dramatic deviations from the general trends of Fig.2.6. These absorption data are generally for some special experimental situation such as small molecules or frequencies that are far from the centers of the absorption bands. One example of this behavior is CF_3D [2.59]. Between about 0.002 and 0.1 photons absorbed per molecule, the number of photons absorbed per molecule increases with about the 2/3 power of the fluence, and at still higher fluence the dependence becomes almost linear. The 2/3 power was the observed dependence in the high-fluence region for the species in Fig.2.6. The authors suggested that low anharmonicity and excitation to higher vibrational levels could both contribute to this observation.

In the previous section, we established that <f> for SF_6 at 300 K is typically 5×10^{-3}, which indicates that the values of <f> in Table 2.2 can be considered to be absolute within the accuracy of the study. Inspection of these results indicates that <f> exceeds unity for the large polyatomics (greater than 7 atoms) SF_5NF_2, *sec*-butyl acetate, CH_3COF_3, and perhaps S_2F_{10}. As we pointed out previously, these particular molecules also dissociate in the fluence range for linear absorption. Under these conditions, the model and the interpretation of <f> ceases to be valid. These large values of <f>, however, may still have a significant meaning. Because of the large number of degrees of freedom in the molecule, several vibrational modes may resonantly interact with the radiation field. Also, vibrational excitation enhances anharmonic coupling between vibrational modes, and all of these species have high levels of thermal vibrational excitation at 300 K, prior to any laser excitation. The net result is that a condition of statistical equilibrium may be established on a rapid time scale and the fraction of molecular population that interacts with the radiation field approaches unity. The fact that $\langle f\rangle$ exceeds unity may indicate that many successive interactions occur between the vibrational modes of the molecule and the radiation field, which is analogous to the absorption of several photons by a single molecule. This observation, and the fact that the large polyatomics dissociate in the linear absorption range, indicates that there is little or no "bottleneck" effect for these species.

The small-molecule/large-molecule distinction has become clearer in recent years [2.29]. For small molecules multiphoton resonances are important. Measurements [2.60,61] of absorption with tunable lasers show large enhancements at two- and three-photon resonances. Effects like Rabi cycling and Stark broadening also contribute to small-molecule absorption. The cumulative effect of these nonlinear and coherent effects is dramatic for small species like CF_2HCl [2.24,62] and D_2CO [2.63]. For the former, a 70-fold increase in laser intensity at constant fluence produces a 300-fold increase in dissociation probability. A rate-equation model, based only on linear absorption, predicts no change in dissociation probability for this experiment.

The coupled two-level model and the general functional form of the MPA measurements shown in Fig.2.6 lead to a number of implications that can be evaluated directly. In particular, it is of interest to explore the functional relationships of gas pressure P, bandwidth of the optical source Δ, temporal pulse shape τ_p, and the dependence on optical flux. In order to gain physical insight into these processes, we consider limiting parameter regimes; an evaluation for arbitrary experimental parameters is possible, however, by using the complete expressions for $\langle f\rangle$ as given by (2.42).

In the fluence range where $(\sigma_0\Phi/\langle f\rangle) \gg 1$, the trends in Fig.2.6 indicate that $n(\Phi)$ can be written in the functional form

$$n(\Phi) \propto \langle f\rangle^{1/3}(\sigma_0\Phi)^{2/3} \quad . \tag{2.45}$$

The MPA process can therefore be completely parametrized in terms of σ_0 and $\langle f\rangle$. The value of σ_0 is determined primarily by the properties of the particular molecule. The effects of most other experimental parameters are manifest in $\langle f\rangle$. From (2.45), $n(\Phi) \propto \langle f\rangle^{1/3}$. This proportionality provides the dominant functional relationship for the dependence of $n(\Phi)$ on τ_p, P, and Δ. In Sect.2.6 we investigate the validity of (2.45), with several quantitative comparisons of experimental measurements in SF_6. Those comparisons are consistent with (2.45).

The results of the comparison of experimental data for several polyatomic molecules in terms of the normalized variables indicate that MPA for almost all polyatomic molecules can be described by a single function of fluence. It appears from this comparison that multiple-photon absorption in poylatomic molecules is a general phenomenon that is qualitatively the same for all molecules; quantitative differences can be related to differences in σ_0 and $\langle f\rangle$.

We conclude that because absorption measurements are, to a large extent, determined by parameters (σ_0 and <f>) that describe the initial interaction with the laser field, absorption measurements are not sensitive tests of an MPE theory. The principal features of the data that do require theoretical explanation are:

1) the high-fluence dependence of absorption on fluence (2/3 slope);
2) the small deviations from the value of 2/3; and
3) the wide range of <f> for large molecules.

A successful MPE theory must, of course, be consistent with absorption data. But verification of a theory will require more sensitive experimental data. We note that the observed 2/3 slope (Fig.2.6) is consistent with and a direct consequence [2.41] of the anharmonic oscillator model for MPE [2.37].

2.4 A Review of MPA Data for SF_6

A great deal of experimental data exist in the literature related to multiple-photon absorption (MPA) of SF_6. In addition, the spectroscopy of the ν_3 ground-state transition is now well understood, both theoretically and experimentally, due to several recent tour-de-force efforts in this area. Consequently SF_6 is a reasonable prototype molecule for studying the MPE processes because a quantitative data base can be established from which a comparison with theoretical models can be effected. A study of the literature indicates that, in many cases, the experimental data differ by factors of 2-3 in absolute value, and in other cases a different functional dependence is reported. This can have important consequences for quantitative comparison with theoretical MPE models. Part of this uncertainty originates from the different methods used to reduce the experimental data. In any MPA experiment it is imperative to take into account the spatial profile of the optical beam and the finite optical depth in the determination of the absorption parameters and the fluence at which these values were determined. Failure to account for these effects leads to inconsistencies in the relation of the number of photons absorbed per molecule $n(\Phi)$, the absorption cross section $\sigma(\Phi)$, and the fluence Φ. Some of the reported measurements apparently neglected these effects completely. The intent of this section is to carry out a critical review and summary of the existing SF_6 absorption data and attempt to determine reasonable values for the absorption parameters.

We first present a quantitative discussion of the low-fluence absorption feature. Next, we consider absorption parameters measured at 300 K for several characteristic output frequencies of the CO_2 laser. A similar evaluation at a reduced temperature of 140 K is then discussed. For these summaries, data are selected from measurements where the conditions of SF_6 number density and the temporal nature of the optical pulse are nearly the same. The magnitude and fluence dependence of the absorption cross section can also depend on the pulse shape. The effects of the temporal character of the optical pulse on MPA are considered in detail for the P(20) line frequency. A discussion and summary of the MPA measurements in SF_6 are contained in the final section.

With one or two exceptions, no error limits are stated for the various measurements. It is, therefore, difficult to assess the merit of one measurement over another. As a basis for evaluation, measurements that are performed over an extended range of fluence ($>10^3$) are considered with greater weight. A second factor is the agreement in functional form as determined by the majority of the measurements. Finally, measurements that include a detailed determination of the optical beam profile and corrections for the propagation effects are considered to be of higher quality (Sect.2.2). In the evaluation of various measurements, the technique for determining the optical beam radius and the absorption parameters is noted, as are any corrections for the nonuniform beam profile and finite optical depth.

The absorption curves in this section were, for the most part, taken directly from the original references. For some measurements, the absorption cross section was converted to $n(\Phi)$ using (2.12). If the original data were refitted by a different curve, this is indicated explicitly. The length of the curve in each figure indicates the approximate extent of the fluence measurement. The actual data points on an individual curve have been deleted for purposes of clarity. The reader is referred to the original references for these details.

Most of the absorption data reported from optoacoustic measurements constitute a relative measurement. In order to compare optoacoustic and optical absorption measurements, the optoacoustic data is normalized in absolute value at one point. This data point is usually chosen so that the function shapes of the two curves overlap over some fluence range.

Absorption measurements on a number of lines of the CO_2 laser are reviewed. The P(16) line at 947.7 cm^{-1} is of particular interest. Radiation at this frequency interacts directly with the Q-branch of the ν_3 vibrational mode. The interaction with hot bands is minimal. Because the spectral structure and intensity of the Q-branch absorption feature are now well understood theoretically, it would seem that any quantitative comparison of experimental data with MPE theoretical models could best be effected at this frequency. As will be apparent in the next section, quantitative data exist at this frequency over a range of 10^5 in fluence. Consequently, a data base exists that allows a quantitative comparison with theory.

2.4.1 Low-Fluence Spectral Absorption

A plot of the SF_6 spectral absorption near 1000 cm^{-1} at a temperature of 300 K is shown in Fig.2.7. This particular spectrum was obtained using a Fourier transform spectrometer with a resolution of 0.06 cm^{-1} at a pressure of 0.3 Torr [2.64]. Also included in the figure are the positions of the output frequencies of the CO_2 laser.

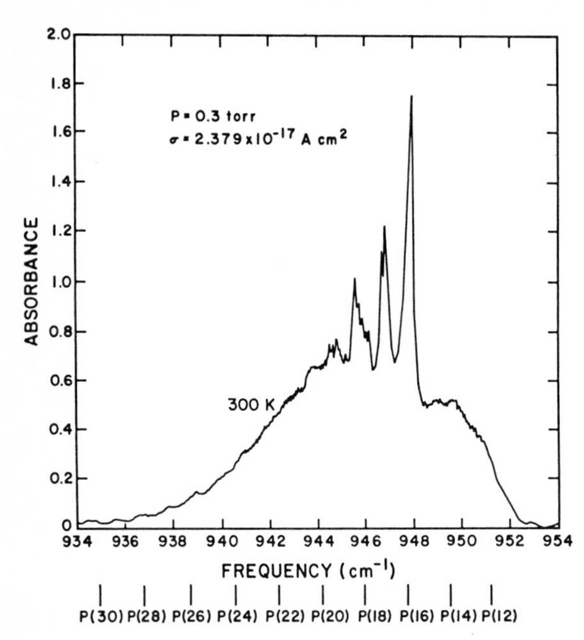

Fig.2.7. Spectroscopic absorption feature of the ν_3 vibrational mode of SF_6 at 300 K. The corresponding output frequencies of the CO_2 laser are shown at the *bottom* of the figure. The quantity "A" in the expression for cross section, σ, is absorbance

Table 2.4. Measured SF_6 absorption cross sections for several CO_2 laser frequencies

CO_2 Laser line	Frequency [cm^{-1}]	Cross section [$10^{-17} cm^2$] T = 300 K	T = 140 K[a]
P(12)	951.19	0.67	0.25
P(14)	949.48	1.24	1.7
P(16)	947.74	2.96	8.3
P(18)	945.98	1.89	1.1
P(20)	944.19	1.58	0.46
P(22)	942.38	1.12	
P(24)	940.54	0.62	0.02
P(26)	938.69	0.31	
P(28)	936.80	0.13	

[a]These values were taken from [2.57] and were derived from computer-generated SF_6 spectra. Because of the detailed spectral structure at this temperature, these values may vary significantly in a given experiment.

The ordinate of the plot is the absorbance A. The absorption cross section can be determined from the relation $\sigma = 2.38 \times 10^{-17} A\ cm^2$. The value of σ_0 for each of the CO_2 laser lines is summarized in Table 2.4.

Higher resolution spectra and the theoretical interpretations are also available in the literature [2.54,55,65-70]. The bandhead of the Q-branch occurs at approximately 948.0 cm^{-1}. The P(16) line of the CO_2 laser is at 947.7 cm^{-1} and interacts directly with the ground-state Q-branch population at a frequency 0.3 cm^{-1} lower than the bandhead. Within this spectral region, most of the Coriolis-split rotational structure has been identified both in intensity and spectral position [2.54]. The spectral character results primarily from the ground-state Q-branch; hot bands contribute only 14% to the absorption cross section [2.55,56]. Absorption at other frequencies of the CO_2 laser is strongly influenced by the hot-band contributions to the intensity. The nature of the spectral structure is more uncertain at these frequencies, although the specific features related to the P- and R-branches of the vibrational ground state have been identified. Quantitative estimates of the specific absorption line frequencies [2.67,70] and the absorption strength [2.69,70] are available for the P(12), P(14), P(18), P(20), and P(22) lines of the CO_2 laser.

A more detailed study of the hot-band structure of the spectral absorption feature has been carried out by *Nowak* and *Lyman* [2.56]. For this study the anharmonicity shift of the absorption is assumed to be proportional to the amount of vibrational energy in the molecule. The absorption spectra predicted by this model are in good agreement with spectral absorption measurements carried out over a temperature range of 400 to 1500 K. The absorption spectrum simplifies considerably at 140 K to a well-defined P-Q-R structure that one would expect at this temperature. Quantitative values of σ_0 at several frequencies of the CO_2 laser are given in Table 2.4.

2.4.2 Fluence-Dependent Absorption at 300 K

In order to compare fluence-dependent absorption data at 300 K, we have selected measurements for which the conditions of pressure and optical pulse characteristics are similar. The CO_2 laser source was multimode for all measurements. Consequently, the 50- to 200-ns envelope of the optical pulse contains a temporal substructure of 5- to 10-ns pulses. Under these conditions, the optical bandwidth is typically 0.01 cm^{-1}. In addition to the quoted pulse length, which refers to the nominal Q-switch spike in the laser output, in many experiments this spike is followed by a long-duration, low-intensity tail. The duration of the tail is typically 1-2 μs and may contain 50% of the quoted pulse energy. As will be discussed later, this leads to additional complications in the interpretation of the absorption data. The presence of a tail on the optical pulse is noted epxlicitly for each meausrement.

The experimental conditions of the various measurements are summarized in Table 2.5. For the purpose of clarity, absorption measurements at each laser frequency are summarized in a single plot (the author's initials given in the table label the curves in the figures).

We first consider absorption at 947.7 cm^{-1} [P(16)], which corresponds to interaction primarily with the ground state Q-branch of the ν_3 vibrational mode. At this frequency, the absorption cross section has been carefully measured over a fluence range of $10^{-6} \leqq \Phi \leqq 0.1$ J cm^{-2} [2.43]. The plot of these data is shown in Fig.2.3. A least-squares fit to the data for $\Phi \geqq 10^5$J cm^{-2} yields a straight line with a slope of 0.38. Two characteristics of this curve should be noted. (a) the deviation from

Table 2.5. SF_6 absorption measurements — 300 K

Symbol	Ref.	Frequency [cm^{-1}]	Pressure [Torr]	Optical pulse width [ns]	Method[a]	Beam diameter [mm]	Optical depth correction
LL	[2.42]	947.7	0.35	150[b]	A	4-10[c]	Yes
LFF	[2.43]	947.7	0.35	150[b]	A	9[d]	Yes
		944.2					
		940.5					
D	[2.71]	947.7	0.25	180[b]	O	1.6[d]	No
		944.2					
		938.7					
HR	[2.72]	951.2-933.0	0.4	50[e]	A	9.5[f]	Yes
RTH	[2.74]	944.2	1.0	50	A	9.5[d]	Yes
BKLL	[2.53]	942.2	0.25	120[b]		3[g]	No
SSK	[2.75]	949.5	0.5	100[b]	A	3[g]	No
		944.2					
		938.7					
		936.8					
AGLM	[2.76]	944.2	0.4	90	A	-[h]	No
BKSY	[2.77]	944.2	4.0	100[b]	O	1.1[c]	Yes
AKSK	[2.78]	947.7	1.0	100[b]	O	-[h]	-[h]
		944.2					
		940.5					

[a] A — optical absorption; O — optoacoustic.
[b] Includes a nominal 1-μs tail in the pulse.
[c] Detailed beam profile measurement performed. Absorption corrected for nonuniform beam.
[d] Beam diameter measurement not specified.
[e] The temporal substructure in this pulse was later measured [2.73] to be less than 1 ns. Consequently, the peak intensity was very high, which may account for the consistently higher values of $\eta(\Phi)$ reported in [2.74].
[f] Beam diameter determined by burn spot.
[g] Focused geometry or beam divergence
[h] Not specified in reference.

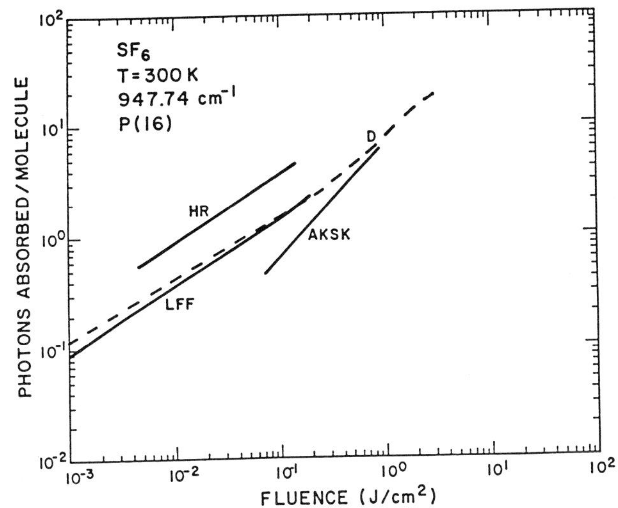

Fig.2.8. Comparison of absorption measurements at a frequency of 947.7 cm^{-1} [P(16)]. T = 300 K

linear absorption occurs at a very low value of the fluence (5 J cm^{-2}); (b) the cross section is a smoothly decreasing function of fluence.

To compare the absorption data and to gain further insight into the MPA process, it is more instructive to consider the number of photons absorbed per molecule, $\eta(\Phi)$, because it is a dimensionless quantity, than the cross section $\sigma(\Phi)$. The two are related by (2.12). The curve in Fig.2.3 is replotted in terms of $\eta(\Phi)$ in Fig. 2.8 (LFF) along with the other absorption measurements for the P(16) line. The slope of the LFF curve is 0.62. The optoacoustic data obtained by Deutsch (D) were normalized to the LFF data at $\Phi = 0.15$ J cm^{-2}. The quoted slope of this curve in the low-fluence range is 0.50 although a value of 0.60 could not be ruled out from a reevaluation of the original data. In general, the two measurements are in good agreement. The magnitude of the HR data appears to be consistently high in all comparisons with other experimental data. As indicated in Table 2.5, the optical-beam diameter was determined by a burn-spot measurement; this technique can be subject to inaccuracies. The related note in Table 2.5 could also account for the larger value of $\eta(\Phi)$. Except for the AKSK curve, the functional dependence of all the other measurements appear to be in good agreement.

Absorption curves at 944.2 cm^{-1} [P(20)] are summarized in Fig.2.9. The slopes of the different curves in the fluence range $\leqq 10^{-1}$ J cm^{-2} are in consistent agreement. The D and LFF measurements have been arbitrarily normalized at 0.06 J cm^{-2} to avoid complete overlap of the curves in a given fluence range. The error bars in the original LFF data appear to be substantial for $\Phi \geqq 0.03$ J cm^{-2}, which raises doubt about the functional behavior indicated above this value. The error limits of the SSK data would accommodate a shift to lower values of η. The HR measurement of η again appears to be consistently larger in magnitude. At $\Phi \geqq 10^{-1}$ J cm^{-2}, the slopes of the different measurements fall into two classes. The BKSY measurements appear to be

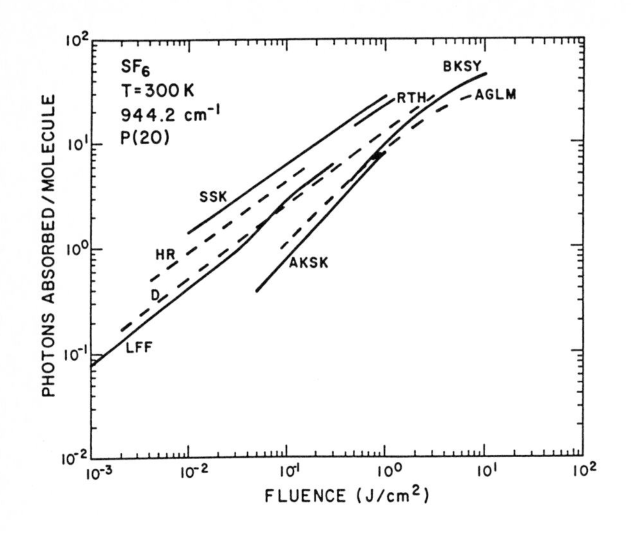

Fig.2.9. Comparison of absorption measurements at a frequency of 944.2 cm^{-1} [P(20)]. T = 300 K

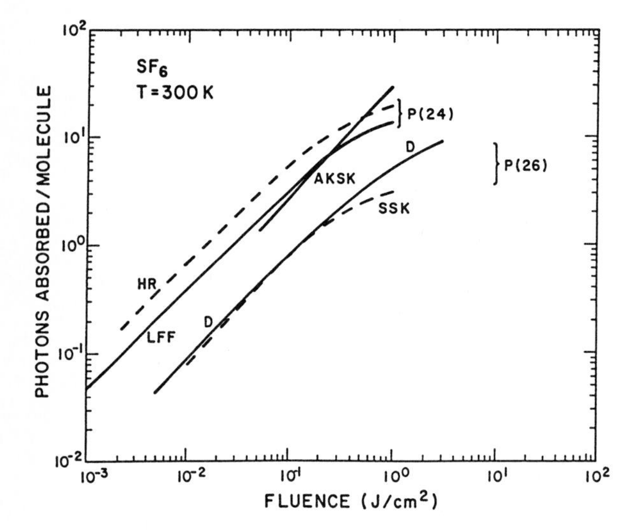

Fig.2.10. Comparison of absorption measurements at frequencies corresponding to the P(24) and P(26) lines of the CO_2 laser. T = 300 K

carefully performed and evaluated. However, a fit to the original data could also be obtained by a straight-line segment (except at the highest fluence values) with a slope that is comparable to the other curves in the plot. A best estimate for $\eta(\Phi)$ may be a slight downward shift of the D curve to agree with LFF at $\Phi \leqq 0.06$ J cm^{-2}. This would also merge with the BKSY curve at high fluence. On physical grounds, it is difficult to account for the steeper slope measured by AKSK and AGLM. A comparison of all the measurements would seem to indicate that absorption on the P(20) line in SF_6 is well described by the relation $\eta(\Phi) \propto \Phi^{\gamma}$ where $\gamma = 2/3$ over the fluence range $10^{-4} \leqq (\Phi) \leqq 3$ J cm^{-2}.

Absorption curves for the P(24) and P(26) lines are shown in Fig.2.10. In general, the functional dependence of $\eta(\Phi)$ at the P(26) frequency is consistent for the two measurements indicated. The functional behavior of $\eta(\Phi)$ at the P(24) line is

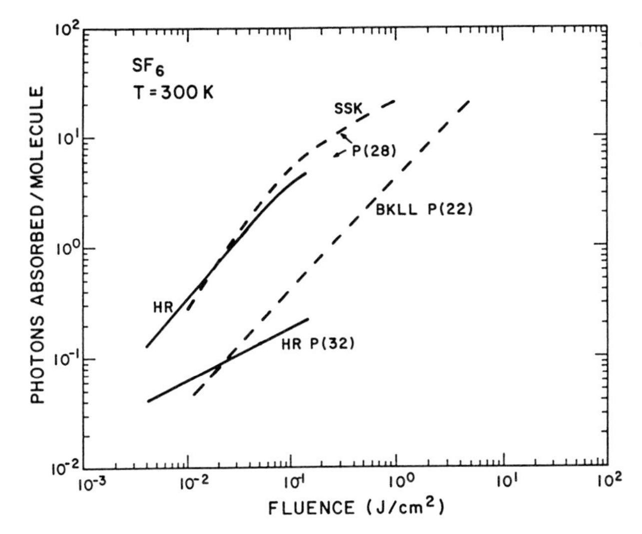

Fig.2.11. Comparison of absorption measurements at frequencies corresponding to the P(22), P(28), and P(32) lines of the CO_2 laser. T = 300 K

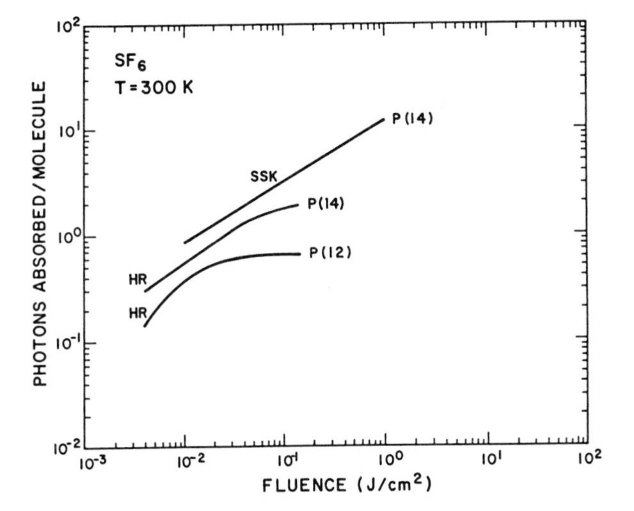

Fig.2.12. Comparison of absorption measurements at frequencies corresponding to the P(14) and P(12) lines of the CO_2 laser. T = 300 K

satisfactory at low fluence. At values of $\Phi \gtrsim 0.2$ J cm^{-2}, there appears to be a disagreement in the functional form of $\eta(\Phi)$. Further measurements will probably be required to establish the correct dependence of $\eta(\Phi)$ at this fluence. There exist only a few quantitative measurements of $\eta(\Phi)$ for other frequencies. Absorption measurements for the P(22), P(28), and P(32) lines are shown in Fig.2.11. The two measurements for the P(28) line appear to be in reasonable agreement within the error limits of the data. A comparison of $\eta(\Phi)$ for the P(22), P(20), P(24), and P(26) lines indicates that the absorption on the P(22) line may be anomalously low. This measurement using P(22) differs from that reported by HR by a factor greater than 10. Measurements for the P(12) and P(14) lines are shown in Fig.2.12.

2.4.3 Fluence-Dependent Absorption at 140 K

Only three quantitative fluence-dependent absorption measurements have been reported at 140 K. The conditions of these experiments are summarized in Table 2.6. The use of a CO_2 laser with a single longitudinal mode in the LAFF measurement resulted in an optical bandwidth that was typically 5×10^{-4} cm^{-1}. Because of the narrow spectral structure of SF_6 at 140 K, it was observed that the measurement was more sensitive to small fluctuations in the laser frequency than measurements at 300 K. Consequently, the error bars on the data are substantial, particularly for $\Phi > 10^{-2}$J cm^{-2}. A comparison of measured absorption at the P(16) line is shown in Fig.2.13. If the optoacoustic measurement D is normalized to the LAFF curve at 10^{-2}J cm^{-2}, a continuous curve is obtained for $10^{-4} \leqq \Phi \leqq 2$ J cm^{-2} provided the LAFF data for $\Phi > 10^{-2}$J cm^{-2} are deleted. The LAFF curves shown in the figure were obtained from a reanalysis of the original data; consequently these curves differ slightly from those derived in the original reference. The AKSK data is in reasonable agreement with both of the other measurements.

Table 2.6. SF_6 absorption measurements – 140 K

Symbol	Ref.	Frequencies [cm^{-1}]	Number [$\times10^{16}$cm^{-3}]	Optical pulse width [ns]	Method[a]	Beam diameter [mm]	Optical depth correction
LAFF	2.57	947.7 944.2 940.5	1.0 1.0 4.5	150[b]	A	9	Yes
D	2.71	947.7 944.2 940.5	0.9 0.9	180[c]	0	1.6[d]	No
AKSK	2.78	947.7	1.0	100[c]	0	-[e]	-[e]

[a] A – optical absorption; 0 – optoacoustic.
[b] Single longitudinal mode.
[c] Includes a nominal 1-μs tail in the pulse.
[d] Beam diameter measurement not specified.
[e] Not specified in reference.

Absorption measurements on the P(20) and P(24) lines are summarized in Fig.2.14. The optoacoustic measurements were normalized to the LAFF data to give a reasonable overlap of the curves. The error bars on the original data of the LAFF curves are sufficiently large that the functional behavior of the D curve could be easily accommodated. In general, the measurements at 140 K are in satisfactory agreement.

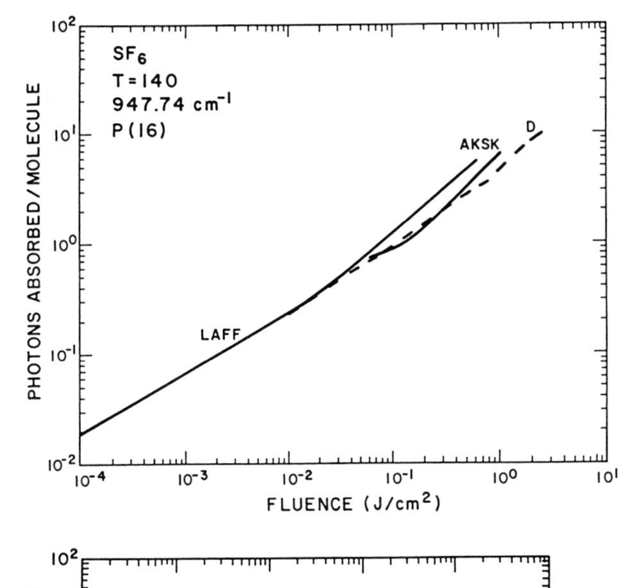

Fig.2.13. Comparison of absorption measurements at 947.7 cm^{-1} [P(16)]. T = 140 K

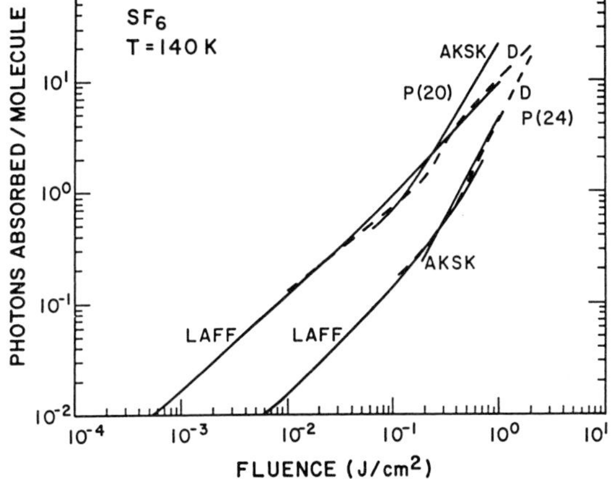

Fig.2.14. Comparison of absorption measurements at frequencies corresponding to the P(20) and P(24) lines of the CO_2 laser. T = 140 K

2.4.4 Effect of Optical Pulse Duration and Shape in MPA Processes

High-fluence absorption experiments in SF_6 have been performed with laser pulses that differed significantly in temporal character. It is instructive to compare the dependence of the absorption on the duration and shape of the optical pulse. Several studies of this effect have been reported. Although the results differ somewhat, the general conclusion appears to be that the pulse width can be varied from 0.5 ns to 1 μs with less than a factor-of-three change in the absorption.

Most absorption experiments are performed with a CO_2 laser operating with a multimode optical output beam. Although the results of the measurements are consistent, it is difficult to account for the effects of the temporal substructure in any theoretical MPE model. This situation is complicated by the existence of a long, low-intensity tail in many experiments. Detailed studies have shown [2.75] that the low-intensity tail modified the fluence dependence of the absorption cross section from that due to a single smooth pulse. This behavior is also strongly pressure dependent. Other experiments indicate, however, that the magnitude of the cross section is essentially independent of pulse duration [2.77,79].

A specific comparison of the fluence dependence of $\eta(\Phi)$ for several different optical pulses is shown in Fig.2.15. A summary of the experimental parameters is contained in Table 2.7. Back-to-back experiments were performed using a CO_2 laser with and without a low-pressure CO_2 gain cell in the optical resonator. The envelope of the pulse was the same in both measurements. Without the gain cell, the envelope contains a substantial amount of temporal substructure in the range of 5-10 ns (unsmoothed). With the gain cell, a smooth envelope results with no temporal substructure (smoothed). The results shown in Fig.2.15 (see Fig.2.1 also) indicate that the absorption is always larger for the unsmoothed pulse than for the smoothed pulse. The fluence behavior of the LFF curves for $\Phi > 0.03$ J cm^{-2} is uncertain for reasons pointed out previously. If the low-fluence curves were extended at a constant slope to high fluence, it would appear that both the U and S curves of LAFF and BKSY would result in a consistent functional dependence for $\eta(\Phi)$. The general result is that the absorption is increased about a factor of 2.3 for unsmoothed pulses over the total fluence range indicated.

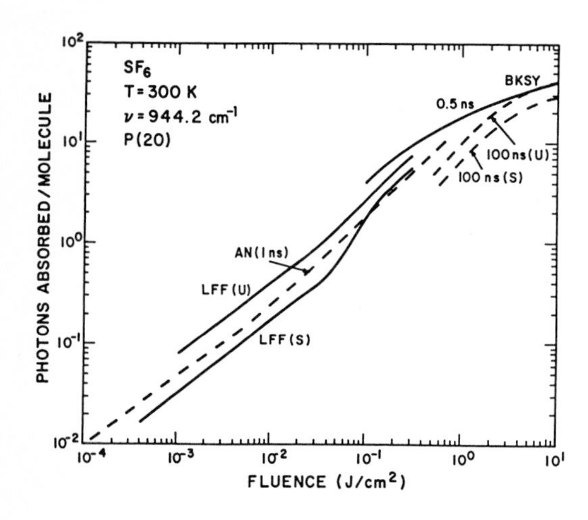

Fig.2.15. Comparison of absorption measurements obtained at 944.2 cm^{-1} [P(20)] for optical pulses with different temporal characteristics. T = 300 K

Also shown are absorption data using 0.5- and 1-ns optical pulses. Although the AN measurement was carefully performed and the data were reduced using a technique identical to that for LAFF, the absolute magnitude of the resulting curve appears to be lower than the LAFF (U). The functional dependence appears to be consistent, however. The fact that the AN data are lower than LAFF (U) is inconsistent on physical grounds and is also inconsistent with the BKSY measurements. Consistency is obtained if the AN curve is increased by a factor of 1.5.

Table 2.7. SF_6 absorption measurements – 300 K: effect of optical pulse width

Symbol	Ref.	Pressure [Torr]	Optical pulse width [ns]	Method[a]	Beam diameter [mm]	Optical depth correction
LFF(U)	2.43	0.35	150[b]	A	9[c]	Yes
LFF(S)	2.43	0.35	150[d]	A		
BKSY	2.77	4.0	0.5	O	1.1[e,f]	Yes
BKSY(U)	2.77	4.0	100[b]	O	1.1[f]	Yes
BKSY(S)	2.77	4.0	100[d]	O	1.1[f]	Yes
AN	2.80	3.12	1.1	A	4.4[c]	No

[a] A – optical absorption; O – optoacoustic.
[b] Includes a nominal 1-μs tail in the pulse.
[c] Beam diameter measurement not specified.
[d] Single longitudinal mode.
[e] Focused geometry or beam divergence.
[f] Detailed beam profile measurement performed. Absorption corrected for nonuniform beam.

From these observations, the effect of differential optical pulse durations on $\eta(\Phi)$ at constant fluence is typically one in magnitude; the functional behavior appears to be the same except at fluence levels in excess of 1 J cm^{-2}. The effect on the magnitude of $\eta(\Phi)$ at a given fluence is typically a factor of 2 to 3.

2.4.5 Discussion and Summary

Some absorption measurements in SF_6 appear to be inconsistent with respect to the functional form of $\eta(\Phi)$. In other measurements, there exists a difference in magnitude; uncertainties of a factor of three are sometimes indicated. From a review of the measurements, however, one can perhaps reduce this factor to 1.5 to 2. In view of the different measurement and data-reduction techniques, this uncertainty is predictable. We have speculated on "best estimates" for $\eta(\Phi)$ in certain cases but have preserved the data in a form that the reader can also effect a "best estimate". A summary of the authors' recommendations for $\eta(\Phi)$ at different frequencies is shown in Fig.2.16.

A comparison of $\eta(\Phi)$ at 300 K and 140 K indicates that the absorption is slightly lower at the lower temperature for the same fluence. At 947.7 cm^{-1} the functional dependence of the absorption is nearly the same at the two temperatures. This is consistent with the fact that the interaction occurs primarily with the ground-state Q-branch of the ν_3 vibrational mode rather than hot-band transitions. At other frequencies the fluence dependence is stronger at 140 K than at 300 K; for the P(24) line, $\eta(\Phi) \propto \Phi^{\gamma}$ where $\gamma = 2.0$ for $\Phi > 0.5$ J cm^{-2}. Because of the collapse of the absorption features at 140 K, the P(24) line falls in the wing of the feature. A similar fluence behavior is observed in the wings of the absorption feature at

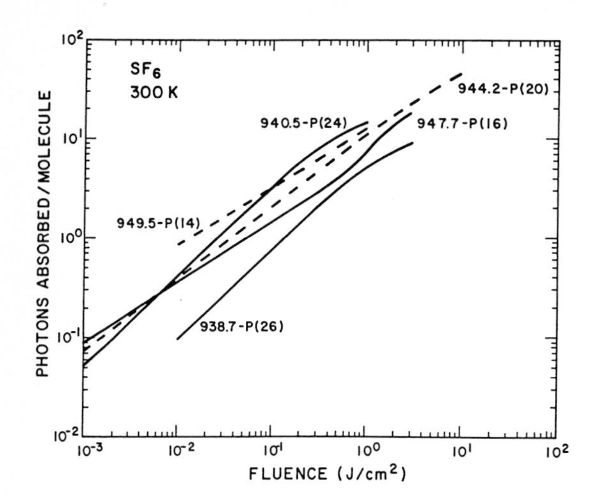

Fig.2.16. Recommended values for the number of absorbed photons per molecule at a given fluence for several SF_6 absorption frequencies

300 K; at the frequency of the P(28) line, $\eta(\Phi)$ increases with a dependence for which $\gamma > 1$. Increased temperature continues to enhance the absorption in this spectral region up to at least 800 K [2.81].

In multimode laser beams, the temporal substructure produces peak electric fields that can be 5-10 times larger than that resulting from a single-mode optical pulse of the same duration. This effect clearly must be taken into account when calculating the Rabi frequency and other field-related parameters in a theoretical model of MPE processes. Experiments have shown, however, that the electric field can vary by factors of 30 to 100 and still produce only a factor of 2 or 3 change in the values of the MPA parameters. The same conclusion results when considering dissociation.

We conclude this section with comments on experimental techniques. In many MPA measurements, the experiment is not sufficiently characterized, the data has not been reduced in a totally satisfactory manner, or the measurements have only been performed over a limited range of the experimental parameters. These limitations preclude any quantitative evaluation of the MPE process or comparison with theoretical models. The techniques to circumvent most of the above limitations are available. More careful attention to experimental technique will be needed for future progress and detailed understanding of the MPE processes.

2.5 A Comparison of Multiple-Photon Dissociation Data

We have treated in detail the absorption of laser radiation by polyatomic molecules and will now turn to a discussion of multiple-photon dissociation (MPD) experiments. Both absorption and dissociation experiments are necessary to understand the multiple-photon process. Dissociation (or isomerization) reactions probe the high-energy tail of the vibrational energy distribution; therefore, quantitative reaction measurements accompanied by accurate absorption measurements could lead to a better understanding of the vibrational energy distribution produced by multiple-photon excitation. Primary reaction product identification and reaction isotopic selectivity aid our understanding of the reaction mechanism by helping to clarify the roles of intramolecular and intermolecular vibrational energy transfer.

In this section we review experimental techniques that researchers have used to study laser-induced dissociation and isomerization reactions; we present a compilation of several molecular properties that we feel have a major effect on MPD; and finally, we assemble the available dissociation and isomerization data in a manner that assists comparing different molecular species.

In our discussion of the dissociation and isomerization data, we attempt to identify the molecular and laser properties that contribute most to the MPD process. Among the properties that we consider are vibrational-state density, bond strength (weakest bond), small-signal cross section, infrared bandwidth, laser frequency (relative to the infrared band center), specific molecular structure, laser intensity, laser fluence, laser bandwidth, and molecular anharmonicity. For some of these properties, insufficient data exist to adequately assess the effect of the property on the MPD process. We stress that our primary purpose is to determine what properties most influence the experimental results; we have not attempted to give a detailed evaluation of various theories. We do, however, investigate the relationship between absorption and reaction, and we reach some general conclusions about the nature of the vibrational energy distribution produced by the laser pulse.

We find that our analysis of the MPD experiments has greatly increased our capability for predicting experimental results.

2.5.1 Experimental Techniques

For the purpose of this discussion we divide the experimental dissociation and isomerization measurement techniques into three categories: stable-species analysis techniques, real-time monitoring techniques, and molecular-beam techniques.

Most of the early experiments, including most of those discussed in this review, employed stable-species analysis techniques. These experiments typically involve irradiating a sample of reactant gas with a pulsed infrared laser in a static reaction cell and following the irradiation with a chemical analysis of the sample. In some cases researchers used focused laser radiation, and in others, unfocused radiation, which gives more easily interpretable results. Analytical techniques that researchers have used include infrared spectroscopic analysis, gas chromatographic analysis, mass spectroscopic analysis, and pressure change upon irradiation.

The stable-species analysis techniques have several advantages over other techniques:

1) The required experimental apparatus is generally simpler and more readily available.
2) These techniques are quite general; most molecular species are amenable to this type of experiment; they require no special optical properties of the reactants or products.
3) These techniques tend to give results that are more quantitative; absolute dissociation probability measurements, for example, are easier with these techniques.

The disadvantages of the stable-species analysis techniques include:

1) Secondary thermal and photolytic reactions, and perhaps even wall reactions, may obscure the primary reaction.
2) Sample-size requirements prevent experiments under truly collisionless conditions.
3) Direct observation of the dynamics of the reaction is impossible with these techniques.

These techniques have provided, and will continue to provide, many of the basic MPD data. Wherever possible in future experiments, researchers should avoid experimental conditions that obscure interpretation, such as focused radiation and long irradiation times.

Several real-time monitoring techniques overcome some of the disadvantages of the stable-species analysis techniques. These techniques allow one to identify the primary photolysis products and to monitor the dynamics of the reaction. Calibration of the signal, however, is generally more difficult.

One technique that is proving to be extremely valuable in following reaction products in real time is laser-induced fluorescence (LIF). Several groups have used this technique to identify the primary photolysis products, to show that most of the reaction occurs rapidly, and even to monitor the vibrational state of products. Table 2.8 summarizes experiments performed with this technique.

In an experiment that used a related technique, *Karl* and *Lyman* [2.97] used chlorine atom resonance radiation to induce vacuum ultraviolet fluorescence of chlorine atoms produced from MPD of SF_5Cl. These experiments showed that this species reacts on a collisionless time scale by cleavage of the weakest bond (S-Cl).

A second real-time technique, that of observing visible or ultraviolet fluorescence from irradiated samples (reactants or products), has been in use for a number of years. In 1973 *Isenor* et al. [2.98] used the technique to demonstrate that a polyatomic molecule (SiF_4) could absorb large amounts of energy from an intense infrared field under collisionless conditions. *Ambartzumian* et al. [2.99] used fluorescence of a secondary product to show that BCl_3 dissociated isotopically selectively when irradiated with CO_2 laser pulses. They also used the technique to obtain data about several other reactions, such as the intensity (fluence) dependence of the photolysis of OsO_4 [2.100].

Table 2.8. Species study by LIF

Reactant species	Product species	Comments	Ref.
NH_3	NH_3	Reacts with off-resonance excitation	[2.82,83]
CF_2Cl_2	CF_3	Measured product T_V	[2.84,85]
CF_2Br_2	CF_2	Measured product T_V	[2.84,85]
CF_2HCl	CF_2	Measured product T_V and reaction rate	[2.33,85]
CF_3Br	CF_2	Secondary reaction produces CF_2	[2.86]
C_2H_4	C_2	Measured product T_V; C_2 is collisionless product; yield enhanced by collisions	[2.87,88]
CH_3OH	OH, CH	Measured product relaxation	[2.89,90]
CH_3CN	CH, CN, C_2	Measured product T_V and T_R	[2.91-93]
CH_3NC	CH, CN, C_2	Higher product T_V and T_R than CH_3CN	[2.94]
CH_3NH_2	NH_2	Reacts by collisional and collisionless mechanisms. Short pulses more efficient	[2.83,95,96]

Groups at the University of Southern California and at Brookhaven National Laboratory have used infrared chemiluminescence from a secondary reaction product (generally HF or HCl) to monitor MPD reactions in near real time. These groups have demonstrated prompt reaction of species such as SF_6 [2.101,102], SF_5Cl [2.102], CF_3Cl [2.102], SeF_6 [2.103], UF_6 [2.104], and several halogenated ethanes and ethenes [2.105]. These researchers determined reaction thresholds for many of these species. This is an excellent technique for obtaining this type of data; again, however, absolute signal calibration is difficult.

One real-time monitoring technique that has found limited application is to follow the growth or decay of a product or reactant with changes in its ultraviolet absorbance. This technique is only applicable to species that have strong absorption bands at convenient frequencies. *Lyman* and *Jensen* [2.106] used the technique to demonstrate the nonthermal nature of the infrared photolysis of N_2F_4 in one of the earliest quantitative laser-induced chemistry experiments. *Proch* and *Schröder* [2.107] have more recently used this method in their MPD experiments with the triatomic molecule O_3.

Several groups have performed molecular-beam MPD studies with pulsed CO_2 lasers. These groups have used a variety of techniques to verify the MPD reaction and to characterize the products. These techniques include monitoring the photolytic depletion of the molecular beam [2.108], electron ionization of the reaction products [2.109-114], reaction product detection by chemionization with Sm [2.115], and the laser-induced fluorescence technique discussed above [2.82].

These researchers reached several conclusions that have significantly augmented the understanding of MPD processes:

1) The MPD reaction may occur under completely collisionless conditions.
2) The dissociation reaction of SF_6 is to SF_5 and F.
3) For sufficient energy fluence the reaction products (SF_5) may undergo further photolytic reactions (to SF_4 and F).
4) The velocity distribution for reaction products is consistent with a statistical distribution of the vibrational energy among the available vibrational states at a given energy.
5) Many species react by MPD, and the reaction channel is virtually always the one with the lowest energy barrier.
6) The distribution over vibrational energy states is nonthermal.
7) A laser beam of sufficiently high fluence can dissociate virtually all of the irradiated molecules.

2.5.2 Molecular Properties

If we are to understand the multiple-photon dissociation of poylatomic molecules, we must certainly have available the molecular properties that may influence the absorption and dissociation process. We, therefore, have tabulated a few of the most critical molecular properties of some of the species for which quantitative measurements of reaction probability or absorption data are available. This tabulation (Appendix 2.8, Table 2.9) contains the energy of activation for the reaction and the information necessary to estimate the density of vibrational states. A third molecular property that strongly influences the absorption process is the low-intensity absorption cross section. This quantity, however, depends on the laser frequency and is not appropriate for this particular table.

Many researchers believe that the density of vibrational states has a greater effect on the multiple-photon absorption process than most other molecular properties. The density of vibrational states defines the quasi-continuum region [2.116], it determines the unimolecular reaction rate in RRKM and other theories [2.117], and it dominates the theory of intermode energy flow [2.118].

An exact computation of the density-of-states function for a given molecule requires a large amount of spectroscopic data such as fundamental vibrational frequencies, anharmonicity constants, and Fermi-resonance parameters. Many of these data are unavailable for most molecules. In many applications one may obtain the density of vibrational states by directly counting states of a molecule that is assumed to be a collection of uncoupled harmonic oscillators. This method gives a reasonably good approximation to the true density-of-states function, and it requires only a knowledge of the fundamental vibrational frequencies. However, a direct count of harmonic oscillator states becomes very time consuming for a large molecule at high levels of excitation.

Approximation methods, such as the Whitten-Rabinovitch method [2.117], work well for our purposes. This method gives

$$N(E_V) = \frac{(E_V + aE_Z)^{s-1}}{(s-1)! \prod_{i=1}^{s} h\nu_i} \left[1 - \beta \frac{dw(E')}{dE'}\right] \qquad (2.46)$$

for the density of vibrational states, $N(E_V)$, at vibrational energy E_V above the ground vibrational state. The quantity E_Z is the zero-point vibrational energy, s is the number of vibrational degrees of freedom, and the ν_i are the set of fundamental vibrational frequencies. The factor a is

$$a = 1 - \beta w(E') \quad \text{where} \qquad (2.47)$$

$$\beta = \frac{s-1}{s} \frac{\langle\nu^2\rangle}{\langle\nu\rangle^2} \quad , \qquad (2.48)$$

and the empirical function w(E') is

$$w(E') = (5.00\, E' + 2.73\, E'^{0.5} + 3.51)^{-1} \quad , \qquad 0.1 < E' < 1.0$$

$$w(E') = \exp(-2.4191\, E'^{0.25}) \quad , \qquad E' > 1.0 \qquad (2.49)$$

and

$$E' = E_V/E_Z \quad . \qquad (2.50)$$

We see that this approximation method required four parameters for each species: the arithmetic mean of the fundamental vibrational frequencies, $\langle\nu\rangle$; the geometrical mean of the fundamental vibrational frequencies, $\bar{\nu}$; the number of vibrational degrees of freedom, s; and the frequency spread parameter β. Equation (2.46) then becomes

$$N(E_V) = \frac{(s\langle\nu\rangle/2\bar{\nu})^{s-1}(E' + a)^{s-1}}{(s-1)!\,h\bar{\nu}} \left(1 - \beta \frac{dw(E')}{dE'}\right) \quad . \qquad (2.51)$$

Table 2.9 of the Appendix 2.8 lists the quantities necessary to calculate the vibrational density-of-states function from the Whitten-Rabinovitch method (2.51) for a number of molecular species. The table also lists the density of vibrational states at $E_V = 10\,500\,cm^{-1}$. We also show in Figs.2.17-19 the density of vibrational states for 15 species for the energy range between 0 and $50\,000\,cm^{-1}$. We see that the density of vibrational states depends very strongly on the number of atoms in the molecule (or the number of vibrational degrees of freedom).

One molecular property that most certainly plays a major role in determining the probability of reaction for a given laser energy fluence is the energy required to promote the reaction. For a simple dissociation reaction, such as the dissociation of SF_6 to SF_5 and F, that energy, the activation energy, is very nearly the

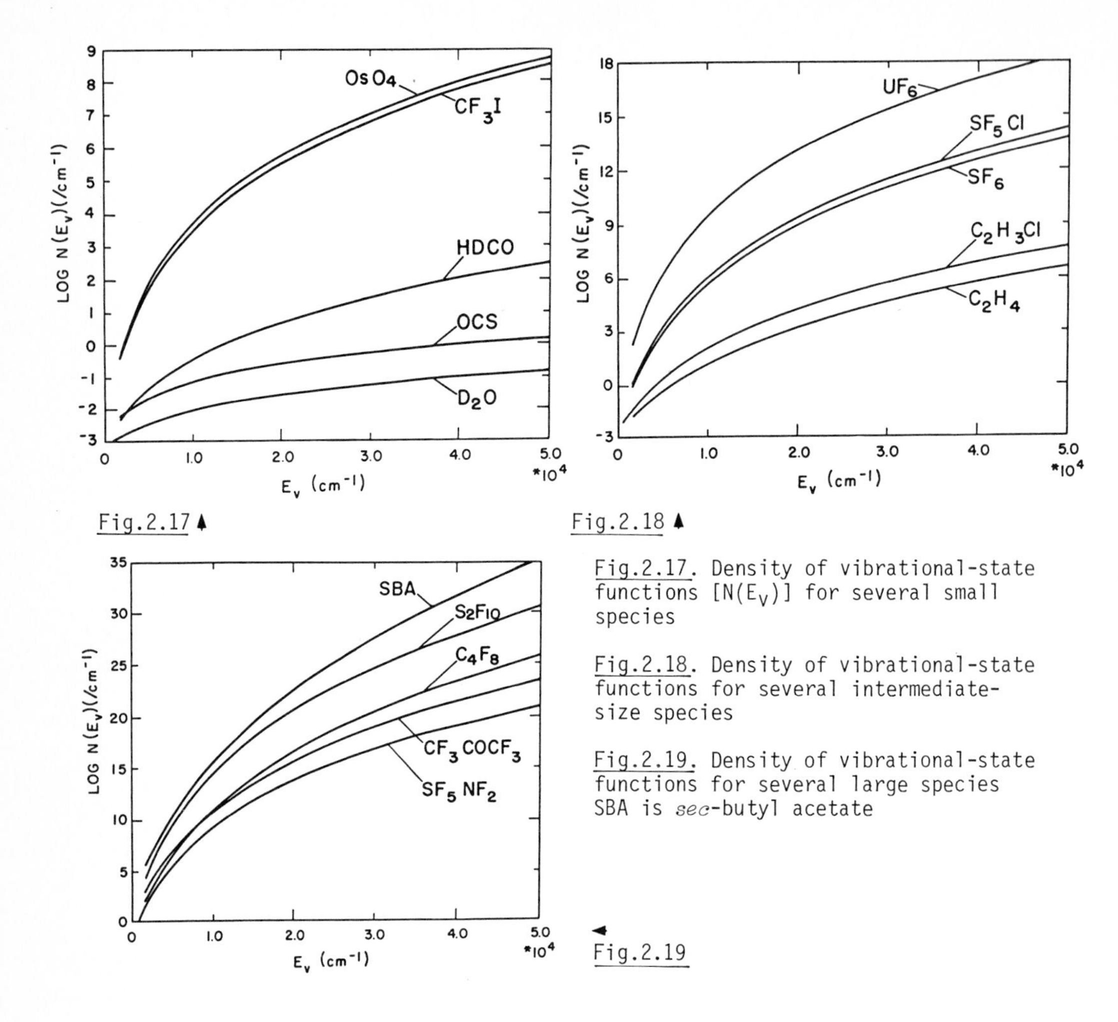

Fig.2.17. Density of vibrational-state functions [$N(E_V)$] for several small species

Fig.2.18. Density of vibrational-state functions for several intermediate-size species

Fig.2.19. Density of vibrational-state functions for several large species SBA is *sec*-butyl acetate

bond strength or the enthalpy change for the observed reaction. For more complex reactions, such as the elimination of a diatomic molecule from an ethane derivative, the activation normally exceeds the change in enthalpy for the reaction by a substantial amount.

Because the activation energy is essential for comparison of multiple-photon dissociation results for different species, we have included this quantity in the tabulation of molecular properties. The value that we have listed in the table is the Arrhenius thermal activation energy (high-pressure value) if it is available. Otherwise, we have listed the enthalpy change for the reaction if the data are available, an estimate of the enthalpy change from data on similar reactions, or an estimate of the activation energy from photochemical quantum-yield studies.

The SF_5-F bond strength, or enthalpy change, for the SF_6 dissociation reaction deserves special comment because of its dominant role in multiple-photon dissociation studies and because of the wide range of values which authors have quoted for this quantity. The primary experimental data for determining the SF_5-F bond strength are from the shock-tube dissociation study of *Bott* and *Jacobs* [2.119], several mass spectroscopic studies [2.120-122], and the recent molecular-beam study of *Kiang* et al. [2.123].

In their analysis of their own experimental data, Bott and Jacobs obtained a value of 75.92 kcal mol^{-1} for the SF_5-F bond strength. The early mass spectroscopic studies [2.120,121], gave a very low threshold for the production of F^- from dissociative electron attachment to SF_6. *Hildenbrand* [2.124] estimated the bond strength to be 78 kcal mol^{-1} from these data; however, more recent electron-attachment experiments set a higher threshold for F^- production [2.83] which in turn gives a stronger SF_5-F bond. Three more recent analyses of the shock-tube data also indicate that the bond is stronger. The RRKM analysis by *Rabinovitch* et al. [2.125] gives 86.0 kcal mol^{-1}, *Benson*'s [2.126] analysis gives 92.6 ± 3 kcal mol^{-1}, and *Lyman*'s [2.127] RRKM analysis confirm Benson's higher value. In the molecular-beam study *Kaing* et al. [2.123] obtained a value of 89.9 ± 3.4 kcal mol^{-1} from chemiluminescent reactions of Ca and Sr with SF_6. The SF_5-F bond strength is probably within 3 kcal mol^{-1} of 90 kcal mol^{-1}. We take *Benson*'s value and use his heat of formation of SF_5 [2.126] to obtain the strengths of other SF_5-X bonds.

2.5.3 Comparison of MPD Results

We feel that great benefits can accrue from comparing results of experiments with a wide range of molecular species. These comparisons would be less cumbersome if we were to define a quantity that is a sensitive measure of the efficiency of the MPD process but, at the same time, could be extracted from the available data. The quantity that we have chosen for this purpose is the laser energy fluence necessary to give a reaction probability of 1% ($\Phi_{1\%}$). This quantity, as we shall see, is sensitive to molecular properties and reaction conditions; it can be calculated, or at least estimated, from the results of experiments on many species.

The MPD data for many species are much too extensive to be summarized by a single quantity. We shall review the experiments performed on these species in greater detail. We have used the methods of the data-analysis section to tabulate $\Phi_{1\%}$ and $n_{1\%}$ for a large number of experiments (Appendix 2.8, Table 2.10).

Many experimental papers do not report quantitative reaction-probability information. For these experiments the list contains only the irradiated species and the laser frequency. For each entry that contains a value of $\Phi_{1\%}$ we also list three spectroscopic quantities: the difference between the laser frequency and the center of the infrared absorption band in that region, $\Delta\nu$; the full width at half height of the absorption band, $\nu_{1/2}$; and the small-signal absorption cross section at the laser frequency, σ_0. We obtained these three quantities from published infrared spectra of these species. Spectra of some of the species were not available to us. In some cases the experimenters also obtained laser absorption data. For these experiments we have listed the average energy per molecule at the fluence necessary to react 1% of the molecules. This energy, $n_{1\%}$, is tabulated as CO_2 laser photons (944.2 cm^{-1}) even if the experiment was performed at another frequency. The final entry in the table is the pressure at which the reaction occurred. This pressure includes the pressure of any diluent present.

We have grouped the entries in the table by the number of atoms in the molecule. The first entries refer to recent demonstrations that triatomic molecules will indeed absorb sufficient infrared laser radiation to undergo multiple-photon dissociation. In the case of O_3 and OCS the authors verified that absorption and reaction occurred under collisionless conditions. These species are the smallest for which this has been demonstrated. On the other hand, from the results of the SO_2 experiments it appears that collisions are required to induce dissociation of that species.

For most of the species listed in the table, the laser-induced reaction is dissociation by simple cleavage of the weakest molecular bond. Dissociation of SF_6 to SF_5 and F is the prototype for this type of reaction. The table refers to several molecular-beam experiments that verify this reaction. The laser-induced reaction of some of the species is elimination of a diatomic molecule from the irradiated parent such as a partially halogenated ethane or ethylene species. One of several examples listed in the table is the elimination of HF from vinyl fluoride, which *Quick* and *Wittig* [2.101] studied by monitoring HF infrared fluorescence. Isomerization is a third type of laser-induced reaction that several groups have studied. An example of this type of reaction is the isomerization of CH_3NC to CH_3CN.

In some cases there may be more than one energetically accessible reaction pathway. Reactions of this type provide an excellent opportunity to study intramolecular vibrational energy transfer dynamics. One example of this type of reaction is the dissociation of ethylvinylether to either stable or radical reaction products. In his study of this reaction, *Brenner* [2.128] found that a shorter laser pulse produced more radical products (the higher energy reaction path) than a longer pulse did. He presented this result as evidence of restriction of the rate of intramolecular randomization of vibrational energy in this species. He noted, however, (and we prefer this explanation) that there are severe entropy restrictions (lower Arrhenius preexponential factor) to the lower energy reaction pathway. The difference in reaction products for two different pulse lengths, then, simply shows that the faster pumping rate of the shorter pulse competes with dissociation by the lower energy reaction pathway.

Many investigators have searched for evidence of bond-selective, or mode-selective, laser-induced chemical reactions. While there are many examples of species that dissociate by the thermally favored pathway, we are aware of only a few claims of mode- or bond-selective MPD. In many of these cases an alternative explanation based on rapid randomization of internal energy is equally (or more) plausible. We feel, however, that the question of how much internal energy is necessary to promote rapid randomization is still unresolved.

For many species listed in the table, reaction data are available for a range of frequencies. These data are useful for determining the effect of frequency on the absorption process. Also, there are some data available on dissociation of a single species irradiated in different infrared absorption bands (CCl_4, CF_2Cl_2, CH_3OH, CH_3CN, C_2H_3Cl, SF_6, SF_5NF_2, and cyclopropane).

As we discussed above, the density of vibrational states is an important parameter in theories of absorption of intense infrared radiation as well as in theories of unimolecular reaction. The vibrational-state density determines the unimolecular reaction rate for statistical theories and it defines the quasi-continuum region. What may we learn about the role of the vibrational-state density on the MPD process from the available reaction-probability data?

We first plot the fluence necessary to give 1% reaction probability (the quantity $\Phi_{1\%}$ from Table 2.10 of the Appendix 2.8) against the density of vibrational states of the species at an energy level of 10 500 cm^{-1} [the quantity N(10 500) from Table 2.9]. This energy level is about 11 CO_2 laser quanta. This is about the average excitation necessary to give 1% reaction probability for thermally distributed vibrational energy in a sample of molecules with moderate activation energy (65 kcal mol^{-1}). This plot (Fig.2.20) contains all of the $\Phi_{1\%}$ entries from Table 2.10 for which the laser excitation frequency is within $\pm 0.5\ \nu_{1/2}$ of the absorption band center. The vertical bars show the range of $\Phi_{1\%}$ from different sources or from different frequencies within an absorption band. The open circles are for data where the infrared absorption band was very weak, i.e., where the peak absorption cross section was less than $10^{-20} cm^2$. These were, for the most part, at the frequencies of combination bands.

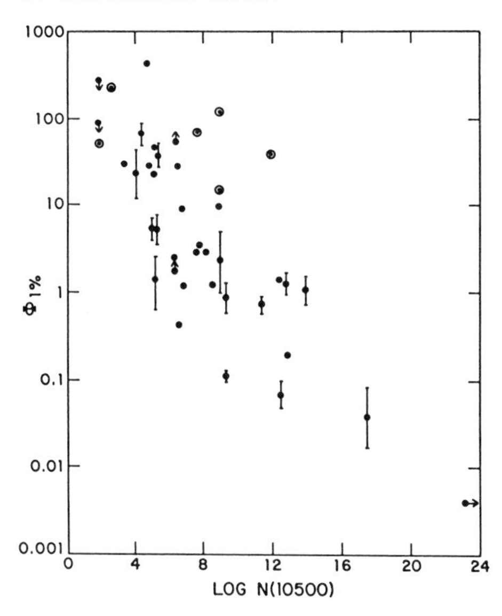

Fig.2.20. Fluence necessary to react 1% ($\Phi_{1\%}$) vs the base-ten logarithm of the density of vibrational states at an excitation level of 10 500 cm^{-1}. ↕ upper or lower limits; ⊙ weak (combination) bands; vertical bars represent range of data for a given absorption band. All data come from Table 2.10 (see Appendix 2.8)

We see from Fig.2.20 that there is a definite trend of decreasing values of $\Phi_{1\%}$ with increasing values of N(10 500). This observation is quite clear even though the spread of $\Phi_{1\%}$ at a given value of N(10 500) may be as wide as three decades. It is apparent that larger molecules react more easily by MPD than smaller molecules. This conclusion will surprise no one who is familiar with the MPD literature.

We hasten to point out that many molecular properties besides the vibrational-state density change from species to species. We see from Table 2.9, for example, that the activation energies for reaction of these species range between 20 and 142 kcal mol^{-1}. The small-signal cross sections (σ_0 from Table 2.10) also vary by several decades. The laser-induced reaction probability will most certainly depend on both of these quantities. Note, for example, that irradiation within weak bands (open circles) tends to give high values of $\Phi_{1\%}$ and that the lighter molecules tend to have higher activation energies (Table 2.9).

In an attempt to normalize the data for these two variables, we plot $\Phi_{1\%}\ \sigma_0/E_a$ versus N(10 500) in Fig.2.21 for data points where σ_0 and E_A are available. The units of the ordinate quantity were chosen to make it dimensionless. We may interpret this quantity as the energy (as a fraction of the activation energy) that would be absorbed at 1% reaction probability if the absorption cross section were independent of fluence. With these two corrections we see that the effect of the density of vibrational states is not quite so obvious, and that the vertical spread is sill nearly as great as in Fig.2.20. A careful examination of this figure and the tables in the Appendix reveals that the lower points in Fig.2.21 tend to come from the irradiation in the broader absorption bands. We would expect absorption of large amounts of laser energy to be easier at frequencies within broad bands because compensation of the anharmonic frequency shift that accompanies vibrational excitation is easier, and because the integrated band intensity (approximately given by $\sigma_0\nu_{1/2}$) is probably a better parameter for measuring interaction with the laser field at high levels of excitation than the small-signal cross section at a single frequency. This is also consistent with the results of absorption studies discussed in Sect.2.3, where the absorbed energy could be specified in terms of the two parameters σ_0 and f_r. Since $f_r \propto 1/\nu_{1/2}$, $(\sigma_0/\langle f\rangle) \leftrightarrow (\sigma_0\nu_{1/2})$.

We make a third modification to the reaction data by multiplying the ordinate quantity in Fig.2.21 by the width of the absorption band at half-peak cross section ($\nu_{1/2}$ from Table 2.10). Figure 2.22 shows a plot of this quantity, $\Phi_{1\%}\sigma_0\nu_{1/2}/E_a$, versus N(10 500). With this correction we see that the vertical spread is reduced to a little more than one decade, which is about the spread in this quantity for a small change in frequency within a single band of a single species (S_2F_{10}, for example). We also see that the trend of decreasing values of $\Phi_{1\%}$ (as modified) with increasing vibrational-state density is again readily apparent. Even after correction for the small-signal cross section, bandwidth, and activation energy, the larger molecules (higher vibrational-state density) still tend to react more easily.

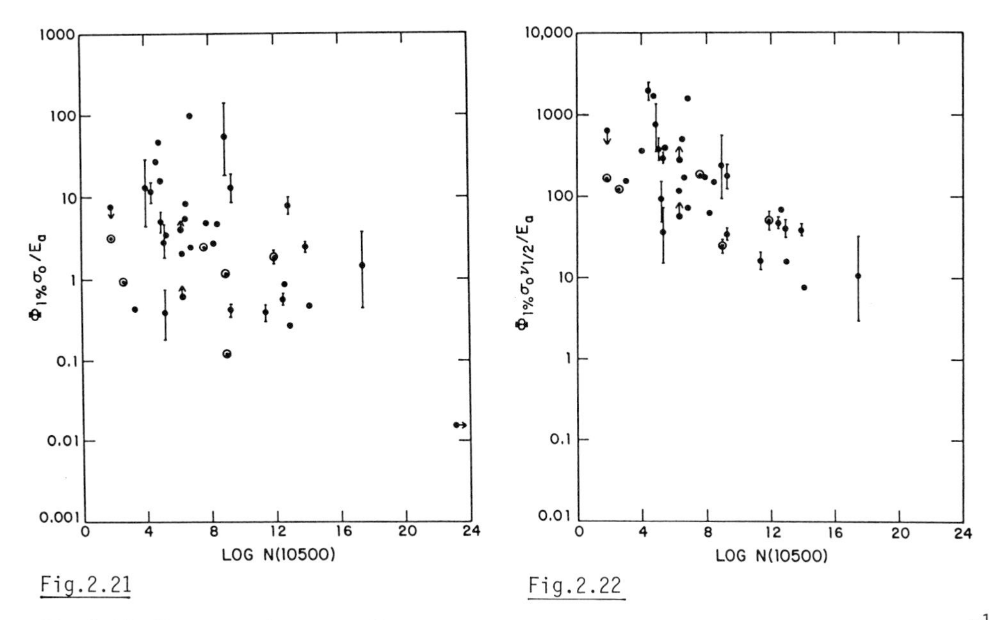

Fig.2.21. The quantity $\Phi_{1\%}\sigma_0/E_a$ vs the density of vibrational states at 10 500 cm^{-1} for all Fig.2.4 points where normalization data are available

Fig.2.22. The quantity $\Phi_{1\%}\sigma_0\nu_{1/2}/E_a$ vs the density of vibrational states at 10 500 cm^{-1} for all Fig.2.5 points where normalization data are available

We suggest several reasons for this behavior:

1) The smaller species tend to have more well-defined, narrow infrared absorption features that saturate at high intensity more easily. A higher fluence is therefore necessary to give the same amount of absorbed energy as in a large molecule. We discuss this aspect in more detail later.

2) The larger species tend to have a higher density of vibrational states than the smaller ones. The larger species will, therefore, tend to have more internal energy. (At thermal equilibrium the amount of vibrational energy in a molecule depends on the vibrational-state density function as well as on the temperature, $E_v \cong sRT$.) One would expect, therefore, that a greater energy fluence would be required to bring a sample of triatomic molecules up to the several-thousand-degree temperatures necessary to initiate reaction than to bring the larger molecules up to the several-hundred-degree temperatures necessary to induce reaction in them.

3) "Bottleneck" effects will be greater for the species with a lower density of vibrational states, which means higher intensity for the same amount of absorbed laser energy.

The rate-equation models [2.24-27] do very well for experiments with large molecules. Linear absorption is so unhindered in these large species that any effects of multiphoton transitions go undetected. Detailed treatments [2.27-30] of the laser interaction with the first few energy levels of the molecule improves agreement with experimental data. These treatments require detailed spectroscopic information and a quantum-mechanical description of excitation in the lower levels.

The technique that *Heller* and *West* [2.129] developed demonstrates that highly excited species display large-molecule characteristics. They prepared highly excited CrO_2Cl_2 molecules with a narrow vibrational energy distribution by electronic excitation. After the initial excitation a rapid intersystem crossing converted the excitation to vibrational energy in the ground electronic state. Absorption by these excited species was nearly linear, and broader in frequency than the absorption band of the unexcited species.

Several groups have explored some interesting phenomena that occur mostly in small molecules where nonlinear and "bottleneck" effects are most pronounced.

The phenomenon of self-focusing could have an impact on the interpretation of the absorption and dissociation experiments. Self-focusing and defocusing occur near infrared absorption bands in species like SF_6 [2.130]. These effects are large for some frequencies. Focusing occurs on the high-frequency side of the band and defocusing on the low-frequency side.

Electric and magnetic fields enhance the absorption at high laser intensity for small molecules [2.131] like CF_2HCl and CF_3D. The authors attribute the enhancement to a breakdown of absorption selection rules.

If absorption (and hence, reaction) depends on small-signal cross section, bandwidth, and anharmonic shift, one would expect the fluence necessary to give 1% reaction to depend on the position of the exciting frequency within the absorption band. We see from the entries in Table 2.10, for a single species with multiple frequencies, that this is indeed so. We graphically portray this dependence by plotting the quantity $\Phi_{1\%}\sigma_0/E_a$ versus $\Delta\nu/\nu_{1/2}$ in Fig.2.23 for the species with multiple-frequency data. The dashed horizontal line is what one would expect for the ordinate quantity if the species adsorbed at the small-signal cross section σ_0 and if the distribution over vibrational states were thermal. We note from this figure that the curves tend to peak near, or slightly above, the band center, and that reaction may occur well below the absorption band center.

For some species (S_2F_{10} and CF_3COCF_3) the fluence necessary to induce a reaction at frequencies well below the band center is much lower than one would expect considering only the quantities σ_0 and E_a. This most certainly represents the effect of a downward anharmonic frequency shift of the absorption frequency upon vibrational excitation.

We show two sets of SF_6 data in this figure. The upper curve is for a 0.016 Torr sample [2.132] and the lower curve is for SF_6 in a collisionless molecular beam [2.133]. We see that the two curves agree at the high and low extremes of the absorption band, but at frequencies well within the absorption band the fluence necessary to induce reaction is much lower for the molecular-beam sample. This large

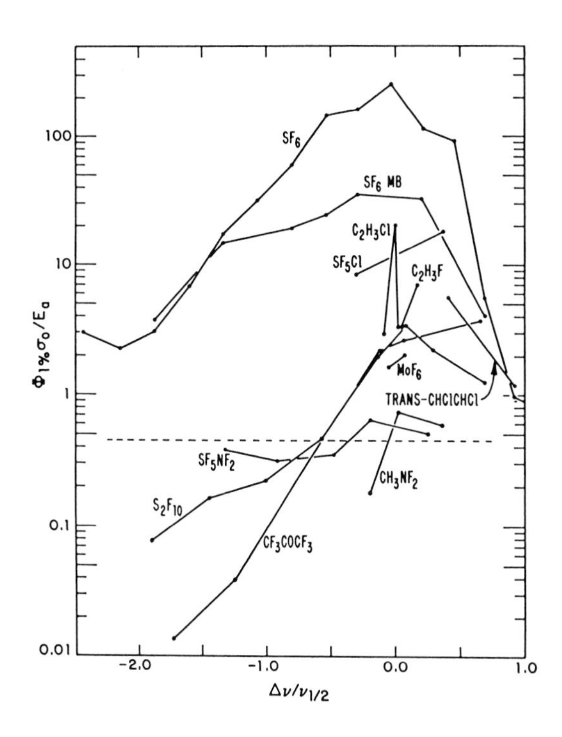

Fig.2.23. The quantity $\Phi_{1\%}\sigma_0/E_a$ vs $\Delta\nu/\nu_{1/2}$ for all multiple-frequency entries in Table 2.10 (see Appendix 2.8)

difference is an effect that requires careful theoretical examination, and perhaps experimental verification.

The SF_5NF_2 curve seems to be somewhat anomalous in that it is very near the thermal dashed line and it does not fall off at lower frequencies as rapidly as the other curves do. The major difference between the infrared spectrum of this and other species is that SF_5NF_2 has a second strong absorption band centered near -1.3 on the $\Delta\nu/\nu_{1/2}$ scale. This band not only provides a higher cross section in this region, but is also plays a major role in compensating for the anharmonic frequency shift.

Two major effects that could contribute to the deviations of the curves in Fig. 2.23 from the thermal value of 0.46 (dashed line) are, first, that the absorbed energy per molecule deviates from $\sigma_0\Phi_{1\%}$, and second, that the absorbed energy is not distributed thermally over the available vibrational states.

To distinguish between these two effects we plot in Fig.2.24 the measured amount of *absorbed* laser radiation necessary to give 1% reaction as a fraction of the activation energy, $\eta_{1\%}/E_a$, versus $\Delta\nu/\nu_{1/2}$. The dashed line is again the value of $\eta_{1\%}/E_a$ for thermal reaction. This figure shows all of the data from Table 2.10, where both absorption and dissociation were measured. This figure shows to some extent how the absorbed laser energy is distributed over the available vibrational states. For the points below the dashed line the distribution over vibrational states is broader than a thermal distribution. That is, the amount of energy ab-

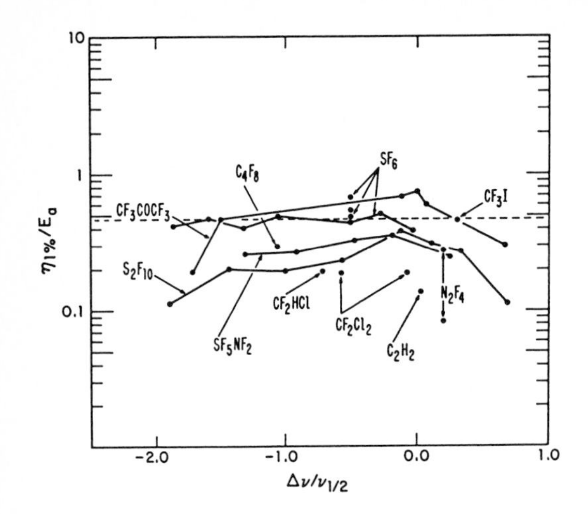

Fig.2.24. The quantity $\eta_{1\%}/E_a$ vs $\Delta\nu/\nu_{1/2}$ for all entries in Table 2.10 that list both absorption and reaction data

sorbed is less than a thermal distribution would give when 1% of the sample molecules have sufficient energy to react. The broad distribution could be due to an enhanced absorption cross section at high levels of vibrational excitation, to the fact that only a fraction of the molecules in the sample interact with the laser field, or to the tendency of the absorbed radiant energy to remain in the absorbing mode. A narrow distribution, which would give points above the dashed line, could result from a diminished absorption cross section at high levels of excitation. Points above the dashed line could also result from a nonnegligible recombination rate. The clustering of most of the points around the dashed line indicates that most of the vertical scatter in Fig.2.23 is due to differences in how much laser energy is actually *absorbed*, and that differences in how that absorbed energy is distributed are smaller, but not negligible.

→

Fig.2.25. Reaction probability vs η for CF_3I at 1074.6 cm^{-1} [2.50]. *Solid line* is the thermal reaction probability calculated by the method discussed in the text from the data of Table 2.9

Fig.2.26. Reaction probability vs η for N_2F_4 at 944.2 cm^{-1}. Point marked "1" is the long-pulse point [2.106,134] and all others are short-pulse (15 ns) points [2.135]. *Solid line* is thermal curve

Fig.2.27. Reaction probability vs η for SF_6 at 944.2 cm^{-1}. "1" = 100-ns single-mode pulse [2.136]; "2" = 100-ns single-mode pulse [2.77]; "3" = 1-ns pulse [2.77]; "4" = threshold measurements for several frequencies with long multimode pulses [2.73]. *Solid line* is thermal curve

Fig.2.28. Reaction probability vs η for SF_5NF_2 [2.46]. "1" = 908.5 cm^{-1}; "2" = 920.8 cm^{-1}; "3" = 934.9 cm^{-1}; "4" = 944.2 cm^{-1}; "5" = 957.8 cm^{-1}. Absorption-band-center frequencies are at 950 and 910 cm^{-1}. *Solid line* is thermal curve

Fig.2.29. Reaction probability vs η for CF_3COCF_3. "1" = 934.9 cm^{-1}; "2" = 940.5 cm^{-1}; "3" = 944.2 cm^{-1}; "4" = 969.1 cm^{-1}; "5" = 970.6 cm^{-1}; "6" = 971.9 cm^{-1}; "7" = 983.3 cm^{-1}; "8" = 985.5 cm^{-1}. All points are from [2.52] except for "5" which are from [2.15]. *Solid line* is thermal curve

Fig.2.30. Reaction probability vs η for S_2F_{10} [2.42]. Frequencies for numbers 1-8 are, respectively, 921.0, 925.1, 929.1, 933.0, 936.8, 938.7, 940.6, and 944.2 cm^{-1}. *Solid line* is thermal curve

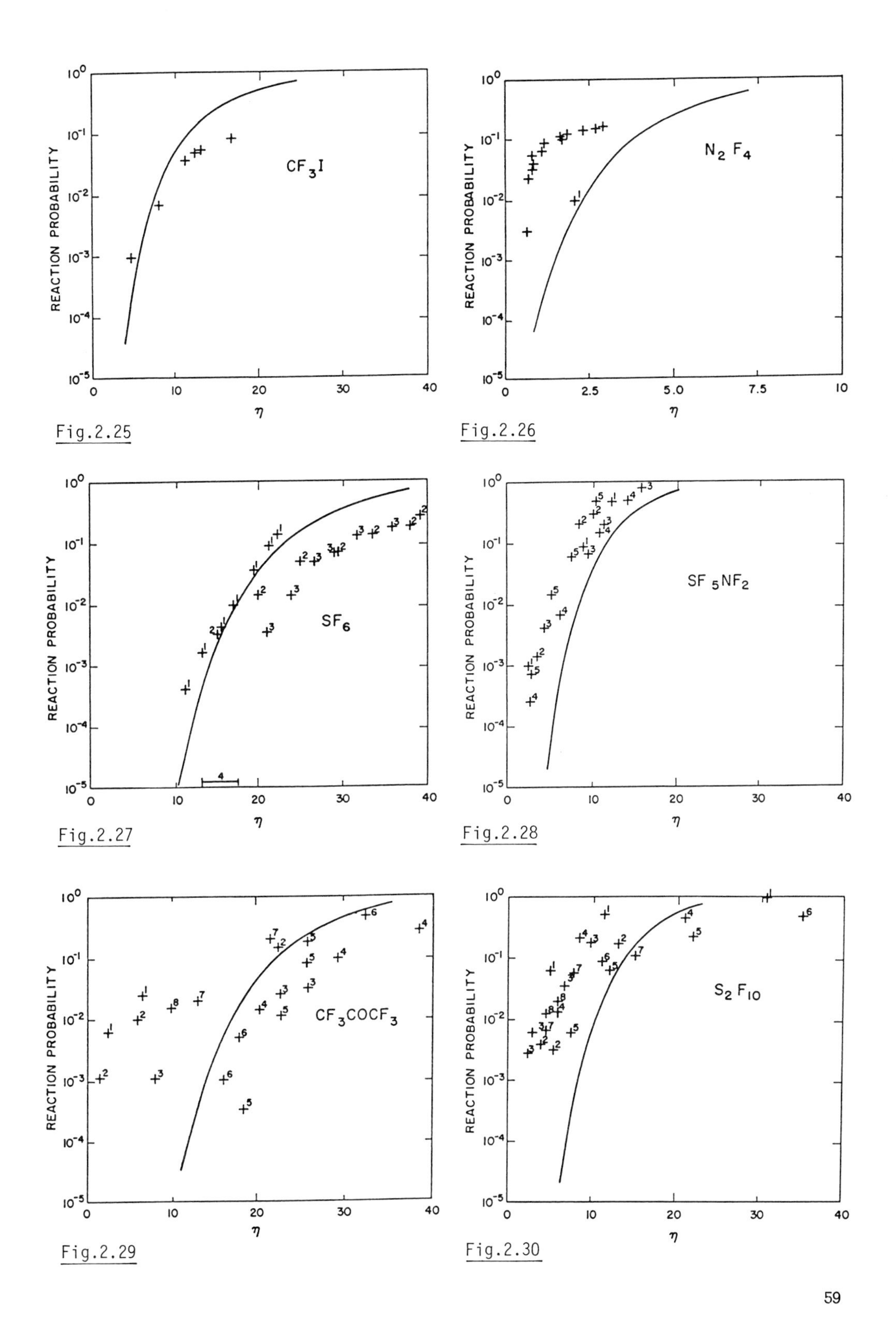

Fig.2.25

Fig.2.26

Fig.2.27

Fig.2.28

Fig.2.29

Fig.2.30

For some of the species in Fig.2.8 more complete absorption and dissociation data are available. For these species we compare (Figs.2.25-30) reaction probability versus energy absorbed with the thermal reaction probability curves. The energy unit η is the CO_2 laser quantum at 944.2 cm^{-1}.

The thermal curve was obtained by first calculating the vibrational-energy distribution function

$$n(E_V) = n_0 N(E_V) \exp(-E_V/kT_V)/Z(E_V) \quad (2.52)$$

for a range of vibrational temperatures. The quantity $n(E_V)$ is the density of molecules with energy E_V per unit energy; n_0 is the total molecular density; $N(E_V)$ is the density of vibrational states calculated from the parameters in Table 2.9; and $Z(E_V)$ is the vibrational partition function. One may easily convert the vibration temperature T_V in (2.52) to absorbed energy η. We assume that the thermal reaction occurs if a molecule has sufficient vibrational energy to react. So by integrating (2.52) for E_V greater than the activation energy (Table 2.9) we obtained the thermal curves in Figs.2.25-30. This estimation of the extent of thermal reaction is clearly not as good as a complete RRKM treatment, but it is sufficient for our purposes, which are to provide points of reference.

We see from these figures that the experimental reaction probabilities generally have shapes that are similar to the thermal curves, but in most cases the experimental curves deviate significantly from the thermal curves. The frequency of the exciting radiation is an important parameter. The reaction probability curves tend to be close to thermal for frequencies near the band centers (for S_2F_{10}, CF_3COCF_3, and SF_5NF_2); however, for frequencies in the high- and low-frequency extremes of the bands, the absorbed laser energy produces reactions much more efficiently. We suggest that in the band extremes the laser interacts with a smaller fraction of the molecules.

The shape (or intensity) of the laser pulse also affects the reaction probability. The SF_6 molecules (Fig.2.27) must absorb significantly more laser energy from the short pulse to produce the same reaction probability than from energy absorbed from the long pulse (a more complete analysis of these data is in [2.77]). For N_2F_4 we see the opposite effect. It is likely that the significant variable in these experiments is frequency bandwidth, not intensity (Sect.2.3).

The distribution of total energy in the CF_2 fragment from CF_2CFCl dissociation 2.24 supports the view of a statistical distribution of internal energy. This is the basic underlying assumption of the RRKM theory of unimolecular reaction rates. The results of MPE experiments [2.24-26] generally support that assumption. *Bagratashvili* et al. [2.137] have used a Raman-spectroscopy technique to set an upper limit of 20 ns for the time (collisionless) for vibrational energy randomization in SF_6. Examples [2.138-140] of restricted randomization involve either small molecules or low levels of excitation. The statistical assumption does not really apply in those situations.

2.5.4 Concluding Remarks

Our primary objective in this review of multiple-photon dissociation experiments was to determine the molecular and laser properties that most strongly influenced the MPD process. We have identified and characterized many of these properties. Those that determine how much energy a sample absorbs are the most important. The previous sections on absorption of laser radiation, therefore, contribute significantly to our understanding of such reactions. Many properties influence not only how much energy a sample of molecules absorbs, but also how that energy is distributed among the molecules of the sample and within a given molecule. The reaction experiments allow us to refine our ability to predict reaction probability because they are sensitive to the details of the vibrational energy distribution.

From this analysis of the reaction experiments we have shown in a general way how vibrational-state density, small-signal cross section, bond strength, infrared bandwidth, laser frequency, and to some extent, laser intensity influence reaction. Given these properties, the analysis we have presented would allow one to predict the result of an MPD experiment with a fairly high degree of certainty, and one could augment this predictive capability by employing the ideas in Sect.2.3 of this review.

Some recent papers report a temperature effect on MPE processes. High temperatures generally enhance absorption and dissociation but degrade isotopic selectivity. For very large molecules increasing the temperature greatly increases the level of vibration excitation. This, in turn, increases the reaction probability [2.141]. For some species [2.7,12,14,18] decreasing the temperature improves the isotopic selectivity. An exception [2.142] is CF_2Cl_2 at frequencies well above the center of the absorption band.

We feel that variation of the collision rate (pressure) is another useful tool for increasing our understanding of the absorption and reaction processes. We treat pressure effects in greater detail in the following section.

2.6 Collisional Effects in Multiple-Photon Absorption and Dissociation Processes

In this section we review the effect of collisions on multiple-photon absorption and dissociation. In general, gas-kinetic collisions promote both a redistribution of the molecular energy and a dephasing of the coherence of the optical interaction. Since the effects can also occur in a collisionless regime, it is not always clear in a given set of experiments which specific contributions arise from collisions and at what pressure a collisionless regime obtains. Experiments indicate pressure effects under conditions where the collision times are much longer than the time scale of the measurement. Some investigators believe that the collisionless regime is only approached under molecular-beam conditions.

Other more practical reasons for studying the effects of increased pressure on the MPA or MPD processes include the development of economical laser-induced chemical processes, such as isotope separation. The effects of diluents on infrared laser energy absorption or the effects of radical scavengers on the MPA or MPD process itself, for example, are serious considerations. All of the 100 or so papers consulted for this portion of the review discussed directly the role of collisions in the outcome of their respective experiments. It will be clear from what follows that a good deal more of the results in this field have perhaps been influenced by collisional effects.

Table 2.11 shows a list of the molecules from which a study was made of the effect of collisions on absorption, dissociation, or isotopic selectivity. The arrow upward (downward) implies that increasing the pressure of the buffer gas listed next to it caused an increase (decrease) in the quantity at the top of the column for the molecule at the left. In some rare cases the arrow is horizontal, indicating no change. In other cases there are two arrows side by side. This implies that as the pressure increased from zero, the quantity of interest first went one way and then the other. Also, R denotes the reactant (the species that is being pumped by the laser).

Some preliminary observations include: First, increasing the pressure is detrimental to the isotopic selectivity in all cases but one. Second, an increase in pressure results in more energy being absorbed. Third, there is no obvious general trend concerning the effect of collisions on the laser-induced reaction. Because of the sensitivity of reaction to the vibrational energy distribution, we expect that the laser-induced reaction would be dependent on the experimental details, such as laser fluence, molecular absorption cross section, and the actual pressure range of the study.

Collisions play a variety of roles in the MPA or MPD process. Each of these roles is a subset of the more general effect of driving the energy distribution of the sample towards the statistically most likely (or thermal) distribution. The following is a short list of specific collisional roles, some of which, along with others, will be discussed in the following sections.

1) Rotational relaxation will couple states that are not being pumped by the laser to those that are (rotational hole filling).

2) Dephasing or phase-interruptive collisions will destroy any coherent interactions between the laser field and the molecule.

3) Collisions can quench the vibrational energy into rotation or translation, and thus inhibit dissociation by a purely vibrational energy route and create a strong possibly for thermally induced effects.

4) Because intermolecular vibration-to-vibration energy transfer collisions tend to conserve the total system vibrational energy, these collisions may either enhance

or diminish the reaction rate. The vibration energy distribution produced by the laser determines which occurs.

5) The effect of collisions that randomize the vibrational energy within a single molecule (intramolecular V-V transfer) will depend on the rate and extent of collisionless randomization processes. If the collisionless processes are rapid and complete, this type of collision will have no effect.

In what follows, we will discuss the effect of collisions on the absorption process, the dissociation process, and the threshold for reaction. There will then be a discussion of laser-induced isomerization and molecular elimination reactions, followed by a discussion of the effects of secondary processes.

2.6.1 Collisional Effects on Multiple-Photon Absorption

Several investigators (Table 2.11) have shown that the addition of a nonabsorbing buffer to the irradiated sample causes an increase in the infrared energy absorbed. This phenomenon has been investigated in detail for SF_6 [2.143,144]. Figure 2.31 shows the results for the dependence of the average number of photons absorbed by SF_6 on the pressure of the added diluent (in this case, N_2) for a number of different values of the CO_2 laser fluence. For low or moderate fluences, there is a dramatic increase in the number of photons per molecule absorbed by the SF_6 as the N_2 pressure is increased. At high fluences, above 1 J cm^{-2}, there is very little effect. The slope of the high-fluence curves becomes steeper as the diluent becomes lighter [2.145]. This indicates that above a certain degree of excitation for SF_6, collisions are simply causing $V \rightarrow T$ relaxation and a cycling through a single transition. The lighter collision partners are more effective than one would expect [2.146,147].

The low-fluence behavior can be explained by the phenomenon of rotational hole filling, treated in Sect.2.3.1. The rotational relaxation by the buffer gas causes an increase in the number of molecules interacting with the field by replenishing the supply in rotational states pumped by the laser. This hole-filling process could also apply to the vibrational states where the laser excitation upsets the equilibrium between the pumped level and the other levels with which it is otherwise in thermal equilibrium. This effect is not as important as the rotational hole filling.

As the intensity of the laser is increased, a greater fraction of the rotational states may interact with the radiation field. One would therefore expect the rotational hole-filling process to enhance absorption at low fluence (or intensity) but not at high fluence. Figure 2.31 shows that the changes in the energy absorbed by SF_6 as the pressure increases are negligible for fluences near 1 J cm^{-2} and above. This observation is consistent with an independent experimental determination [2.148] that the fraction of molecules interacting with the laser pulse approaches unity near this fluence.

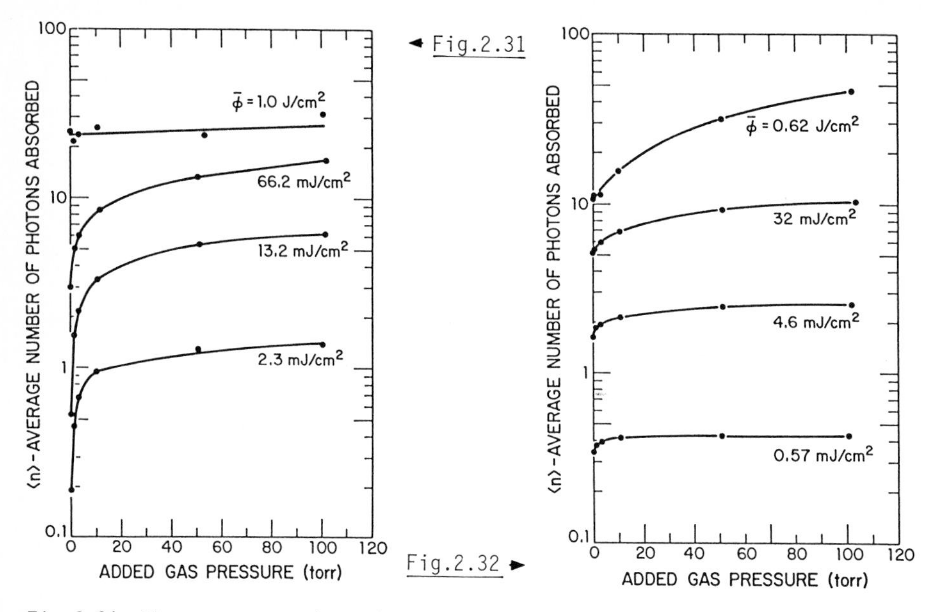

Fig.2.31. The average number of photons absorbed by SF_6 (0.1 Torr) as a function of added diluent (N_2) pressure for different values of CO_2 laser fluence. Laser operating on the P(20) 001-100 transition

Fig.2.32. The average number of photons absorbed by S_2F_{10} (0.1 Torr) as a function of added He pressure for different values of CO_2 laser fluence. The laser line was P(26) of the 001-100 band

As a second example of collisional effects in the MPA process, consider Fig.2.32. This shows the average number of photons absorbed per S_2F_{10} molecule as a function of added He buffer-gas pressure (experiments with Xe buffer gave identical results). The behavior here is somewhat different from the SF_6 case. As the fluence increases, the magnitude of the collisional effect increases. This is opposite to the effect for SF_6. In S_2F_{10}, there is only a slight contribution to the increased absorption at the lower fluences from rotational relaxation. The dissociation probability becomes significant at a fluence between 10^{-2} and 10^{-1}J cm^{-2}, and for a fluence of 0.6 J cm^{-2}, 50% or more of the S_2F_{10} is dissociated [2.42]. In an absorption measurement, this loss of S_2F_{10} during the course of the laser pulse would actually show up as an apparent decrease in absorption. Thus, we suggest that the addition of a buffer gas quenches the dissociation, and thus maintains a higher concentration of absorbers during the laser excitation resulting in an increase in the absorbed energy. The laser apparently produces a vibrational energy distribution that is broader than a thermal distribution with the high-energy tail above the dissociation limit. Collisions tend to deactivate highly excited species and bring the distribution closer to a thermal one. The fact that Xe and He gave the same pressure dependence reflects the lack of mass effect for V-T processes with large highly excited species [2.149].

For D_2CO, collisions are apparently necessary [2.63] to induce absorption to dissociation. The nonlinear and coherent processes are not sufficient for dissociation, at least for the laser fluences used in the experiments of *Orr* and *Haub* [2.63]. They found that collisional redistribution of both rotational and vibrational energy were necessary for dissociation. However, these processes occurred at rates that were faster than gas kinetic for the former and one-third gas kinetic for the latter.

The pressure of a diluent gas may influence absorption, dissociation, and isotopic selectivity. Increased pressure generally increases absorption and dissociation [2.1,10,11,58,150-154] for small molecules (mainly due to rotational relaxation) in the low-pressure range. For higher pressures and for larger molecules [2.26,141,155] vibrational deactivation begins to dominate. Consequently, pressure increases result in a reduction of dissociation probability. In one set of experiments [2.156] substituting three deuterium atoms for hydrogen atoms in CH_2ClCH_3 changed the absorption characteristics from that of a small molecule to a large molecule.

Generally, pressure increases degrade isotopic selectivity [2.1,2,157,158]. However, a pressure increase may actually improve the isotopic selectivity. In one example [2.159] the authors suggested that the higher collision rate suppressed radical products from chloroethane while favoring stable molecular products. Increased diluent pressure sometimes improves the isotopic selectivity for frequencies on the low [2.3,160] side of the absorption band.

Some molecules show no collision effects. For example, absorption by *sec*-butyl acetate [2.145] shows no dependence on pressure at fluences up to about 2 J cm^{-2}. Intramolecular energy redistribution in this large molecule is expected to be much faster than any collisional effects at the usual pressures considered in these experiments.

There are many possible explanations for a given collisional effect. For example, in [2.48] the increased absorption in vinyl chloride in the presence of a helium buffer is attributed to the pressure broadening of absorption lines, which together with power broadening, overcomes the saturation of the ground-state absorbing transitions and permits more of the incident laser energy to be coupled into the system. No mention is made of the possible contributions from rotational relaxation, which are probably nonnegligible.

An interesting experiment reported by *Akhmanov* et al. [2.79] investigates the apparent role of rotational relaxation in their absorption process. This was done by measuring the energy absorbed SF_6 at a constant fluence as a function of the laser pulse duration (15-ns to 1.5-μs duration). At an SF_6 pressure of 0.25 Torr, the rotational relaxation time was given as 160 ns [2.161,162], and therefore the laser pulse-length variation corresponded to a transition from a collisionless regime (15 ns) to a regime where rotational relaxation was extremely efficient

(1.5 μs). The fact that the energy absorbed was constant (within 30%) over the entire range of pulse widths was taken as evidence that rotational relaxation does not contribute significantly to absorption. Similar experiments were conducted in C_2H_4 and CH_3F. The claim was that above approximately 1 mJ cm^{-3} there is a purely radiative depletion of a large number of rotational states even though the Rabi field-broadening effect is much less than the width of the vibrational band. These results would appear to negate the previous discussion, particularly that which is relevant to the SF_6 results.

If we consider the coupled-two-level-oscillator model presented in Sect.2.3.1, we see that absorbed energy for the conditions of these experiments is proportional to $\langle f \rangle$ (Table 2.1, high-flux case). The effective fraction of molecules interacting with the laser, $\langle f \rangle$, is in turn a function of the quantity $f_r\tau_p/\tau$, where f_r is the fraction of molecules interacting with the laser field, τ_p is the optical pulse length, and τ is the collisional equilibration time. If, in the Akhmanov experiment, the field interaction bandwidth is not determined by the Rabi bandwidth but is instead dependent on the actual bandwidth of the laser, then f_r is proportional to $1/\tau_p$, which exactly compensates for the increase in $\langle f \rangle$ due to the pulse length τ_p. Under these conditions, one would not expect to see a change in the absorption with changing pulse width.

A detailed treatment of measurements of the type shown in Fig.2.31 in the context of the coupled-two-level absorber described in Sect.2.3.2 is given in [2.41]. This treatment gives

$$\eta(\Phi) \propto \langle f \rangle^{1/3}(\sigma_0\Phi)^{2/3} = \left[1 - \exp\left(-\frac{f_r}{1-f_r}\frac{\tau_p}{\tau}\right)\right]^{1/3}(\sigma\Phi)^{2/3} \tag{2.53}$$

for the dependence of $\eta(\Phi)$ on f_r, τ, and τ_p for the conditions of high fluence and low to moderate pressure. The derivation of (2.53) from (2.44) involved taking the high-fluence limit of (2.42) (Table 2.1) with the observation that G(x) in (2.44) is $x^{2/3}$ at high fluence. According to (2.53) the dominant effect of collisions is to determine the number of molecules that interact with the radiation field. If the gas pressure is sufficiently low that $f_r\tau_p/\tau \ll 1$, then $\eta(\Phi) \propto p^{1/3}\Phi^{2/3}$. The validity of these results is subject to the condition that $\tau_p/\tau \gtrsim 1$.

Measurements of $\eta(\Phi)$ as a function of the Φ and the pressure of the xenon buffer gas for the P(16) and P(20) lines of the CO_2 laser are reported in [2.144]. The most direct quantitative comparison can be effected for the P(16) line since f_r has been calculated for the Q-branch of the ν_3 vibrational mode (Table 2.3) and the effects of hot-band contributions are minimal. The rotational equilibration time for SF_6-Xe collisions has not been measured. As a first approximation, we consider it to be approximately the same as the SF_6-SF_6 collision time of 100 ns Torr. At 0.5 Torr and a laser pulse duration of 150 ns, $\tau_p/\tau \cong 1$. The value of f_r relevant to the experiment is $f_r = 5 \times 10^{-3}$. Consequently, at a fluence $\Phi > 10^{-2}$ mJ cm^{-2}, the conditions for the validity of (2.53) are satisfied and a comparison of this expression with the SF_6 measurements is justified.

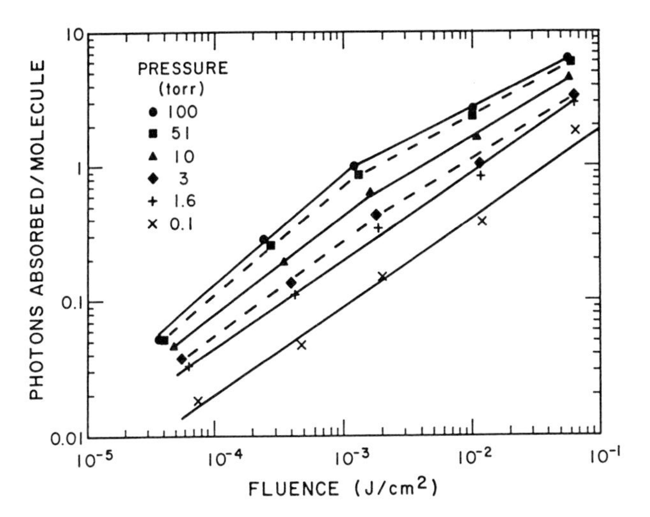

Fig.2.33. Dependence of $\eta(\Phi)$ on fluence for SF_6 at several pressures of xenon buffer gas. The pressure of SF_6 is 0.1 Torr. The irradiation frequency is 947.7 cm^{-1}, which is near the peak of the ground-state Q-branch of the ν_3 vibrational mode

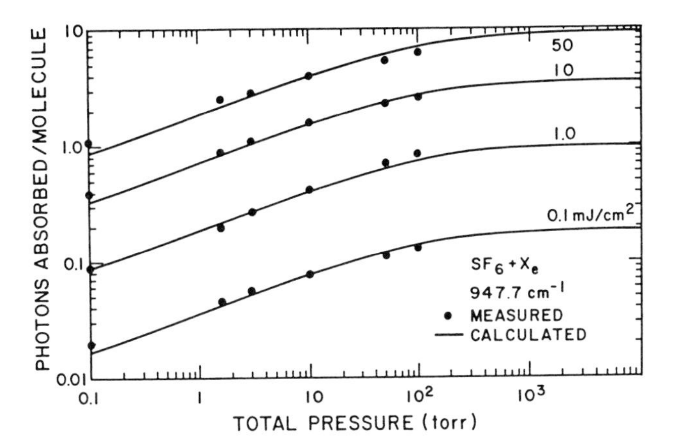

Fig.2.34. Dependence of $\eta(\Phi)$ on pressure for a constant fluence. The points are the experimental data; the *solid curves* are the calculated dependences using (2.42,44)

Figures 2.33,34 show the original data in a form that displays the functional dependence on fluence and xenon gas pressure. The measured data in Fig.2.33 indicates that $\eta(\Phi)$ is proportional to $\Phi^{2/3}$ for $0.1 \leqq P \leqq 10$ Torr. In the pressure range $10 \leqq P \leqq 100$ Torr, there is a small departure from the $\Phi^{2/3}$ dependence. The $p^{1/3}$ dependence for $\eta(\Phi)$ is clearly evident in Fig.2.34 for $P < 10$ Torr. The solid (and dashed) curves in Figs.2.33,34 represent calculations of η from (2.53). We see that this approach provides an excellent fit to the experimental data.

In view of the above discussion, it appears that rotational relaxation may be a dominant collisional effect for infrared laser absorption in polyatomic molecules, particularly when pumping ground-state molecules. For molecules already excited, collisional vibrational energy transfer processes may also play a significant role by altering the population distribution among excited states.

2.6.2 The Influence of Collisions on Multiple-Photon Dissociation

A simple scheme that is often used to describe the MPD process [2.77,163] is to divide the mechanism into three regions. Region I is characterized by multiple-photon excitation in discrete levels where the density of vibrational states is small. The vibrational quasicontinuum (region II) is characterized by a high density of vibrational states. In region III, the molecule has absorbed enough energy to react. Further optical excitation above the reaction threshold may occur, and statistical unimolecular reaction theories indicate that the rate constants for reaction will be strongly dependent on energy content. De-excitation processes will compete with the optical excitation and unimolecular decay. Thus, the intensity of the laser field is important in region III because it determines the optical excitation rate, which in turn determines, under certain circumstances, the unimolecular decay path to be followed as well as the energy content of the products. A discussion of these ideas and experiments to test them appear in [2.164].

The effect of collisions on the MPD process can be broken down into their effects on the three regions. In region I, collisions will couple the discrete levels, those which interact directly with the laser field, with those which do not, by means of rotational relaxation as discussed before in relation to the MPA process. Also, the increased collision frequency will destroy any coherent interaction at sufficiently high pressure. Thus, intensity effects in the absorption process have been shown [2.43,136] to be small for SF_6, and therefore the coherent processes may not be significant. Results for hexafluoracetone [2.165] suggest that more careful experiments may reveal the subtleties of the coherent processes, an area which needs more refined and careful experiments.

The rate of energy flow among the internal degrees freedom will influence the role of collisions in region II. There are large molecules that have extremely high densities of vibrational states even at the room-temperature level of excitation. Such a molecule is *sec*-butyl acetate, which has 54 normal modes of vibration. There is no pressure-induced change in the absorption of infrared laser photons for this molecule [2.145]. There is apparently no restriction to absorption in this region which can be altered by collisions. Two theoretical papers [2.33,166] address the effect of rapid intramolecular energy transfer on absorption in this region.

Except for the case of a very narrow vibrational energy distribution, vibration-to-translation energy transfer will adversely affect the MPD process in all regions. It is probably most destructive in region III, near the dissociation threshold, where optical-pumping rates may be insufficient to give a large unimolecular dissociation rate. Chemical-activation studies [2.167] indicate that V-T processes in highly excited molecules are extremely rapid. One collision may remove an amount of energy corresponding to one CO_2 laser photon. Thus, although collisions may assist the MPA process by rotational relaxation, they will undoubtedly destroy the MPD process by $V \rightarrow T$ relaxation away from the reaction threshold at sufficiently high pressure.

As can be seen from Table 2.11, the measured dependence of collisional effects on the reaction probability depends on the molecule irradiated and the collision partner. If the molecular/laser parameters are such that the molecules are near the reaction threshold, then collisions will probably quench the excitation. Thus one finds, under certain circumstances, that as the pressure increases, there is an increased dissociation rate as more of the molecules are coupled to the field by collisional processes; but as the pressure increases beyond the point where the interaction is optimized, the dissociation rate decreases due to collisional quenching.

Experiments indicate that increased pressure is not conducive to increased isotopic selectivity (Table 2.11). When the reactant pressure is increased, intermolecular V - V transfer between the laser-excited isotope and the unexcited ones enhance excitation of the unwanted isotope. Also, V - T quenching processes by any buffer gas will increase the probability of thermal dissociation, which is inherently nonselective. These processes plus intermode V - V energy transfer make bond-selective chemistry extremely unlikely [2.168,169].

2.6.3 Collisional Effects in Laser-Induced Isomerization Reactions and Molecular Elimination

The laser-induced isomerization reaction has been shown to be selective (i.e., nonthermal) in a number of experiments (for example, [2.154,170-172] and the references contained therein). Collisions have not been beneficial in any of the cases studied, but the increased pressure causes two notable effects: (1) The collisional increase in the absorbed energy increases the likelihood of the molecular dissociation channel [2.171], which has a higher activation energy. (2) The increased V - T relaxation rate increases the importance of the thermal-isomerization route [2.154,170] which destroys the selectivity of the laser process.

In [2.154], for example, at sufficiently low pressure, the laser-induced isomerization of methyl and ethyl isocyanide to the corresponding nitrite was shown to be *isotopically* selective at sufficiently low pressure. However, at pressures above 10 Torr there is almost complete isomerization due to the thermal process (with a single laser pulse) and a resultant loss in selectivity.

The isomerization rate of *trans*- to *cis*-dichloroethylene decreased as a function of increased buffer-gas pressure even though the absorbed energy increased [2.170]. One could explain this in two ways. Either the increased pressure quenched the laser-excited process or it caused an increase in the heat capacity, resulting in a lower temperature and smaller thermally induced isomerization. Both processes probably contribute; however, the buffer-gas efficiencies for quenching this isomerization do follow the same sequence, $H_2 > N_2 > He > Xe$, as the V - T deactivation rates.

The majority of the laser-induced unimolecular elimination reactions studied to date have been with the halogenated ethanes and ethylenes (Table 2.10) where the dissociation channel with the lowest activation energy is the collisionless elimination of the hydrogen-halide molecule. Several other cases of interest involve the photodecomposition of ethylvinylether (EVE) [2.128,173], the dehydration of alcohols [2.174], and the decomposition of allylmethylether, ethyl acetate, and isopropyl bromide [2.175].

The results for the HF elimination [2.105] show that the addition of an inert buffer gas (Ar, N_2, Xe, H_2, and He) produces an enhancement in the number of HF molecules generated. At low pressure this effect can be attributed to rotational hole filling. However, at the highest pressures, there will undoubtedly be a contribution in all of these elimination reactions due to the thermal processes [2.174-176]. If more than one elimination route is available in the dissociation process, then collisions will have a strong effect on the branching ratio. In [2.128], it was shown that at sufficiently high fluence the higher energy pathway for the dissociation of EVE was favored over the lower one. At higher pressures the decrease in yield and branching ratio implied that $V \rightarrow T$ quenching near the high-energy reaction threshold is important.

2.6.4 Secondary Processes

A number of investigators have considered the role of intermolecular vibrational energy transfer in the dissociation process [2.48,175,177-189]. In the MPD process a large fraction of the energy absorbed is stored as vibrational energy in unreacted molecules. In fact, it has been shown in some cases [2.183,185,186] that the total dissociation yield can be dominated by these collisional-energy pooling processes where two highly vibrational excited molecules collide resulting in the dissociation of one and the deexcitation of the other. For example, under the proper conditions of reactant pressure, virtually all of the SiF_4 [2.183] and over 90% of the tetramethyldioxetane [2.185] that reacted did so as a result of binary collision processes that occurred after the laser pulse.

This channel for dissociation can still be isotopically selective. However, this intermolecular V-V energy exchange process is the first step toward the loss of selectivity because vibrational energy builds up in the unwanted isotope by V-V exchange and the translational temperature increases due to the fast V-T processes. The result is a higher probability of dissociation in the unwanted isotope.

It is important to reemphasize this point by stating these ideas in a different way. Certainly, the energy *absorbed* by the molecule depends on the collisional process occurring *during* the laser pulse. But while a fraction of the molecules may react in a collisionless sense during this excitation phase, there is a high probability that collisions occurring after the laser pulse will contribute to the overall reaction yield even at low pressure. One can get around these restrictions by making the experimental observations on a real-time basis as is done in the experiments that use the laser-induced fluorescence-probe technique (see Sect.2.5.1) or in carefully considered (for collisional effects) multiphoton-induced electronic-luminescence experiments. Another obvious solution is to do the experiment in a truly collisionless molecular beam.

2.7 Conclusion

We have attempted to meet several objectives in this chapter. We feel that we have succeeded in the following areas.

First, we have compiled a fairly complete review of the results of experiments that are relevant to the multiple-photon excitation process for single-frequency excitation. We have presented three categories of experimental results: multiple-photon absorption experiments, multiple-photon dissociation experiments, and experiments dealing with the effect of pressure on MPE processes. A major part of this compilation was a critical analysis of the SF_6 MPA experiments.

Second, we have suggested several ways to improve MPE data-reduction techniques. These suggestions deal particularly with the inevitable spatial variation of fluence or intensity in any laser experiment. More attention to data-reduction techniques will improve the reliability of experimental results.

Third, the chapter identifies many molecular properties, laser-beam properties, and experimental conditions that influence the absorption and reaction processes. Some of these include the spectroscopic absorption cross section, the laser fluence, the fraction of molecules absorbing the radiation, the molecular size (or density of vibrational states), the reaction activation energy, the laser frequency relative to the center of an infrared absorption band, the width of the infrared absorption band, the frequency width of the exciting laser, the gas temperature (or degree of internal excitation), and the pressure of the absorber or diluent. These are certainly not all independent properties; many are interrelated. For example, for many species, the pressure, through rotational relaxation, determines to a large extent the fraction of molecules that are excited by the radiation field.

Fourth, we have shown, in a heuristic manner, in what way many of these properties influence the excitation and dissociation of poylatomic molecules. From these considerations, one can predict with reasonable accuracy how much laser energy a molecular sample will absorb and what the reaction probability will be. We have identified several general trends, such as how absorbed laser radiation depends on a normalized fluence parameter.

Finally, our treatment of the experimental data allows one to better determine what experimental observables are basic to the MPE process and require theoretical consideration. For example, we find that although absorption experiments are not really sensitive diagnostics of the multiple-photon excitation process, the number of photons absorbed does depend on the 2/3 power of the fluence. Another theoretical issue relates to why the fraction of molecules absorbing laser radiation changes as it does with molecular size and the degree of internal excitation.

2.8 Appendix

We list here the three large tables referred to in the text of Sects.2.5.6, namely, Table 2.9 on molecular properties, Table 2.10 on molecules dissociated by MPD, and Table 2.11 on the effect of pressure on MPE processes.

In Table 2.9 we list several molecular properties of a selection of molecular species for which researchers have reported quantitative multiple-photon reaction or absorption measurements.

The quantity E_a is the activation energy in kcal mol^{-1} of the unimolecular reaction that is thermally most favorable [2.117,190]. The activation energy is very nearly the enthalpy change for the reaction for many simple bond-cleavage reactions. Therefore, for species that react by this simple mechanism we have listed the reaction enthalpy change. For other types of reactions we have attempted to obtain measured activation energies.

The three quantities $\langle\nu\rangle$, $\bar{\nu}$, β are necessary to calculate the density of vibrational states by the Whitten-Rabinovitch method (Sect.2.5.2). The first $\langle\nu\rangle$ is the arithmetic mean of the fundamental vibrational frequencies (in cm^{-1}) of the species. The second, $\bar{\nu}$, is the geometrical mean of the same set of fundamental frequencies. The third, β, is the frequency-spread parameter (2.48).

The final column in this table, N(10 500), is the density of vibrational states per cm^{-1} at an energy level 10 500 cm^{-1} above the ground vibrational state. We calculated this quantity from the three spectroscopic parameters listed in the table (2.51).

Researchers have reported many successful MPD experiments with a wide range of species. These we summarize in Table 2.10. This table lists the molecular species grouped according to the number of atoms in the molecule. The other columns in the table are: the laser frequency ν in cm^{-1}; the offset of the laser frequency from the center of the absorption band center, $\Delta\nu$, in cm^{-1}; the width of the absorption band at half its peak cross section, $\nu_{1/2}$, also in cm^{-1}; the small-signal cross section at the laser frequency, σ_0, in units of cm^2; the fluence necessary to give 1% reaction, $\Phi_{1\%}$, in J cm^{-2}; the average number of photons absorbed per molecule at that fluence, $n_{1\%}$, as CO_2 laser photons at 944.2 cm^{-1}; and the pressure of the absorbing gas mixture, P, in Torr. We list only the species, the laser frequency, and in some cases the pressure, where the researchers report no quantitative dissociation-probability data.

The footnotes in the laser frequency column refer to the published reports of the experiments, and those in the small-signal cross-section column refer to the sources of the spectroscopic information.

Table 2.11 summarizes experimental investigations of the effect of pressure on absorption, reaction, and isotopic selectivity. The first column lists the investigated molecule (or reactant), and the last column lists the reference to the report of the experiment. The other columns show the effect of pressure on one of the three processes. An upward pointing arrow indicates a positive effect of pressure, a downward pointing arrow indicates a negative effect, and a horizontal arrow indicates no effect. Two arrows indicate a reversal of the effect as the pressure increased. An entry in a column implies that the pressure effects on that process were studied by increasing the pressure of the species noted. The entry R implies that the reactant molecule pressure was changed. The entry TM implies that the pressure of a thermal monitor was increased. The entry D implies that the effect of the pressure of several diluents was studied and the results are the same for all. Because of the vastness of the information contained in the original works, this table cannot be expected to give any more than a qualitative description. For additional details, consult the references.

Table 2.9. Molecular properties

Species	E_a	$\langle\nu\rangle$	$\bar{\nu}$	β	N(10 500)	References[a]
D_2O	...	2284	2127	0.74	1.00×10^1	...,[2.191]
OCS	73.3	1149	976	0.89	8.02×10^1	[2.192,192]
O_3	25.5	953	935	0.69	8.71×10^1	[2.192,192]
SO_2	132.1	1010	933	0.75	8.87×10^1	[2.192,192]
BCl_3	105	567	483	1.08	1.75×10^5	[2.192,192]
HDCO	72	1684	1592	0.94	4.48×10^2	[2.193,194]
NH_3	103	2458	2203	1.08	2.56×10^3	[2.192,192]
CCl_4	78.2	463	408	1.11	4.78×10^7	[2.192,192]
CF_2Cl_2	81	642	565	1.11	4.09×10^6	[2.192,195]
CF_2HCl	56	1106	915	1.30	1.55×10^5	[2.192,196]
CF_3I	53.9	674	569	1.15	4.18×10^6	[2.192,192]
HCOOH	>35	1589	1345	1.22	1.27×10^4	[2.197,198]
OsO_4	73	611	531	1.12	6.59×10^6	[2.100,199]
SiF_4	142	622	537	1.13	6.10×10^6	[2.192,192]
CDF_3	69	1050	947	1.09	1.03×10^5	[2.200,200]
CH_3OH	91	1812	1495	1.19	4.85×10^4	[2.198,201]
CH_3NC	38.4	1588	1239	1.25	2.41×10^5	[2.202,203]

Table 2.9 (cont.)

Species	E_a	$<\nu>$	$\bar{\nu}$	β	N(10 500)	References[a]
Trans-dichlorethylene	41	1193	870	1.47	4.54×10^{6}	[2.170,204]
C_2H_3Cl	72	1512	1238	1.29	1.91×10^{5}	[2.205,206]
C_2H_3F	70	1564	1316	1.25	1.09×10^{5}	[2.105,207]
C_2H_4	71	1799	1586	1.15	2.36×10^{4}	[2.198,208]
N_2F_4	19.8	494	414	1.18	2.39×10^{9}	[2.192,192]
CH_3NH_2	79	1826	1533	1.20	1.75×10^{5}	[2.201,209]
SF_5Cl	61	577	546	1.04	2.06×10^{9}	[2.126,210]
SF_6	92.6	624	591	1.03	8.50×10^{8}	[2.66,126[b]]
SeF_6	90	512	475	1.06	1.05×10^{10}	...[c],[2.211]
MoF_6	94	423	345	1.25	6.77×10^{11}	[2.192,192]
UF_6	68	347	286	1.28	6.41×10^{12}	[2.212,213]
CF_3CDCl_2	79	1233	935	1.44	1.48×10^{8}	...[c],...[c]
CF_3CH_3	69	1251	979	1.41	7.41×10^{7}	[2.176,214]
C_2H_5F	60	1604	1338	1.24	2.38×10^{6}	[2.176,215]
C_2H_5NC	38.2	1412	1007	1.44	7.00×10^{8}	[2.216],...[c]
C_2H_5OH	91	1605	1238	1.31	4.15×10^{7}	[2.201],...[c]
SF_5NF_2	50	566	480	1.12	2.60×10^{12}	[2.126],..[c]
Cyclopropane	66	1647	1431	1.25	2.85×10^{6}	[2.117,209]
CF_3COCF_3	82	710	491	1.40	8.68×10^{13}	...[c],[2.217]
Hexafluorobutene	47	732	546	1.39	9.06×10^{12}	[2.117,218]
S_2F_{10}	58	524	390	1.23	3.07×10^{17}	[2.126,219]
Perfluorocyclobutane	74	722	575	1.31	1.13×10^{14}	[2.220,221]
Ethylvinylether	44	1393	1007	1.43	2.74×10^{11}	[2.173],...[c]
Ethyl acetate	48	1361	952	1.48	3.90×10^{12}	[2.222],...[c]
Tetramethyldioxetane	25	1321	897	1.53	4.03×10^{16}	[2.223],...[c]
Sec-butyl acetate	47	1276	796	1.59	4.09×10^{18}	[2.216],...[c]
UO_2 $(hfacac)_2$.THF[d]	30	1100	700	1.60	$\sim10^{30}$	[2.169],...[c]

[a] The first reference is the source for E_a and the second is for the fundamental vibrational frequency.
[b] See the discussion in Sect.2.5.2.
[c] Estimated.
[d] Uranyl-bis-hexafluoroacetylacetonate·tetrahydrofuran.

Table 2.10. Molecules dissociated by MPD

Three-atom molecules

Species	ν	$\Delta\nu$	$\nu_{1/2}$	σ_0	$\Phi_{1\%}$	$\eta_{1\%}$	P	References[a]
OCS	1043.2	- 5	48	3.1×10^{-20}	50	...	~ 0	[2.107,224]
O_3	1037.4	...	...	...	< 85	...	0.5	[2.107],...
SO_2	1084.6	- 65	80	2.8×10^{-20}	< 250	...	2	[2.224,225]

[a] The first reference, or group of references, is the source for ν, and the second for σ_0.

Four-atom molecules

Species	ν	$\Delta\nu$	$\nu_{1/2}$	σ_0	$\Phi_{1\%}$	$\eta_{1\%}$	P	References[a]
BCl_3	944.2	- 11	11	1.2×10^{-18}	29	...	3	[2.226[b],230]
H_2CO	944.2	...	...	...	...	...	4.7	[2.230],...
HDCO	944.2	- 97	120	1.4×10^{-21}	218	...	5-20	[2.194,230-232]
D_2CO	944.2	...	...	...	...	...	3	[2.230],...
HN_3	945.9	...	...	...	...	...	...	[2.233],...
DN_3	945.9	...	...	...	...	...	...	[2.233],...
NH_3	933.0	...	...	...	...	...	...	[2.82],...

[a] The first reference, or group of references, is the source for ν, and the second for σ_0.
[b] See also [2.99,227-229].

Five-atom molecules

Species	ν	$\Delta\nu$	$\nu_{1/2}$	σ_0	$\Phi_{1\%}$	$\eta_{1\%}$	P	References[a]
CCl_4	979.7	- 22	80	1.8×10^{-20}	70	...	...	[2.234,234]
	971-988	...	...	...	...	...	...	[2.235],...
	780.5	- 13	12	2.1×10^{-18}	1.3	...	...	[2.236[b],224]

Table 2.10 (cont.)

Species	ν	$\Delta\nu$	$\nu_{1/2}$	σ_0	$\Phi_{1\%}$	$\eta_{1\%}$	P	References[a]
$CFCl_3$	1078.6	...	...	...	...	...	...	[2.237],...
	1074.6	...	...	...	...	...	...	[2.111,113][c],...
$CFHCl_2$	1055.6	...	...	...	...	...	...	[2.111,113][c],...
	1082.3	...	...	...	...	...	0.001-0.1	[2.84,85],...
	1081.1	...	...	...	...	...	...	[2.238],...
	1084.6	...	...	...	...	...	...	[2.111,113][c],...
	944.2	22	29	7.7×10^{-19}	21	...	0.5	[2.226,224]
	929.0	...	...	...	...	...	...	[2.84,184,238,239],...
	925.1	...	...	...	...	...	...	[2.111,113][c],...
	1089.0	...	...	...	...	...	...	[2.111,113][c],...
	920.8	- 1	29	8.4×10^{-18}	...	5.6	12	[2.240,224]
	1087.9	-15	28	3.0×10^{-18}	...	5.6	12	[2.240,224]
	1082.3	...	...	...	...	...	0.001-0.1	[2.84,84,237],...
	1081.1	...	...	...	...	...	2	[2.241],...
	931.0	9	29	8.0×10^{-18}	1.8	...	0.5	[2.195,224]
	1085.8	-17	28	2.5×10^{-18}	1.1	...	0.5	[2.195,224]
CF_2HCl	1082.3	...	...	...	...	...	...	[2.85,113],...
	1087.9	-20	28	1.1×10^{-18}	0.65	4.0	1.6	[2.196,224]
CF_2HCl	1085.8	-22	28	7.7×10^{-19}	2.4	...	0.004	[2.33,224]
	...	...	...	...	...	...	...	[2.242]
CF_2H_2	1077.3	...	...	...	...	...	...	[2.243],...
	1082.3	...	...	...	...	...	...	[2.111,113][c],...
CF_3Br	1040-1060	...	...	...	...	...	...	[2.244-246],...
	1082.3	...	...	...	...	...	...	[2.86],...

	1078.6	...	...	...	...	...	...	[2.111,113][c],...
CF_3Cl	1090.0	...	...	...	...	...	...	[2.111,113][c],...
CF_3I	1073.3	...	...	...	...	...	...	[2.111,112][c],...
	1074.6	6	20	7.2×10^{-18}	0.42	9.0	0.50	[2.50[d],50]
CrO_2Cl_2	969-984	...	...	...	...	...	...	[2.248],...
HCOOH	3593.8	23	60	7.5×10^{-20}	12	...	1.6	[2.180,224]
	3577.5	7	60	1.3×10^{-19}	42	...	1.0	[2.224,249]
OsO_4	947.8	-13	30	1.1×10^{-18}	1.1	...	0.03	[2.100,199,250,251]
	927.0	-33	30	5.3×10^{-21}	0.45	...	0.03	[2.100,199,250,251]
SiF_4	944.2	...	...	...	...	...	...	[2.98,183,252],...
	1031.5	0	16	1.1×10^{-17}	8.9	...	0.5	[2.224,226]
SiH_4	944.2	...	...	...	...	...	72.0	[2.189],...
CDF_3	979.7	5	27	3.4×10^{-19}	21.6	...	0.065	[2.200,200][e]

[a]The first reference, or group of references, is the source for ν, and the second for σ_0.
[b]See also [2.188].
[c]Molecular-beam experiments.
[d]See also [2.247].
[e]See also [2.154].

Six-atom molecules

Species	ν	$\Delta\nu$	$\nu_{1/2}$	σ_0	$\Phi_{1\%}$	$\eta_{1\%}$	P	References[a]
CH_3OH	3693.5	15	60	3.9×10^{-20}	440	...	1.3	[2.253,224]
	1046.9	...	...	...	...	...	...	[2.89,90],...
	1048.7	...	...	...	...	...	...	[2.89,90],...
CH_3CN	933.0	...	...	...	...	...	...	[2.91-93],...
	1046.9	...	...	...	...	...	...	[2.91-93],...

Table 2.10 (cont.)

Species	ν	$\Delta\nu$	$\nu_{1/2}$	σ_0	$\Phi_{1\%}$	$\eta_{1\%}$	P	References[a]
CH_3NC	931.0	...	...	...	45.3	...	1.5	[2.154][b],...
$CHClCF_2$	967.7	...	...	...	...	...	...	[2.113,254]
Trans-CHClCHCl	934.9	37	90	5.7×10^{-20}	28	...	0.4-1.0	[2.170,204]
	979.7	82	90	6.5×10^{-21}	48	...	0.4-1.0	[2.170,204]
	934.9	...	...	...	...	...	...	[2.254,255],...
C_2HCl_3	929.1	...	...	...	...	...	...	[2.113],...
C_2H_3Cl	933.0	- 9	100	5.6×10^{-20}	27	...	10	[2.205[c],224]
	942.4	0	100	9.7×10^{-20}	107	...	10	[2.205[c],224]
	944.2	2	100	7.1×10^{-20}	26	...	10	[2.205[c],224]
	1046.9	13	35	4.0×10^{-20}	46	...	10	[2.205[c],224]
CH_2CCl_2	...	...	...	...	...	...	10	[2.254],...
C_2H_3F	934.9	4	78	4.5×10^{-19}	4	...	< 0.5	[2.105,224,256]
	944.2	13	78	4.8×10^{-19}	7	...	< 0.5	[2.105,224,256]
$CF_2=CH_2$	944.2	...	...	...	...	...	0.05-40.0	[2.256],...
C_2H_4	952.7	2.9	170	7.8×10^{-20}	85	3.6	1.0	[2.224,257]
	947.8	- 3.8	170	8.6×10^{-20}	50	...	< 0.2	[2.87,224]
	949.5	...	...	...	...	...	200	[2.208],...
	951.2	...	...	...	...	...	0.003-8.0	[2.88],...
N_2F_4	944.2	-16	80	5.3×10^{-19}	0.12	2.0	2.7	[2.106,134],...[d]
	944.2	-16	80	5.3×10^{-19}	0.10	0.6	10	[2.135][e],...[d]
	975.9	...	...	...	...	...	...	[2.111][f],...

[a] The first reference, or group of references, is the source for ν, and the second for σ_0. [b] See also [2.94]. [c] See also [2.48,254]. [d] Unpublished data. [e] 1.5-ns pulse. [f] Molecular-beam experiment.

Seven-atom molecules

Species	ν	$\Delta\nu$	$\nu_{1/2}$	σ_0	$\Phi_{1\%}$	$\eta_{1\%}$	P	References[a]
CH_3NH_2	1031.5	-14	75	2.8×10^{-20}	3.5	...	0.03	[2.96,224]
	1043.2	- 2	75	5.4×10^{-20}	7.4	...	0.03	[2.96,224]
	1070.5	26	75	3.7×10^{-20}	8.7	...	0.03	[2.96,224]
	1046.9	2	75	5.6×10^{-20}	2.5	...	0.1	[2.95,224]
	1031-1044	...	...	...	...	...	0.5	[2.83],...
CH_3NO_2	920-1060	...	...	...	...	...	1.0	[2.258],...
SF_5Cl	904.9	- 4	14	6.1×10^{-18}	0.60	...	0.8	[2.259,259]
	914.4	5	14	6.0×10^{-18}	1.25	...	0.8	[2.259,259]
	904.9	...	...	...	...	...	...	[2.97],...
SF_6	952.9	5	7	4.8×10^{-19}	5.4	...	~0	[2.133][b],...[c]
	949.5	2	7	1.24×10^{-17}	1.7	...	~0	[2.133][b],...[c]
	946.0	- 2	7	1.89×10^{-17}	1.2	...	~0	[2.133][b],...[c]
	944.2	- 4	7	1.58×10^{-17}	1.0	...	~0	[2.133][b],...[c]
	942.4	- 6	7	1.12×10^{-17}	1.1	...	~0	[2.133][b],...[c]
	938.7	- 9	7	3.10×10^{-18}	3.1	...	~0	[2.133][b],...[c]
	934.9	-13	7	6.2×10^{-19}	3.8	...	~0	[2.133][b],...[c]
	956.2	8.4	7	5.5×10^{-20}	9.9	...	0.016	[2.132][b],...[c]
	954.5	6.5	7	7.5×10^{-20}	8.4	...	0.016	[2.132][b],...[c]
	952.9	4.9	7	4.8×10^{-19}	7.7	...	0.016	[2.132][b],...[c]
	951.2	3.2	7	6.68×10^{-18}	7.93	...	0.016	[2.132][d],...[c]
	949.5	1.5	7	1.24×10^{-17}	5.75	...	0.016	[2.132][d],...[c]
	947.7	- 0.3	7	2.96×10^{-17}	5.61	...	0.016	[2.132][d],...[c]
	946.0	- 2.0	7	1.89×10^{-17}	5.81	...	0.016	[2.132][d],...[c]
	944.2	- 3.8	7	1.58×10^{-17}	4.64	...	0.016	[2.132][d],...[c]
	942.4	- 5.6	7	1.12×10^{-17}	3.90	...	0.016	[2.132][d],...[c]

Table 2.10 (cont.)

Species	ν	$\Delta\nu$	$\nu_{1/2}$	σ_0	$\Phi_{1\%}$	$\eta_{1\%}$	P	References[a]
	940.5	- 7.5	7	6.21×10^{-18}	3.69	...	0.016	[2.132][d],...[c]
	938.7	- 9.3	7	3.10×10^{-18}	3.41	...	0.016	[2.132][d],...[c]
	936.8	-11.2	7	1.31×10^{-18}	3.11	...	0.016	[2.132][d],...[c]
	934.9	-13.1	7	6.2×10^{-19}	3.16	...	0.016	[2.132][d],...[c]
	933.0	-15.0	7	5.0×10^{-19}	2.87	...	0.016	[2.132][d],...[c]
	931.0	-17.0	7	6.0×10^{-19}	3.19	...	0.016	[2.132][d],...[c]
	944.2	- 3.5	7	1.58×10^{-17}	2.10	18.6	0.121	[2.77][e],...[c]
	944.2	- 3.5	7	1.58×10^{-17}	2.10	23.4	0.121	[2.77][f],...[c]
	944.2	- 3.5	7	1.58×10^{-17}	2.1	...	0.08	[2.260][g],...[c]
	944.2	- 3.5	7	1.58×10^{-17}	2.3	...	0.15	[2.38][g],...[c]
	944.2	- 3.5	7	1.58×10^{-17}	2.8	16.9	0.2	[2.136][h],..[c]
	944.2	- 3.5	7	1.58×10^{-17}	2.5	...	1.0	[2.261],...[c]
	944.2	...	...	...	2.5	...	0.2	[2.262][i],...
	944.2	...	...	...	3.2	...	0.2	[2.262][i],...
	944.2	...	...	...	5.2	...	0.2	[2.262][i],...
	947.7	- 0.3	7	2.96×10^{-17}	2.9	...	1.0	[2.261],...[c]
	947.7	- 0.3	7	2.96×10^{-17}	2.9	...	...	[2.235],...[c]
	985.5	6	20	5.0×10^{-20}	14.3	...	...	[2.235],...[c]
	1055.6	0	200	6.4×10^{-22}	109	...	...	[2.235],...[c]
	934.9	-13.1	7	6.2×10^{-19}	...	14.3	1.0	[2.73][j],...[c]
	936.8	-11.2	7	1.31×10^{-18}	...	16.0	1.0	[2.73][j],...[c]
	938.7	- 9.3	7	3.10×10^{-18}	...	13.6	1.0	[2.73][j],...[c]
	940.5	- 7.5	7	6.21×10^{-18}	...	16.6	1.0	[2.73][j],...[c]
	944.2	- 3.8	7	1.58×10^{-17}	...	15.4	1.0	[2.73][j],...[c]
	946.0	- 2.0	7	1.89×10^{-17}	...	17.7	1.0	[2.73][j],...[c]

	947.7	- 0.3	7	2.96×10^{-17}	...	12.8	1.0	[2.73][j],...[c]
	930-948	...	...	...	...	...	...	[2.39,101,102,108-110, 112,164,182,263-267],...
SeF_6	780.5	3	20	2.0×10^{-18}	69	...	0.9	[2.103,211,268]
MoF_6	1050.4	- 2	29	3.0×10^{-20}	34	...	3.0	[2.269],...[c]
	1055.6	3	29	3.0×10^{-20}	44	...	3.0	[2.269],...[c]
UF_6	615	-11	14	3.0×10^{-18}	1.0	...	0.1-0.2	[2.104,212]
	615	-11	14	3.0×10^{-18}	1.5	...	3.5	[2.212,270,271[k]]

[a] The first reference, or group or references, is the source for ν, and the second for σ_0.
[b] Molecular-beam experiment, listed fluence is for 2% dissociation.
[c] Unpublished data.
[d] Listed fluence is threshold of detectable enrichment.
[e] 100-ns pulse.
[f] 1-ns pulse.
[g] See also 2.39 .
[h] See also 2.39,77 .
[i] Temperatures were 223, 293, and 343 K for the three entries.
[j] Threshold for reaction.
[k] Listed fluence is threshold value.

Eight-atom molecules

Species	ν	$\Delta\nu$	$\nu_{1/2}$	σ_0	$\Phi_{1\%}$	$\eta_{1\%}$	P	References[a]
CF_2ClCH_3	954.2	...	...	...	...	...	...	[2.113],...
	973.3	...	...	...	...	...	...	[2.272],...
CF_3CF_2Cl	978.5	...	...	...	...	...	...	[2.113],...
CF_3CHCl_2	937.8	...	...	5.0×10^{-21}	26	...	0.12	[2.2,2][b]
CF_3CDCl_2	937.8	- 2	24	5.0×10^{-19}	2.9	...	0.12	[2.2[b],215]
CF_3CH_3	973.3	0	35	7.1×10^{-19}	3.3	...	0.6	[2.215,176]
	973.3	...	...	...	...	...	...	[2.272],...
CH_2DCH_2Cl	973.3 & 944.2	...	...	...	...	...	...	[2.273],...

Table 2.10 (cont.)

Species	ν	$\Delta\nu$	$\nu_{1/2}$	σ_0	$\Phi_{1\%}$	$\eta_{1\%}$	P	References[a]
CH_2FCH_2Br	...	...	...	...	...	...	...	[2.176],...
CH_2FCH_2Cl	1035.6	...	...	...	1.4	...	0.01	[2.274],...
CHF_2CH_3	944.2	...	...	...	...	...	0.05-40.0	[2.256],...
CH_3CCl_3	1073.3	...	...	...	...	...	...	[2.113],...
C_2H_5Cl	973.3	...	...	...	...	...	...	[2.272],...
C_2H_5F	1048.7	1	55	3.5×10^{-19}	2.4	...	0.6	[2.215,176]
	1046.9	...	...	...	...	...	0.05-40.0	[2.256],...

[a]The first reference, or group of references, is the source for ν, and the second for σ_0.
[b]See also [2.245].

Nine-atom molecules

Species	ν	$\Delta\nu$	$\nu_{1/2}$	σ_0	$\Phi_{1\%}$	$\eta_{1\%}$	P	References[a]
C_2H_5NC	987.6	...	...	...	9.72	...	1.5	[2.154],...
C_2H_5OH	1039.4	-28	60	2.42×10^{-19}	3.0	...	0.5	[2.224,275]
SF_5NF_2	908.5	- 2/-42	32	2.66×10^{-18}	0.050	4.8	0.1	[2.46,46]
	920.8	11/-29	32	1.45×10^{-18}	0.076	4.9	0.1	[2.46,46]
	934.9	25/-15	32	1.53×10^{-18}	0.081	6.1	0.1	[2.46,46]
	944.2	34/- 6	32	2.19×10^{-18}	0.100	6.6	0.1	[2.46,46]
	957.8	48/8	32	1.25×10^{-18}	0.141	4.6	0.1	[2.46,46]
Cyclopropane	1046.9	...	...	...	...	...	...	[2.181,255],...
	1029.4	...	...	...	...	...	...	[2.255],...
	1050.4	22	60	3.7×10^{-20}	>60	...	1.0	[2.171,224]
	3106	0	80	1.7×10^{-19}	> 2	...	1.0	[2.171,171]

Propylene	931.0	...	...	...	...	...	...	[2.255],...
CH_3OCH_3	1030-1085	...	...	...	...	...	0.1-2.0	[2.276],...

[a]The first reference, or group of references, is the source for ν, and the second for σ_0.

Ten-atom molecules

Species	ν	$\Delta\nu$	$\nu_{1/2}$	σ_0	$\Phi_{1\%}$	$\eta_{1\%}$	P	References
CF_3COCF_3	970.6	0	18	1.8×10^{-18}	1.3	22.5	0.35	[2.51,52]
	940.5	-31	18	$\sim 4.4 \times 10^{-21}$	1.76	5.8	0.38	[2.52,52]
	944.2	-27	18	$\sim 1 \times 10^{-20}$	2.21	14.4	0.38	[2.52,52]
	969.1	- 2	18	1.7×10^{-18}	0.74	20.9	0.38	[2.52,52]
	971.9	1	18	1.7×10^{-18}	0.89	18.0	0.38	[2.52,52]
	983.3	12	18	8.0×10^{-19}	2.7	9.0	0.38	[2.52,52]
Hexafluoro-cyclobutene	952.9	-23	40	4.3×10^{-19}	0.2	...	0.5	[2.172,218]

Twelve-atom molecules

Species	ν	$\Delta\nu$	$\nu_{1/2}$	σ_0	$\Phi_{1\%}$	$\eta_{1\%}$	P	References[a]
$CH_3(CH_2)_2OH$	949.5	...	...	...	...	...	...	[2.275],...
S_2F_{10}	920.8	-17	9	3.7×10^{-19}	0.085	2.4	0.10	[2.42,42]
	925.0	-13	9	1.3×10^{-18}	0.052	4.4	0.10	[2.42,42]
	929.0	- 9	9	4.0×10^{-18}	0.022	4.2	0.10	[2.42,42]
	933.0	- 5	9	1.04×10^{-17}	0.018	5.0	0.10	[2.42,42]
	936.8	- 1	9	1.72×10^{-17}	0.070	8.1	0.10	[2.42,42]
	938.7	1	9	1.75×10^{-17}	0.080	6.9	0.10	[2.42,42]
	940.5	3	9	1.09×10^{-17}	0.082	5.8	0.10	[2.42,42]
	944.2	6	9	3.2×10^{-18}	0.160	4.0	0.10	[2.42,42]

Table 2.10 (cont.)

Species	ν	$\Delta\nu$	$\nu_{1/2}$	σ_0	$\Phi_{1\%}$	$\eta_{1\%}$	P	References[a]
Perfluorocyclo-butane	949.5	-17	16	2.6×10^{-19}	0.96	8.0	1.0	[2.220,277]
C_6F_5H	944.2	...	...	...	1.4	...	0.41	[2.278],...

[a]The first reference, or group of references, is the source for ν, and the second for σ_0.

Molecules with more than twelve atoms

Species	ν	$\Delta\nu$	$\nu_{1/2}$	σ_0	$\Phi_{1\%}$	$\eta_{1\%}$	P	References[a]
Ethylvinylether	...	...	...	...	...	...	...	[2.173][b],...
	1041.2	0	40	1.6×10^{-19}	0.6	...	0.003	[2.128[c],279[d]]
	1041.2	0	40	1.6×10^{-19}	0.9	...	0.003	[2.128[e],279[d]]
Allylmethylether	1081.1	...	...	...	3.0	...	0.3	[2.175][b],...
Ethyl acetate	1081.1	21	80	2.1×10^{-19}	1.4	...	0.3	[2.175[f],224]
	1035-1051	...	...	...	...	...	...	[2.222,242],...
Sec-butyl acetate	1029.4	- 5	60	2.6×10^{-19}	0.35	...	0.05	[2.275,275[g]]
Tetramethyl-dioxetane	940-980	...	...	...	~100	...	0.35	[2.185,223,280][g],...
$U(OCH_3)_6$	924-941	...	...	...	...	...	...	[2.281][h],...
$UO_2(hfacac)_2$.THF	956.2	...	...	7.2×10^{-19}	0.004	0.15	...	[2.169,282,169,282][i]

[a]The first reference, or group of references, is the source for ν, and the second for σ_0. [b]13 atoms. [c]2-μs pulse. [d]Estimated from liquid spectrum. [e]0.2-μs pulse. [f]14 atoms. [g]20 atoms. [h]31 atoms. [i]44 atoms.

Table 2.11. Effect of pressure on MPE process

Molecule	Absorption	Reaction	Isotopic selectivity	Ref.
O_3		↑R		[2.107]
NH_3	↓R			[2.283]
H_2CO		↑R	↓R	[2.249]
HDCO		↑R		[2.232]
HDCO		↑R		[2.230]
D_2CO		↑R		[2.230]
BCl_3		↑,↓ Air,↓R		[2.284]
BCl_3		↓O_2	↓R,↑O_2	[2.179]
BCl_3			↑↓H_2S/BCl_3	[2.285]
BCl_3		↑R	↓R	[2.226]
CH_3F	→Ar,↑R			[2.286]
CF_3I			↓R	[2.50]
CF_2Cl_2	↑D	↓D		[2.240]
CF_2Cl_2		↑R	↓R	[2.226]
CF_2Cl_2		↑R,↑↓He		[2.241]
CF_2Cl_2			↓R	[2.184]
CF_2HCl	↑D	↓D		[2.196]
CF_2HCl		↑Ar		[2.33]
CDF_3		↑Ar		[2.200]
CCl_4		↓↑R,↑O_2,C_2,H_4	↓O_2,C_2H_4	[2.236]
CCl_4		↓↑R		[2.188]
CF_3Br		↑R		[2.86]
SiH_4			↓R	[2.244]
SiH_4		↑R		[2.287]
SIH_4		↑R		[2.189]
SiF_4		↑R	↓R	[2.226]
HCOOH		↑R,↓N_2O>CO_2		[2.180]
CH_3OH	↑R	↑R,↓D,↑↓Xe	↓R	[2.253]
CH_3OH		↑R		[2.288]
C_2H_4		↑H_2,D_2He,Ar,Xe		[2.257]
C_2H_4		↑R		[2.88]
C_2H_4		↑He		[2.208]
C_2H_3F		↑Ar,He		[2.105]
C_2H_3F		↑Ar,N_2,Xe,H_2		[2.256]
$C_2H_2F_2$		↑A,N_2,Xe,H_2		[2.256]
C_2H_3Cl	↑R,↑He	↑R,↓He		[2.205]
CH_3NC		↑R		[2.154]
$C_2H_2Cl_2$	↑R	↓H_2>N_2>He>Xe		[2.170]

Table 2.11 (cont.)

Molecule	Absorption	Reaction	Isotope selectivity	Ref.
SF_6	$\uparrow H_2$		$\downarrow R$	[2.76]
SF_6			$\uparrow, \downarrow R$	[2.289]
SF_6			$\downarrow H_2 + SF_6$	[2.290]
SF_6			$\downarrow R, \downarrow Ar, O_2, NH_3$	[2.178]
SF_6		$\downarrow H_2$	$\downarrow H_2$	[2.38]
SF_6	$\uparrow, \downarrow R, \uparrow H_2$			[2.182]
SF_6		$\downarrow H_2$		[2.260]
SF_6		$\downarrow R$		[2.101]
SF_6		$\downarrow HF, HCl, HBr, HI, Kr$		[2.291]
SF_6		$\downarrow Ar < N_2 < CF_4$		[2.266]
SF_6		$\downarrow H_2, \downarrow {}^{32}SF_6, \rightarrow {}^{34}SF_6$	$\downarrow H_2$	[2.186]
SF_6		$\downarrow Ar > He > C_3F_8$		[2.164]
SF_6	$\uparrow H_2, N_2, Xe$			[2.143-145]
SF_6	$\uparrow Ar$			[2.109]
UF_6	$\uparrow\downarrow R$			[2.271]
UF_6	$\uparrow N_2$			[2.45]
UF_6	$\downarrow\uparrow R$			[2.292]
CH_3CF_3		$\downarrow R + TM$		[2.176]
CH_3CH_2F		$\downarrow R + TM$		[2.176]
CH_3CH_2F		$\uparrow R, Ar, N_2, Xe, H_2$		[2.256]
CH_2FCH_2Br		$\downarrow R + TM$		[2.176]
CH_3CHF_2		$\uparrow R, Ar, N_2, Xe, H_2$		[2.256]
C_2H_4FCl		$\uparrow R$		[2.274]
Cyclo-C_3H_6		$\uparrow Ar$		[2.171]
Cyclo-C_3H_6		$\uparrow R, NO, O_2, \downarrow He$		[2.181]
$(CH_3)_2O$			$\downarrow R$	[2.276]
CF_3COCF_3		$\uparrow R, C_2F_6, C_2F_4, CO_2$		[2.51]
Cyclo-C_4F_8		$\downarrow Ar$		[2.220]
S_2F_{10}	$\uparrow He, Xe$			[2.145]
EVE[a]		$\rightarrow R$		[2.173]
EVE[a]		$\downarrow R$		[2.128]
Allylmethyl-ether		$\downarrow R$		[2.175]
Ethyl acetate		$\downarrow R$ Approached the		[2.175]
Ethyl acetate		$\downarrow R$ thermal reaction		[2.187]
SBA[b]	$\rightarrow$Nonabsorb-diluents			[2.145]
HFCB[c]		$\downarrow He$		[2.172]

Table 2.11 (cont.)

Molecule	Absorption	Reaction	Isotope selectivity	Ref.
$C_3F_6^+$		↑Ar,$N_2$$SF_6$,↑↓$C_3F_6$		[2.293]
$[(C_2H_5)_2O]H^+$		↓R		[2.294]

[a]Ethylvinylether.
[b]*Sec*-butyl acetate.
[c]Hexafluorocyclobutene.

References

2.1 D.K. Evans, R.D. McAlpine, M.O. Poroshina: Chem. Phys. Lett. **65**, 226 (1979)
2.2 J.B. Marling, I.P. Herman: Appl. Phys. Lett. **34**, 439 (1979)
2.3 S.K. Sarkar, V. Parthasarathy, A. Pandy, K.V.S. Rama Rao, J. Mittal: Chem. Phys. Lett. **78**, 479 (1981)
2.4 S.A. Tuccio, A. Hartford, Jr.: Chem. Phys. Lett. **65**, 235 (1979)
2.5 J. Moser, P. Morand, R. Duperrex, H. van den Bergh: Chem. Phys. **79**, 277 (1983)
2.6 Y. Makide, S. Kato, T. Tominaga, K. Takeuchi: Appl. Phys. B**32**, 33 (1983)
2.7 K. Takeuchi, O. Kurihara, S. Satooka, Y. Makide, I. Inque, R. Nakane: J. Nucl. Sci. Technol. **18**, 68 (1981)
2.8 Y. Makide, T. Tominaga, K. Takeuchi, O. Kurihara, R. Nakane: ACS National Meeting. Symposium "Laser Isotope Separation", Las Vegas, USA, American Chemical Society, Washington, D.C. March 31, 1982
2.9 F. Magnotta, I.P. Herman, F.T. Aldridge: Chem. Phys. Lett. **92**, 600 (1982)
2.10 M. Gauthier, C.G. Cureton, P.A. Hackett, C. Willis: Appl. Phy. B**28**, 43 (1982)
2.11 G.I. Abdushelishvili, O.N. Avatkov, V.N. Bagratashvili, V.Yu. Baranov, A.B. Bakhtadze, E.P. Velikhov, V.M. Vetsko, I.G. Gverdtsiteli, V.S. Dolzhikov, G.G. Esadze, S.A. Kazakov, Yu.R. Kolomiiskii, V.S. Letokhov, S.V. Pigulskii, V.D. Pismennyi, E.A. Ryabov, G.I. Tkeshelashvili: Sov. J. Quantum Electron. **12**, 459 (1982)
2.12 G.I. Abushelishvili, O.N. Avatkov, A.B. Bakhtadze, V.M. Vetsko, G.I. Tkeshelashvili, V.I. Tomilina, V.N. Fedoseev, Yu.R. Kolomiiskii: Sov. J. Quantum Electron. **11**, 326 (1981)
2.13 M. Cauchetier, O. Croix, M. Luce, S. Tistchenko: Note CEA-N-2348, Centre D'Etudes Nucleaires De Saclay (May 1983)
2.14 V.Yu. Baranov, E.P. Velikhov, S.A. Kazakov, Yu.R. Kolomiiskii, V.S. Letokhov, V.D. Pismennyi, E.A. Ryabov, A.I. Starodubtsev: Sov. J. Quantum Electron **9**, 486 (1979)
2.15 V.Yu. Baranov, E.P. Velikhov, Yu.R. Kolomiiskii, V.S. Letokhov, Ev Niz, V.D. Pismennyi, E.A. Ryabov: Sov. J. Quantum Electron **9**, 621 (1979)
2.16 B.B. McInteer, J.L. Lyman, G.P. Quigley, A.C. Nilsson: In *Synthesis and Applications of Isotopically Labeled Compounds*, Proc. Int. Conf., Kansas City, USA, June 6-11, 1982, ed. by W.F. Duncan, A.B. Susan (Elsevier, Amsterdam 1982) p. 397
2.17 W. Fuss, W.E. Schmid: Ber. Bunsenges. Phys. Chem. **83**, 1148 (1979)
2.18 J.A. Horsley, D.M. Cox, R.B. Hall, A. Kaldor, E.T. Maas, Jr., E.B. Priestley, G.M. Kramer: J. Chem. Phys. **73**, 3660 (1980)
2.19 D.M. Cox, E.T. Maas, Jr.: Chem. Phys. Lett. 71, 330 (1980)
2.20 P. Ravinowitz, A. Kaldor, A. Gnauck: Appl. Phys. B**28**, 187
2.21 F.S. Becker, K.L. Kompa: Nucl. Technol. **58**, 329 (1982)
2.22 S.S. Alimpiev, N.V. Karlov, Sh.Sh. Naviev, S.M. Nikiforov, A.M. Prokhorov, B.G. Sartakov: Sov. J. Quantum Electron. **11**, 35 (1981)

2.23 E.A. Cuellar, S.S. Miller, T.J. Marks, E. Weitz: J. Am. Chem. Soc. **105**, 4580 (1983)
2.24 D.S. King: In *Dynamics of the Excited State*, ed. by K.P. Lawley (Wiley, Chichester 1982) p.105
2.25 V.S. Letokhov: *Nonlinear Laser Chemistry*, Springer Ser. Chem. Phys., Vol.22 (Springer, Berlin, Heidelberg 1983)
2.26 W.C. Danen, J.C. Jang: In *Laser-Induced Chemical Processes*, ed. by J.I. Steinfeld (Plenum, New York 1981) p.45
2.27 M. Quack: In *Dynamics of the Excited State*, ed. by K.P. Lawley (Wiley, Chichester 1982) p.395
2.28 H.W. Galbraith, J.R. Ackerhalt: In *Laser-Induced Chemical Processes*, ed. by J.I. Steinfeld (Plenum, New York 1981) p.1
2.29 J.L. Lyman, H.G. Galbraith, J.R. Ackerhalt: Los Alamos Sci. **3**(**1**), 66 (1982)
2.30 M. Quack: Chimia **35**, 463 (1981)
2.31 R.G. Harrison, R.G. Butcher: Contem. Phys. **21**, 19 (1980)
2.32 K.L. Kompa: In *Developments in High-Power Lasers and Their Applications*, ed. by C. Pellegrini, Proc. Int. School of Phys. (Enrico Fermi) (North Holland, Amsterdam 1981) p.274
2.33 J.C. Stephenson, D.S. King, M.F. Goodman, J. Stone: J. Chem. Phys. **70**, 4496 (1979)
2.34 M. Quack, G. Seyfang: J. Chem. Phys. **76**, 955 (1982)
2.35 R. Duperrex, H. van den Bergh: J. Chem. Phys. **71**, 3613 (1979)
2.36 P. Kolodner, H.S. Kwok, J.G. Black, E. Yablonovitch: Opt. Lett. **4**, 38 (1979)
2.37 T.P. Cotter: "Analysis of Measurements of Absorption of Laser Pulses in Gases", Res. Rpt. LA-UR-77-2055, Los Alamos Scientific Laboratory (1977)
2.38 J.L. Lyman, S.D. Rockwood, S.M. Freund: J. Chem. Phys. **67**, 4545 (1977)
2.39 C.D. Cantrell, S.M. Freund, J.L. Lyman: In *Laser Handbook*, Vol. IIIb, ed. by M.L. Stitch (North-Holland, Amsterdam 1979) p.485
2.40 S. Speiser, J. Jortner: Chem. Phys. Lett. **7**, 19 (1970)
2.41 O.P. Judd: J. Chem. Phys. **71**, 4515 (1979)
2.42 J.L. Lyman, K.M. Leary: J. Chem. Phys. **69**, 1858 (1978)
2.43 J.L. Lyman, B.J. Feldman, R.A. Fisher: Opt. Commun. **25**, 391 (1978)
2.44 K. Fox, W.B. Person: J. Chem. Phys. **64**, 5218 (1976)
2.45 R.A. Lucht, J.S. Beardall, R.C. Kennedy, G.W. Sullivan, J.P. Rink: Opt. Lett. **4**, 216 (1979)
2.46 J.L. Lyman, W.C. Danen, A.C. Nilsson, A.V. Nowak: J. Chem. Phys. **71**, 1206 (1978)
2.47 G.P. Quigley: Unpublished data, Los Alamos Scientific Laboratory
2.48 F.M. Lussier, J.I. Steinfeld, T.F. Deutsch: Chem. Phys. Lett. **58**, 277 (1978)
2.49 V.N. Bagratashvili, V.S. Doljikov, V.S. Letokhov, E.A. Ryabov: Appl. Phys. **20**, 231 (1979)
2.50 S. Bittenson, P.L. Houston: J. Chem. Phys. **67**, 4819 (1977)
2.51 P.A. Hackett, M. Gauthier, C. Willis: J. Chem. Phys. **69**, 2924 (1978)
2.52 W. Fuss, K.L. Kompa, F.M.G. Tablas: Faraday Discuss. Chem. Soc. 180 (1979)
2.53 V.N. Bagatashvilli, I.N. Knyazev, V.S. Letokhov, V.V. Lobko: Opt. Commun. **18**, 525 (1976)
2.54 R.S. McDowell, H.W. Galbraith, C.D. Cantrell, N.G. Nereson, E.D. Hinkley: J. Mol. Spectrosc. **68**, 288 (1977)
2.55 R.S. McDowell: "Laser Diode Spectra of Spherical-Top Molecules", in *Laser Spectroscopy III*, ed.. by J.L. Hall, J.L. Carlsten, Springer Ser. in Opt. Sci., Vol.7 (Springer, Berlin, Heidelberg 1977) p.102
2.56 A.V. Nowak, J.L. Lyman: J. Quant. Spectrosc. Radiat. Transfer **15**, 945 (1975)
2.57 J.L. Lyman, R.G. Anderson, R.A. Fischer, B.J. Feldman: Opt. Lett. **3**, 238 (1978)
2.58 E. Borsella, R. Fantoni, A. Giardini Guidoni: Nuovo Cimento **63**B, 83 (1981)
2.59 D.K. Evans, R.D. McAlpine, H.M. Adams: J. Chem. Phys. **77**, 3551 (1982)
2.60 S.S. Alimpiev, N.V. Karlov, S.M. Kikiforov, A.M. Prokhorov, B.G. Sartakov, E.M. Khoklov, A.L. Shtarkov: Opt. Commun. **31**, 309 (1979)
2.61 E. Borsella, R. Fantoni, A. Giardini-Guidoni, D. Masci, Ruess: Chem. Phys. Lett. **93**, 523 (1982)
2.62 D.S. King, J.C. Stephenson: Chem. Phys. Lett. **66**, 33 (1979)
2.63 B.J. Orr, J.G. Haub: Opt. Lett. **6**, 236 (1981)
2.64 H. Flicker: Private communication, Los Alamos Laboratory (1979)

2.65 J.P. Aldridge, H. Filip, H. Flicker, R.F. Holland, R.S. McDowell, N.G. Nereson, K. Fox: J. Mol. Spectrosc. **58**, 165 (1975)
2.66 R.S. McDowell, J.P. Aldridge, R.F. Holland: J. Phys. Chem. **80**, 1203 (1976)
2.67 C.D. Cantrell, H.W. Galbraith: J. Mol. Spectrosc. **58**, 158 (1975)
2.68 R.S. McDowell, H.G. Galbraith, B.J. Krohn, C.D. Cantrell, E.D. Hinkley: Opt. Commun. **17**, 178 (1976)
2.69 K. Fox: Opt. Commun. **19**, 397 (1976)
2.70 R.S. McDowell, H.W. Galbraith, C.D. Cantrell, N.G. Nereson, P.F. Moulton, E.D. Hinkley: Opt. Lett. **2**, 97 (1978)
2.71 T. Deutsch: Opt. Lett. **1**, 25 (1977)
2.72 D.O. Ham, M. Rothschild: Opt. Lett. **1**, 28 (1977)
2.73 D.O. Ham: Private communication (1977)
2.74 M. Rothschild, W. Tsay, D.O. Ham: Opt. Commun. **24**, 327 (1978)
2.75 H. Stafast, W.E. Schmidt, K.L. Kompa: Opt. Commun. **21**, 121 (1977)
2.76 R.V. Ambartzumian, Y.A. Gorohkov, V.S. Letokhov, G.N. Makarov: Sov. Phys.-JETP **42**, 993 (1976)
2.77 J.G. Black, P. Kolodner, M.J. Shultz, E. Yablonovitch, N. Bloembergen: Phys. Rev. A**19**, 704 (1979)
2.78 S.S. Alimpiev, N.V. Karlov, B.G. Sartakov, E.M. Khokhlov: Opt. Commun. **26**, 45 (1978)
2.79 A.S. Akhmanov, V.Y. Baranov, V.D. Pismenny, V.N. Bagratashvili, Y.R. Kolomilisky, V.S. Letokhov, E.A. Ryabov: Opt. Commun. **23**, 357 (1977)
2.80 A. Nowak: Private communication, Los Alamos Scientific Laboratory
2.81 W.S. Tsay, C. Riley, D.O. Ham: J. Chem. Phys. **70**, 3558 (1979)
2.82 J.D. Campbell, G. Hancock, J.B. Halpern, K.H. Welge: Opt. Commun. **17**, 38 (1976)
2.83 G. Hancock, R.J. Hennessy, T. Vissis: J. Photochem. **9**, 197 (1978)
2.84 D.S. King, J.C. Stephenson: Chem. Phys. Lett. **51**, 48 (1977)
2.85 J.C. Stephenson, D.S. King: J. Chem. Phys. **69**, 1485 (1978)
2.86 E. Wurzberg, L.J. Kovokenko, P.L. Houston: Chem. Phys. **35**, 317 (1978)
2.87 J.H. Hall, Jr., M.L. Lesiecki, W.A. Guillory: J. Chem. Phys. **68**, 2247 (1978)
2.88 N.V. Chekalin, V.S. Letokhov, V.N. Lokhman, A.N. Shibanov: Chem. Phys. **36**, 415 (1979)
2.89 S.E. Bialkowski, W.A. Guillory: J. Chem. Phys. **67**, 2061 (1977)
2.90 S.E. Bialkowski, W.A. Guillory: J. Chem. Phys. **68**, 3339 (1978)
2.91 M.L. Lesiecki, W.A. Guillory: Chem. Phys. Lett. **49**, 92 (1977)
2.92 M.L. Lesiecki, W.A. Guillory: J. Chem. Phys. **66**, 4239 (1977)
2.93 M.L. Lesiecki, W.A. Guillory: J. Chem. Phys. **69**, 4572 (1978)
2.94 K.W. Hicks, M.L. Lesiecki, S.M. Riseman, W.A. Guillory: J. Phys. Chem. **83**, 1936 (1979)
2.95 S.E. Bialkowski, W.A. Guillory: "Dynamic Processes Generated by the IR Photolysis of CH_3NH_2", preprint (unpublished)
2.96 M.N.R. Ashford, G. Hancock, G. Ketley: Faraday Discuss. Chem. Soc. 204 (1979)
2.97 R.R. Karl, J.L. Lyman: J. Chem. Phys. **69**, 1196 (1978)
2.98 N.R. Isenor, V. Merchant, R.S. Hallsworth, M.C. Richardson: Can. J. Phys. **51**, 1281 (1973)
2.99 R.V. Ambartzumian, V.S. Letokhov, E.A. Ryabov, N.V. Chekalin: JETP Lett. **20**, 273 (1974)
2.100 R.V. Ambartzumian, G.N. Makarov, A.A. Puretzky: Opt. Commun. **27**, 79 (1978)
2.101 C.R. Quick, Jr., C. Wittig: Chem. Phys. Lett. **48**, 420 (1977)
2.102 J.M. Preses, R.E. Weston, Jr., G.W. Flynn: Chem. Phys. Lett. **48**, 425 (1977)
2.103 J.J. Tiee, C. Wittig: Appl. Phys. Lett. **32**, 236 (1978)
2.104 J.J. Tiee, C. Wittig: Opt. Commun. **27**, 377 (1978)
2.105 C.R. Quick, Jr., C. Wittig: Chem. Phys. **32**, 75 (1978)
2.106 J.L. Lyman, R.J. Jensen: Chem. Phys. Lett. **13**, 421 (1972)
2.107 D. Proch, H. Schröder: Chem. Phys. Lett. **61**, 426 (1979)
2.108 F. Brunner, T.P. Cotter, K.L. Kompa, D. Proch: J. Chem. Phys. **67**, 1547 (1977)
2.109 M.J. Coggiola, P.A. Schultz, Y.T. Lee, Y.R. Shen: Phys. Rev. Lett. **38**, 17 (1977)
2.110 E.R. Grant, P.A. Schultz, Aa.S. Sudbo, M.J. Coggiola, Y.T. Lett, Y.R. Shen: "Multi-Photon Dissociation of Polyatomic Molecules Studied with a Molecular

Beam", in *Laser Spectroscopy III*, ed. by J.L. Hall, J.L. Carlsten, Springer Ser. in Opt. Sci., Vol.7 (Springer, Berlin, Heidelberg 1977) p.94
2.111 Aa.S. Sudbo, P.A. Schultz, E.R. Grant, Y.R. Shen, Y.T. Lee: J. Chem. Phys. **70**, 912 (1979)
2.112 E.R. Grant, M.J. Coggiola, Y.T. Lee, P.A. Schultz, Aa.S. Sudbo, Y.R. Shen: Chem. Phys. Lett. **52**, 595 (1977)
2.113 Aa.S. Sudbo, P.A. Schultz, E.R. Grant, Y.R. Shen, Y.T. Lee: J. Chem. Phys. **67**, 1306 (1978)
2.114 E.R. Grant, P.A. Schultz, Aa.S. Sudbo, Y.R. Shen, Y.T. Lee: Phys. Rev. Lett. **40**, 115 (1978)
2.115 G.J. Diebold, F. Engelke, D.M. Lubman, J.C. Whitehead, R.N. Zare: J. Chem. Phys. **67**, 5407 (1977)
2.116 N. Bloembergen, E. Yablonovitch: Phys. Today **31**, No. 5, 23 (1978)
2.117 P.J. Robinson, K.A. Holbrook: *Unimolecular Reactions* (Wiley, London 1972)
2.118 M.F. Goodman, J. Stone, D.A. Dows: J. Chem. Phys. **65**, 5052 (1976)
2.119 J.F. Bott, T.A. Jacobs: J. Chem. Phys. **50**, 3870 (1969)
2.120 R.K. Curran: J. Chem. Phys. **34**, 1069 (1961)
2.121 P. Harland, J.C. Thynne: J. Phys. Chem. **73**, 4031 (1969)
2.122 C.L. Chen, P.J. Chantry: Bull. Am. Phys. Soc. **15**, 418 (1970)
2.123 T. Kiang, R.C. Estler, R.N. Zare: J. Chem. Phys. **70**, 5925 (1979)
2.124 D.L. Hildenbrandt: J. Phys. Chem. **77**, 897 (1973)
2.125 B.S. Rabinovitch, D.G. Keil, J.F. Burkhalter, G.B. Skinner: In *Modern Developments in Shock Tube Research*, Proc. 10th Int. Shock Tube Symp., ed. by G. Kaminoto (Japan 1975) p.579
2.126 S.W. Benson: Chem. Rev. **78**, 23 (1978)
2.127 J.L. Lyman: J. Chem. Phys. **67**, 1868 (1977)
2.128 D.M. Brenner: Chem. Phys. Lett. **57**, 357 (1978)
2.129 D.F. Heller, G.A. West: Chem. Phys. Lett. **69**, 419 (1980)
2.130 A.V. Nowak, D.O. Ham: Opt. Lett. **6**, 185 (1980)
2.131 P. Gozel, D. Braichotte, H. van den Bergh: J. Chem. Phys. **79**, 4924 (1983)
2.132 M.E. Gower, K.W. Billman: Opt. Commun. **20**, 123 (1977)
2.133 F. Brunner, D. Proch: J. Chem. Phys. **68**, 4936 (1978)
2.134 J.L. Lyman: Ph.D. Dissertation, Brigham Young University, Provo (1973)
2.135 P. Lavigne, J.L. Lachambre, G. Otis: Opt. Commun. **25**, 75 (1977)
2.136 P. Kolodner, C. Winterfield, E. Yablonovtich: Opt. Commun. **21**, 112 (1977)
2.137 V.N. Bagratashvili, Yu.G. Vainer, V.S. Doljikov, S.F. Koliakov, A.A. Makarov, L.P. Malyavkin, E.A. Ryabov, E.G. Silkis, V.D. Titov: Appl. Phys. **22**, 101 (1980)
2.138 H. Frei, G.C. Pimentel: J. Chem. Phys. **78**, 3698 (1983)
2.139 D.M. Brenner, K. Brezinsky: Chem. Phys. Lett. **67**, 36 (1979)
2.140 X. De Hemptinne, D. De Keuster: J. Chem. Phys. **73**, 3170 (1980)
2.141 D.M. Cox, J.A. Horsley: J. Chem. Phys. **72**, 864 (1980)
2.142 J.-S.J. Choo, E.R. Grant: J. Chem. Phys. **74**, 5679 (1981)
2.143 G.P. Quigley. "Collisional Effects in Multiple Photon IR Absorption", in *Advances in Laser Chemistry*, ed. by A.H. Zewail, Springer Ser. Chem. Phys., Vol.3 (Springer, Berlin, Heidelberg 1978) p.374
2.144 G.P. Quigley: Opt. Lett. **3**, 106 (1978)
2.145 G.P. Quigley, J.L. Lyman: "Collisional Effects in the Multiple-Photon Infrared Laser Pumping of Polyatomic Molecules", in *Laser-Induced Processes in Molecules*, ed. by K.L. Kompa, S.D. Smith, Springer Ser. Chem. Phys., Vol.6 (Springer, Berlin, Heidelberg 1979) p.134
2.146 J.D. Lambert: *Vibrational and Rotational Relaxation in Gases* (Clarendon, Oxford 1977)
2.147 J.I. Steinfeld, I. Burak, G.D. Sutton, A.V. Nowak: J. Chem. Phys. **52**, 5421 (1970)
2.148 V.N. Bagratashvili, V.S. Dolzhikov, V.S. Letokhov: Sov. Phys. JETP **49**, 8 (1979)
2.149 D.C. Tardy, B.S. Rabinovitch: Chem. Rev. **77**, 369 (1977)
2.150 A.C. Baldwin, H. van den Bergh: J. Chem. Phys. **74**, 1012 (1981)
2.151 R. Duperrex, H. van den Bergh: J. Chem. Phys. **75**, 3371 (1981)
2.152 K. Takeuchi, I. Inque, R. Nakane, Y. Makide, S. Kato, T. Tominaga: J. Chem. Phys. **76**, 398 (1982)

2.153 A. Makide, S. Hagiwara, T. Tominaga, K. Takeuchi, R. Nakane: Chem. Phys. Lett. **82**, 18 (1981)
2.154 A. Hartford, Jr., S.A. Tuccio: Chem. Phys. Lett. **60**, 431 (1979)
2.155 P.A. Hackett, V. Malatesta, W.S. Nip, C. Willis, P.B. Corkum: J. Phys. Chem. **85**, 1152 (1981)
2.156 J.S. Francisco, Z. Qingshi, J.I. Steinfeld: J. Chem. Phys. **78**, 5339 (1983)
2.157 E. Borsella, R. Fantoni, A. Giardini-Guidoni, G. Sanna: Chem. Phys. Lett. **72**, 25 (1980)
2.158 A. Hason, P. Gozel, H. van den Bergh: Helv. Phys. Acta **55**, 187 (1982)
2.159 A. Gandini, C. Willis, R.A. Back: Can. J. Chem. **55**, 4156 (1977)
2.160 V.S. Doljikov, Yu R. Kolomisky, E.A. Ryabov: Chem. Phys. Lett. **80**, 433 (1981)
2.161 T.F. Deutsch, S.R.J. Brueck: Chem. Phys. Lett. **54**, 258 (1978)
2.162 P.F. Moulton, D.M. Larson, J.N. Walpole, A. Mooradian: Opt. Lett. **1**, 51 (1977)
2.163 S. Mukamel, J. Jortner: J. Chem. Phys. **65**, 5204 (1976)
2.164 R. Duperrex, H. van den Burgh: Chem. Phys. **40**, 275 (1979)
2.165 R. Naaman, R.N. Zare: (unpublished)
2.166 J.R. Ackerhalt, H.W. Galbraith: J. Opt. Soc. Am. **70**, 598 (1980)
2.167 P.J. Marcoux, D.W. Setser: J. Phys. Chem. **82**, 97 (1978)
2.168 D.F. Denver, E. Grunwald: J. Am. Chem. Soc. **98**, 5055 (1976)
2.169 A. Kaldor, R.B. Hall, D.M. Cox, J.A. Horsley, P. Rabinowitz, G.M. Kramer: J. Am. Chem. Soc. **101**, 4465 (1979)
2.170 R.V. Ambartzumian, N.V. Chekalin, V.S. Doljikov, V.S. Letokhov, V.N. Lokhman: Opt. Commun. **18**, 400 (1976)
2.171 R.B. Hall, A. Kaldor: J. Chem. Phys. **70**, 4027 (1979)
2.172 A. Yogev, R.M.J. Benmair: Chem. Phys. Lett. **46**, 290 (1977)
2.173 R.N. Rosenfeld, J.I. Bauman, J.R. Barker, D.M. Golden: J. Am. Chem. Soc. **99**, 8063 (1977)
2.174 W.C. Danen: J. Am. Chem. Soc. **101**, 1187 (1979)
2.175 D. Gutman, W. Braun, W. Tsang: J. Chem. Phys. **67**, 4291 (1977)
2.176 T.H. Richardson, D.W. Setser: J. Phys. Chem. **81**, 2301 (1977)
2.177 V.M. Akulin, S.S. Alimpiev, N.V. Karlov, B.G. Sartakov, L.A. Shelepin: Sov. Phys.-JETP **44**, 239 (1976)
2.178 R.V. Ambartzumian, Yu.A. Gorokhov, V.S. Letokhov, G.N. Makarov, A.A. Puretskii: Sov. Phys.-JETP **44**, 231 (1976)
2.179 R.V. Ambartzumian, Yu.A. Gorokhov, V.S. Letokhov, G.N. Makarov, E.A. Ryabov, N.V. Chekalin: Sov. J. Quantum Electron. **6**, 437 (1976)
2.180 R. Corkum, C. Willis, R.A. Back: Chem. Phys. **24**, 13 (1977)
2.181 M.L. Lesiecki, W.A. Guillory: J. Chem. Phys. **66**, 4317 (1977)
2.182 N.G. Basov, V.T. Galochkin, V.G. Kartyshov, A.G. Lyapin, I.M. Mazurin, A.N. Oraevskii, N.F. Starodubtsev: Zh. Eksp. Teor. Fiz. **72**, 918 (1977) [English transl.: Sov. Phys.-JETP **45**, 479 (1977)]
2.183 V.E. Merchant: Opt. Commun. **25**, 259 (1978)
2.184 R.E. Huie, J.T. Herron, W. Braun, W. Tsang: Chem. Phys. Lett. **56**, 193 (1978)
2.185 G. Yahav, Y. Haas: Chem. Phys. **35**, 41 (1978)
2.186 S.T. Lin, S.M. Lee, A.M. Ronn: Chem. Phys. Lett. **53**, 260 (1978)
2.187 Tsang, J.A. Walker, W. Braun, J.T. Herron: Chem. Phys. Lett. **59**, 487 (1978)
2.188 R.V. Ambartzumian, B.I. Vasil'ev, A.Z. Grasyuk, A.P. Dyadkin, V.S. Letokhov, N.P. Furzikov: Sov. J. Quantum Electron. **8**, 1015 (1978)
2.189 T.F. Deutsch: J. Chem. Phys. **70**, 1187 (1979)
2.190 H.M. Frey, R. Walsh: Chem. Rev. **69**, 103 (1969)
2.191 W.F. Libby: J. Chem. Phys. **11**, 101 (1943)
2.192 D.R. Stull, H. Prophet: *JANAF Thermochemical Tables*, 2nd ed. (National Bureau of Standards, Washington, D.C. 1971)
2.193 H.G. Schecker, W. Jost: Ber. Bunsenges. Phys. Chem. **73**, 521 (1969)
2.194 D.W. Davidson, B.P. Stoicheff, H.J. Bernstein: J. Chem. Phys. **22**, 289 (1954)
2.195 P. Fettweis, M. Neve de Mevengnies: J. Appl. Phys. **49**, 5699 (1978)
2.196 E. Grunwald, K.J. Olszyna, D.F. Dever, B. Knishkowy: J. Am. Chem. Soc. **99**, 6515 (1977)
2.197 G.M. Schwab, A.M. Watson: Trans. Faraday Soc. **60**, 1833 (1964)
2.198 T.Shimanouchi: *Tables of Molecular Vibrational Frequencies Consolidated*, Vol. 1, NSRDS NBS 39 (National Bureau of Standards, Washington, D.C. 1972)
2.199 R.S. McDowell, M. Goldblatt: Inorg. Chem. **10**, 625 (1971)

2.200 I.P. Herman, J.B. Marling: Chem. Phys. Lett. **64**, 75 (1979)
2.201 J.A. Kerr: Chem. Rev. **66**, 465 (1966)
2.202 F.W. Schneider, B.S. Rabinovitch: J. Am. Chem. Soc. **84**, 4215 (1962)
2.203 M.G. Krishna Pillai, F.F. Cleveland: J. Mol. Spectrosc. **5**, 212 (1960)
2.204 H.J. Bernstein, D.A. Ramsay: J. Chem. Phys. **17**, 556 (1949)
2.205 F.M. Lussier, J.I. Steinfeld: Chem. Phys. Lett. **50**, 175 (1977)
2.206 S. Enomoto, M. Asahina: J. Mol. Spectrosc. **19**, 117 (1966)
2.207 B. Bak, D. Christensen: Spectrochim. Acta **12**, 355 (1958)
2.208 N.C. Peterson, R.G. Manning, W. Braun: J. Res. Natl. Bur. Stand. **83**, 117 (1978)
2.209 T. Shimanouchi: *Tables of Molecular Vibrational Frequencies*, Part 2, NSRDS NBS 11 (National Bureau of Standards, Washington, D.C. 1967)
2.210 J.E. Griffiths: Spectrochim. Acta **23**A, 2145 (1967)
2.211 J. Gaunt: Trans. Faraday Soc. **49**, 1122 (1953)
2.212 D.L. Hildenbrand: J. Chem. Phys. **66**, 4788 (1977)
2.213 R.S. McDowell, L.B. Asprey, R.T. Paine: J. Chem. Phys. **61**, 3571 (1974)
2.214 J.R. Nielsen, H.H. Claassen, D.C. Smith: J. Chem. Phys. **18**, 1471 (1950)
2.215 D.C. Smith, R.A. Saunders, J.R. Nielsen, E.E. Ferguson: J. Chem. Phys. **20**, 847 (1952)
2.216 K.M. Maloney, B.S. Rabinovitch: J. Phys. Chem. **73**, 1652 (1969)
2.217 E.L. Pace, A.C. Plausch, H.V. Samuelson: Spectrochim. Acta **22**, 993 (1966)
2.218 J.R. Nielsen, M.Z. El-Sabban, M. Alpert: J. Chem. Phys. **23**, 324 (1955)
2.219 J.K. Wilmhurst, H.J. Bernstein: Can. J. Chem. **35**, 191 (1957)
2.220 J.M. Preses, R.E. Weston, Jr., G.W. Flynn: Chem. Phys. Lett. **46**, 69 (1977)
2.221 R.P. Bauman, B.J. Bulkin: J. Chem. Phys. **45**, 496 (1966)
2.222 W.C. Danen, W.D. Munslow, D.W. Setser: J. Am. Chem. Soc. **99**, 6961 (1977)
2.223 Y. Haas, G. Yahav: Chem. Phys. Lett. **48**, 63 (1977)
2.224 *The Sadtler Standard Spectra, Gases and Vapors* (Sadtler Research Laboratories, Philadelphia 1972)
2.225 S.E. Bialkowski, S.W. Guillory: Chem. Phys. Lett. **60**, 429 (1979)
2.226 J.L. Lyman, S.D. Rockwood: J. Appl. Phys. **47**, 595 (1976)
2.227 R.V. Ambartzumian, N.V. Chekalin, V.S. Doljikov, V.S. Letokhov, E.A. Ryabov: Chem. Phys. Lett. **25**, 515 (1974)
2.228 R.V. Ambartzumian, V.S. Doljikov, V.S. Letokhov, E.A. Ryabov, N.V. Chekalin: Sov. Phys.-JETP **42**, 36 (1975)
2.229 N.V. Bourimov, V.S. Letokhov, E.A. Ryabov: J. Photochem. **5**, 49 (1976)
2.230 G. Koren, M. Okon, U.P. Oppenheim: Opt. Commun. **22**, 351 (1977)
2.231 G. Koren, U.P. Oppenheim, D. Tal, M. Okon, R. Weil: Appl. Phys. Lett. **29**, 40 (1976)
2.232 G. Koren, U.P. Oppenheim: Opt. Commun. **26**, 449 (1978)
2.233 A. Hartford, Jr.: Chem. Phys. Lett. **57**, 352 (1978)
2.234 R.V. Ambartzumian, Yu.A. Gorokhov, V.S. Letokhov, G.N. Makarov, A.A. Puretzki: Phys. Lett. **56**A, 183 (1976)
2.235 R.V. Ambartzumian, Yu.A.. Gorokhov, V.S. Letokhov, A.A. Puretskii: JETP Lett. **22**, 177 (1977)
2.236 R.V. Ambartzumian, N.P. Furzikov, V.S. Letokhov, A.P. Dyad'kin, A.Z. Grasyuk, B.I. Vasil'yev: Appl. Phys. **15**, 27 (1978)
2.237 J.W. Hudgens: J. Chem. Phys. **68**, 777 (1978)
2.238 J.J. Ritter: J. Am. Chem. Soc. **100**, 2441 (1978)
2.239 J.J. Ritter, S.M. Freund: J. Chem. Soc., Chem. Commun. 812 (1976)
2.240 G.A. Hill, E. Grunwald, P. Keehan: J. Am. Chem. Soc. **99**, 6521 (1977)
2.241 G. Folcher, W. Braun: J. Photochem. **8**, 341 (1978)
2.242 W. Braun, J.T. Herron, W. Tsang, K. Churney: Chem. Phys. Lett. **59**, 492 (1978)
2.243 S.T. Lin, A.M. Ronn: Chem. Phys. Lett. **49**, 255 (1977)
2.244 M. Gauthier, P.A. Hackett, M. Drouin, R. Pilon, C. Willis: Can. J. Chem. **56**, 2227 (1978)
2.245 J.B. Marling, I.P. Herman, S.J. Thomas: J. Chem. Phys. **72**, 5603 (1980)
2.246 V.N. Bagratashvili, V.S. Doljikov, V.S. Letokhov, E.A. Ryabov: "Study of Primary Characteristics of Multiple IR Photon Excitation and Dissociation of CF_3I", in *Laser-Induced Processes in Molecules*, ed. by K.L. Kompa, S.D. Smith, Springer Ser. Chem. Phys., Vol.6 (Springer, Berlin, Heidelberg 1979) p.179

2.247 I.N. Knyazev, Yu.A. Kudriavtzev, V.S. Letokhov, A.A. Sarkisian: Appl. Phys. **17**, 427 (1978)
2.248 Z. Karny, A. Gupta, R.N. Zare, S.T. Lin, T. Neiman, A.M. Ronn: Chem. Phys. **37**, 15 (1979)
2.249 D.K. Evans, R.D. McAlpine, F.K. McClusky: Chem. Phys. **32**, 81 (1978)
2.250 R.V. Ambartzumian, Yu.A. Gorokhov, G.N. Makarov, A.A. Puretzki, N.P. Furzikov: Chem. Phys. Lett. **45**, 231 (1977)
2.251 R.V. Ambartzumian, V.S. Letokhov, G.N. Makarov, A.A. Puretzki: Opt. Commun. **25**, 69 (1978)
2.252 N.R. Isenor, M.C. Richardson: Appl. Phys. Lett. **18**, 224 (1971)
2.253 R.V. McAlpine, D.K. Evans, F.K. McClusky: Chem. Phys. **39**, 263 (1979)
2.254 C. Reiser, F.M. Lussier, C.C. Jensen, J.I. Steinfeld: J. Am. Chem. Soc. **101**, 350 (1979)
2.255 Z. Karny, R.N. Zare: Chem. Phys. **23**, 321 (1977)
2.256 C.R. Quick, C. Wittig: J. Chem. Phys. **69**, 4201 (1978)
2.257 O.N. Avatkov, V.N. Bagratashvili, I.N. Knyazev, Yu.R. Kolomiiskii, V.S. Letokhov, V.V. Lobko, E.A. Ryabov: Kvantovaya Electron. **4**, 741 (1977) [English transl.: Sov. J. Quantum Electron. **7**, 412 (1977)]
2.258 A. Hartford, Jr.: Chem. Phys. Lett. **53**, 503 (1978)
2.259 K.M. Leary, J.L. Lyman, L.B. Asprey, S.M. Freund: J. Chem. Phys. **68**, 1671 (1978)
2.260 W. Fuss, T.P. Cotter: Appl. Phys. **12**, 265 (1977)
2.261 J.D. Campbell, G. Hancock, K.H. Welge: Chem. Phys. Lett. **43**, 581 (1976)
2.262 R. Duperrex, H. van den Burgh: J. Chem. Phys. **70**, 5672 (1979)
2.263 D. Tal, U.P. Oppenheim, G. Koren, M. Okon: Chem. Phys. Lett. **48**, 67 (1977)
2.264 H.N. Rutt: J. Phys. D**10**, 869 (1977)
2.265 V.N. Bagratashvili, Yu.R. Kolomisky, V.S. Letokhov, E.A. Ryabov, V.Hu. Baranov, S.A. Kazakov, V.G. Nizijev, V.D. Pismenny, A.I. Starodubtser, E.P. Velokhov: Appl. Phys. **14**, 217 (1977)
2.266 P. Bado, H. van den Burgh: J. Chem. Phys. **68**, 4188 (1978)
2.267 G.F. Nutt, B.J. Orr: Opt. Commun. **29**, 57 (1979)
2.268 J.J. Tiee, C. Wittig: J. Chem. Phys. **69**, 4756 (1978)
2.269 S.M. Freund, J.L. Lyman: Chem. Phys. Lett. **55**, 435 (1978)
2.270 P. Rabinowitz, A. Stein, A. Kaldor: Opt. Commun. **27**, 381 (1978)
2.271 D.M. Cox, A. Stein, P. Rabinowitz, R. Brickman, J.A. Horsley, A. Kaldor: Proc. Tech. Programm-Electro Optics/Laser 1978 Conf. Expo. 1978, 140. Ind. Sci. Conference Manage. Inc., Chicago, Ill.
2.272 G.A. West, R.E. Weston, Jr., G.W. Flynn: Chem. Phys. **35**, 275 (1978)
2.273 A.J. Calussi, S.W. Benson, R.J. Hwang, J.J. Tiee: Chem. Phys. Lett. **52**, 349 (1977)
2.274 A.V. Baklanov, A.K. Petrov, Yu.N. Molin: Dokl. Phys. Chem. **242**, 808 (1978)
2.275 W.C. Danen: Opt. Eng. **19**, 21 (1980)
2.276 V.V. Vishin, Yu.N. Molin, A.K. Petrov, A.R. Sorokin: Appl. Phys. **17**, 385 (1978)
2.277 H.H. Claasen: J. Chem. Phys. **18**, 543 (1950)
2.278 V.V. Vizhin, A.K. Petrov, Yu.N. Molin: Dokl. Phys. Chem. **243**, 907 (1978)
2.279 H.A. Szymanski: *Interpreted Infrared Spectra*, Vol.3 (Plenum, New York 1967)
2.280 Y. Haas, G. Yahav: J. Amer. Chem. Soc. **100**, 4885 (1978)
2.281 S.S. Miller, D.D. DeFord, T.J. Marks, E. Weitz: J. Am. Chem. Soc. **101**, 1036 (1979)
2.282 D.M. Cox, R.B. Hall, J.A. Horsley, G.M. Kramer, P. Rabinowitz, A. Kaldor: Science **205**, 390 (1979)
2.283 V. Starov, C. Steel, S. Butcher, R.G. Harrison, P. John, R. Leo: "Excitation of Ammonia in the Megawatt Region Using a CO_2 Laser", in *Laser-Induced Processes in Molecules*, ed. by K.L. Kompa, S.D. Smith, Springer Ser. Chem. Phys., Vol.6 (Springer, Berlin, Heidelberg 1979) p.201
2.284 R.V. Ambartzumian, Yu.A. Gorokhov, V.S. Ketokhov, G.N. Makarov, E.A. Ryabov, N.V. Chekalin: Sov. J. Quantum Electron. **5**, 1196 (1976)
2.285 S.M. Freund, J.J. Ritter: Chem. Phys. Lett. **32**, 255 (1975)
2.286 R.E. McNair, S.F. Fulghum, G.W. Flynn, M.S. Feld, B.J. Feldman: Chem. Phys. Lett. **48**, 241 (1977)

2.287 A.N. Oraevskii, A.V. Pankratov, A.N. Skachkov, V.M. Shabarshin: High Energy Chem. **12**, 48 (1978)
2.288 R. Bhatnagar, P.E. Dyer, G.A. Oldershaw: "Laser-Induced Decomposition of Methanol: A Comparative Study Using Pulsed HF and CO_2 Lasers", in *Laser-Induced Processes in Molecules*, ed. by K.L. Kompa, S.D. Smith, Springer Ser. Chem. Phys., Vol.6 (Springer, Berlin, Heidelberg 1979) p.195
2.289 J. Dupré, P. Pinson, J. Dupré-Maquaire, C. Meyer, P. Barchewitz: C.R. Acad. Sci. **283**, 311 (1976)
2.290 G. Hancock, J.D. Campbell, K.H. Welge: Opt. Commun. **16**, 177 (1976)
2.291 Yu.I. Arkhangel'skii, V.D. Klimov, V.A. Kuz'menko, V.A. Legasov, S.L. Nedoseev: Dokl. Phys. Chem. **235**, 781 (1977)
2.292 A. Kaldor, P. Rabinowitz, D.M. Cox, J.A. Horsley, R. Brickman: J. Opt. Soc. Am. **68**, 684 (1978)
2.293 R.L. Woodin, D.S. Bomse, J.L. Beauchamp: Chem. Phys. Lett. **63**, 630 (1979)
2.294 D.S. Bomse, R.L. Woodin, J.L. Beauchamp: J. Am. Chem. Soc. **101**, 5503 (1979)

3. Molecular-Beam Studies of Laser-Induced Multiphoton Dissociation

A. S. Sudbø, P. A. Schulz, Y. R. Shen, and Y. T. Lee*

With 18 Figures

We review the evidence from molecular-beam experiments establishing that laser-induced multiphoton dissociation is a collisionless unimolecular process. We interpret the experimental results using a model in which stepwise incoherent multiphoton absorption excites the molecule through the quasi continuum and above the dissociation threshold, and in which dissociation is described by RRKM theory. We find good agreement between the measured and calculated translational energy distributions.

3.1 Overview

Laser-induced multiphoton dissociation (MPD) of molecules is a collisionless unimolecular process. In the early studies, the collisionless nature of the process was inferred by the observation of instantaneous luminescence from the dissociation products following laser excitation and by the observation of linear pressure dependence of the dissociation yield at sufficiently low gas pressures [3.1-4]. The experiments were done in gas cells in which excitation may be assumed collisionless if the laser pulse is much shorter than the mean free time between collisions of the molecules. Even so, after the laser pulse is over, molecular collisions in the gas cell are still unavoidable, leading to possible collisional dissociation of excited molecules and chemical reactions among dissociation products and excited parent molecules. Thus, it is generally recognized that the primary dissociation fragments cannot be unambiguously identified in the gas cell experiments.

The best way to study a collisionless process is, of course, in a collisionless experiment. A molecular beam method is therefore most appropriate for the study of MPD. Indeed, observation of infrared MPD in a molecular beam provides the most direct evidence that the process is collisionless [3.5]. The use of a mass spectrometer to detect the dissociation fragments from the beam allows us to identify the primary dissociation products in a straightforward way [3.5,6]. The latter information is in fact of great fundamental importance for the understanding of MPD,

*This work was supported by the Division of Advanced Systems Materials Production, Office of Advanced Isotope Separation, U.S. Department of Energy under contract No. W-7405-Eng-48.

because it reveals whether there is any correlation between the pattern of molecular dissociation and the vibrational mode through which the initial excitation is attained.

With a sophisticated molecular beam apparatus, which we shall describe later in this chapter, the angular and velocity distributions of the fragments can also be measured. From these measurements, much additional information about the dynamics of MPD can be deduced. There are a number of important questions, aside from major dissociation fragments and channels, that need to be answered before a reasonable physical understanding of MPD can be achieved.

1) What is the excitation mechanism? How does a molecule get excited through the discrete levels into the quasi-continuum states and then through the quasi continuum into the dissociation continuum states? Does the excitation energy remain in the vibrational mode being excited or get quickly randomized into other modes as the level of excitation increases? How many photons does each molecule absorb before dissociation, or equivalently, what is the average excitation level from which a molecule will dissociate? What eventually limits the level of excitation?

2) Through which channel (or channels) does the molecule dissociate and how does this depend on the laser excitation? Is the major dissociation channel in MPD different from that in thermal decomposition? What is the dissociation rate of a molecule and how does it depend on the level of excitation? Can competing dissociation channels exist with dissociation rates depending on laser excitation?

3) What is the dynamics of dissociation? How does the energy available to the fragments distribute itself among the various degrees of freedom (translation, rotation, and vibration) of the fragments? Can the fragments absorb more laser energy and undergo a secondary MPD?

Most of these questions are usually difficult to answer from the analysis of final products in the gas cell experiments. However, as we shall see in this chapter, they can be and have been answered by measuring angular and velocity distributions of the fragments in molecular-beam experiments. In addition to the molecular-beam experiments, the newly developed laser-induced fluorescence technique for detecting a small number of molecules has been used to detect dissociation fragments from collisionless MPD in a low-pressure gas cell [3.7-12]. Because of the good spatial and temporal resolution of the probing pulse, this technique can also yield information on the dissociation dynamics of the fragments. In principle, this detection method is sometimes even superior to the usual mass spectrometric detection method used in most molecular-beam experiments, in the sense that it can also measure the rotational and vibrational energy distributions in the fragments. In practice, however, this technique is limited to some smaller fragments by the fact that the optical transitions of many larger dissociation fragments either are not known, are too complicated, or cannot be reached by the available probe laser.

As has been stressed in other chapters, MPD is isotopically selective. It has also been proposed that MPD is mode-selective, that is, the excitation energy should remain to a large extent in the vibrational mode being excited. If this were true, the dissociation products could be different from those expected in thermal decomposition, and the application of MPD to chemical synthesis could lead to a revolutionary change in the field. So far, however, aside from some erroneous conclusions, no concrete evidence of mode-selective MPD has been reported. The molecular beam experiments on many molecules described here have shown that in the infrared MPD process the rate of intramolecular energy transfer of dissociating molecules is faster than the rate of dissociation, such that the statistical theory of unimolecular decomposition [3.13] can be used to describe the dissociation of excited molecules satisfactorily. This is not really surprising, as will be discussed later, in view of the fact that the energy deposition rate as well as the rate of dissociation is rather slow compared with intramolecular energy transfer rates. Indeed, the statistical theory, used convincingly to explain our molecular-beam experiments, is the key to answering most of the questions concerning the dynamics of MPD. It also allows us to establish a simple and reliable phenomenological model which enables us to understand and predict the MPD process more quantitatively.

In this chapter, our emphasis is on the study of MPD in a molecular beam with high-power infrared lasers. We shall first review the theory of unimolecular dissociation and the theory of multiphoton excitation of molecules. We shall then describe the experimental apparatus and the experimental results, followed by a thorough discussion of the results, the interpretation, the various aspects of the problem, and our understanding of the MPD process.

3.2 Theory

The infrared multiphoton dissociation of molecules can be considered as a multiphoton excitation process followed by a dissociation process. In the multiphoton excitation, the energy deposition in the molecule, after initial excitation over discrete energy levels, is mainly through stepwise incoherent one-photon transitions [3.14]. A simple phenomenological model using a set of rate equations can describe the time-dependent excitation of the molecules to and beyond the dissociation level. Prior to dissociation, the energy deposited is likely to be randomly distributed among all the vibrational modes of the molecule. Consequently, a simple statistical theory of unimolecular dissociation can be used to describe the dynamics of dissociation. Here, for the later discussion of the experimental results, we shall briefly review the theory of incoherent multiphoton excitation and the subsequent unimolecular dissociation.

We begin with a review of some basic ideas of the theory of thermal unimolecular reactions [3.13]. The reaction scheme used in the well-known RRKM theory is

$$A + M \underset{k_2}{\overset{\delta k_1(E^* \rightarrow E^* + \delta E^*)}{\rightleftarrows}} A^*(E^*) + M$$

$$A^*(E^*) \xrightarrow{k_a(E^*)} A^+ \xrightarrow{k^+} \text{products} \quad . \tag{3.1}$$

The reactant molecules A are energized and deenergized by collisions to and from the energy range between E^* and $E^* + \delta E^*$ with the rate constants δk_1 and k_2 respectively. The energized molecules A^* are converted to activated complexes A^+ with a rate constant k_a. The activated complexes subsequently dissociate with a rate constant k^+. Assuming the processes of activation and deactivation to be essentially single-step processes (the so-called strong collision assumption), the quantity $\delta k_1/k_2$ may be equated to the equilibrium Boltzmann distribution of molecules in the given energy range,

$$\frac{\delta k_1(E^* \rightarrow E^* + \delta E^*)}{k_2} = \frac{N^*(E^*)\, e^{-E^*/kT}}{Q_2} \cdot \delta E^* \quad , \tag{3.2}$$

where $N^*(E^*)$ is the density of quantum states at energy E^* and Q_2 is the molecular partition function for all the active modes of A. Since relatively large amounts of energy are generally transferred in molecular collisions of highly excited polyatomic molecules (5 kcal mol^{-1} or more) and since the average excitation energies in thermal reactions are typically 5-15 kcal mol^{-1} above the critical energy E_0, the strong collision assumption is reasonably good for most thermal reactions. The "equilibrium hypothesis," which assumes that the steady-state concentration of activated complexes A^+ is equal to the equilibrium concentration which would be present if no reaction were occurring, is really implicit in the strong collision assumption.

The second major assumption of the RRKM theory is the energy randomization hypothesis, which states that the excitation energy in the active vibrational and rotational degrees of freedom is subject to rapid statistical redistribution. With this assumption, the rate of formation of the "critical configuration" (or activated complex) from the energized molecules, $k_a(E^*)$, is just proportional to the ratio of the sum of all vibrational and rotational quantum states available to the critical configuration with vibrational-rotational energy in the range from $E^+ = 0$ to $E^* - E_0$ to the density of quantum states in the energized molecule at energy E^*,

$$k_a(E\ ) \propto \frac{\sum\limits_{E^+=0}^{E^*-E_0} N^+(E^+)\Delta E^+}{N^*(E^*)} \tag{3.3}$$

This is the quantum-mechanical equivalent of taking the ratio of the volumes in phase space corresponding to the critical configuration and the energized molecule, respectively. The energized molecules $A^*(E^*)$ will then dissociate in time with a probability distribution of the form

$$p(t) \propto e^{-t/\tau^*} \tag{3.4}$$

where $\tau^*(E^*) = 1/k_a(E^*)$ represents an average dissociation lifetime.

The rate of formation of activated complexes with translational energy, in the reaction coordinate, between E_t and $E_t + \Delta E_t$ is

$$k_a(E^*, E_t)^+ \propto \frac{{}^+N(E^* = E - E_0 - E_t)\Delta E_t}{N^*(E)^*} \quad . \tag{3.5}$$

This expression can be used to describe correctly the translational energy distribution of the dissociation products in the center-of-mass coordinates only if there are no interactions between the fragments after the critical configuration is passed. When this condition is met for a moderately large polyatomic molecule without excessive angular momentum, the translational energy distribution is expected to peak near zero and fall off roughly exponentially, leaving most of the excess energy in the internal degrees of freedom of the products.

In the high-pressure limit, the rate constant of thermal unimolecular dissociation can be approximated by

$$k = -\frac{1}{[A]}\frac{d[A]}{dt} = \int_{E^*=E_0}^{\infty} k_a(E^*) \frac{\delta k_1(E^* \rightarrow E^* + \delta E^*)}{k_2} \quad , \tag{3.6}$$

which reduces to the well-known Arrhenius form

$$k_\infty = Ae^{-E_0/kT} \quad . \tag{3.7}$$

The Arrhenius-type dependence of the high-pressure rate constant is simply a consequence of averaging the microscopic rate constant $k_a(E^*)$ over an equilibrium Boltzmann distribution of energized molecules with $E^* > E_0$.

The two basic assumptions of RRKM theory mentioned above are quite different in nature. The energy randomization hypothesis is directed toward single isolated molecules and permits calculation of the microscopic rate constant k_a as a function of excess energy in the molecule. It assumes that the intramolecular dynamics is ergodic on the time scale of the dissociation process so that statistical considerations alone can predict the dissociation lifetimes and product energy distributions. On the other hand, the strong collision assumption is related to the special case of thermal energization and deenergization of molecules by molecular collisions. It leads to a Boltzmann distribution of energized molecules above the critical energy and predicts an Arrhenius dependence of the rate constant on temperature at high pressure.

In MPD experiments, the absorption and the stimulated emission of infrared photons replace molecular collisions as the means of excitation and deexcitation. Molecules which are translationally cold but vibrationally very hot are prepared. In some ways the multiphoton excitation (MPE) process resembles chemical activation where highly vibrationally excited molecules or radicals are prepared by a bimolecular addition reaction of atoms or radicals. In both cases a nonequilibrium situation is produced in which molecules can acquire internal energies well above the average thermal energy. Chemical activation and MPE also differ in several important respects [3.15]. The energy distribution of chemically activated molecules is usually very narrow, most of the excited molecules having energies within 10% (or less) of the average. Also, because of the extremely rapid energy deposition which occurs in a chemical activation experiment, molecules with dissociation lifetimes in the subpicosecond range may often be studied. In MPE, the energy distribution of the excited molecules depends on the size of the molecule, the magnitude of the critical energy, and the laser pulse conditions. The molecule continues to absorb IR photons until it reaches levels where the dissociation rate becomes comparable to or larger than the up-pumping rate due to the laser field. As will be seen later, for the typical CO_2 TEA laser with 50- to 100-ns pulse duration, 10^6-10^8W cm^{-2} intensity, and reasonable absorption cross sections of $\sim 10^{-20}cm^2$, the average dissociation lifetime should fall in the 1- to 50-ns range, independent of the nature of the molecule or the type of dissociation process occurring. This places most MPD experiments on a much slower timescale than the intramolecular relaxation time of highly excited poylatomic molecules, 10^{-11}-10^{-12}s, derived from a series of chemical activation experiments [3.16]. This also implies the applicability of a statistical theory in describing the dissociation of multiphoton excited molecules.

In the multiphoton excitation, the width of the energy distribution is often fairly broad because of the stochastic nature of the excitation process. The distribution of molecules above the dissociation energy level also depends on how rapidly the dissociation rate constant increases with increased excitation. The RRKM theory predicts that the rate constant for large molecules with a high density of states near the dissociation level will increase rather slowly with excitation, leading to a distribution of molecules over a broad range of levels from which dissociation can occur. For small molecules and molecules with low dissociation energies, the rate constant increases more rapidly with excitation energy and then the molecules can only be excited into a narrow range of levels just above the dissociation level. The RRKM rate constants for CF_3I and SF_6 are shown in Fig.3.1. If levels with lifetimes between 1 and 10 ns contribute to the dissociation yield, the spread in excess energy is around 3 kcal mol^{-1} for CF_3I and 10 kcal mol^{-1} for SF_6.

In order to get a more quantitative picture of the population distribution of excited molecules during and after the laser pulse, some assumptions must be made concerning the dynamics of the MPE process. The simplest assumption is that the

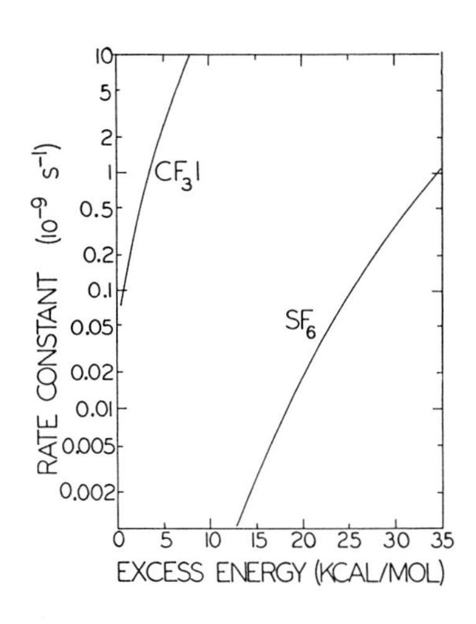

Fig.3.1. RRKM dissociation rates as functions of excess energy for CF_3I and SF_6

laser excitation is equivalent to thermal heating and that the thermal population distribution is not appreciably perturbed by dissociation. This is equivalent to making the strong collision assumption in thermal unimolecular reaction theory. Then, the population distribution at any time will be a Boltzmann distribution whose temperature is determined by the average number of photons absorbed per molecule. This approach has been used by the Harvard group to interpret their measurements of the dissociation yield and the average vibrational excitation of SF_6 as a function of laser energy fluence [3.14].

A somewhat more sophisticated model utilizes a set of phenomenological rate equations to describe the multiphoton excitation [3.17,18]. Such a model has also been considered by *Lyman* [3.19]. In our model we assume that: (1) The multiphoton excitation of a molecule may be described by stepwise incoherent one-photon transitions among a set of equally spaced levels in the quasi continuum. (2) The degeneracy of each level is given by the corresponding molecular density of states. (3) The ratio of the emission and absorption cross sections for transitions between two adjacent levels is given by the ratio of the level degeneracies. (Spontaneous emission is neglected.) (4) The dissociation rates of molecules from levels above the dissociation threshold are given by RRKM theory. With these assumptions the rate equations may be written as

$$\frac{dN_m}{dt} = \frac{I(t)}{h\nu}\left[\sigma_{m-1}N_{m-1} + \frac{g_m}{g_{m+1}}\sigma_m N_{m+1} - \left(\frac{g_{m-1}}{g_m}\sigma_{m-1} + \sigma_m\right)N_m\right] - k_m N_m \tag{3.8}$$

where N_m is the normalized population in level m with energy $mh\nu$, $I(t)$ is the laser intensity, g_m is the density of states of level m, σ_m is the cross section for absorption from level m to m+1, and $k_m = k_a(E^* = mh\nu)$ is the RRKM dissociation rate constant for level m. With given $I(t)$, g_m, σ_m, and k_m, the population distribution N_m as a function of time can be calculated from (3.8).

The density of states g_m is calculated with the Whitten-Rabinovitch approximation [3.13]. Use of this approximation tacitly assumes that the molecules in the quasi-continuum are randomly distributed over all the vibrational states at a given energy. It is not obvious whether the rotational degrees of freedom should be included in the calculation of the density of states. This was investigated for SF_6 by calculating the density of states with the rovibrational Whitten-Rabinovitch approximation [3.18]. The ratio of density of states for adjacent levels changed by about 4% at the bottom of the quasi continuum ($m = 3,4$) and by only 1% around the dissociation threshold. These changes cause at most a 1% difference in the population of any level. Thus, the question of the inclusion of the rotational degrees of freedom in the calculation of the density of states is not an important consideration in the rate-equation model.

The absorption cross section and its dependence on excitation energy are difficult to estimate from an ab initio calculation. The line shape of the ν_3 transition of SF_6 has been observed to broaden and shift with increasing temperature. Thus, at the peak of the linear absorption spectrum, the absorption cross section should decrease with excitation, as has been observed by *Nowak* and *Lyman* [3.20] and *Bott* [3.21]. In our numerical calculations on SF_6 the absorption cross section is assumed to have the form

$$\sigma_m = \sigma_0 \exp(-\gamma m) \tag{3.9}$$

where $\sigma_0 = 8 \times 10^{-19} cm^2$ and $\gamma = 5 \times 10^{-2}$. These parameters were chosen to fit the experimental results of *Black* et al. [3.14], as will be discussed later. A comparison of the laser-excited population distribution and a thermal distribution is shown in Fig.3.2. The laser-excited distribution with $\langle n \rangle = 20$ was obtained from our model calculation for a 20-ns 200-MW cm^{-2} rectangular laser pulse. The thermal distribution with $T = 2200$ K has an average excitation energy of $\langle n \rangle h\nu \cong 20\ h\nu$. We notice

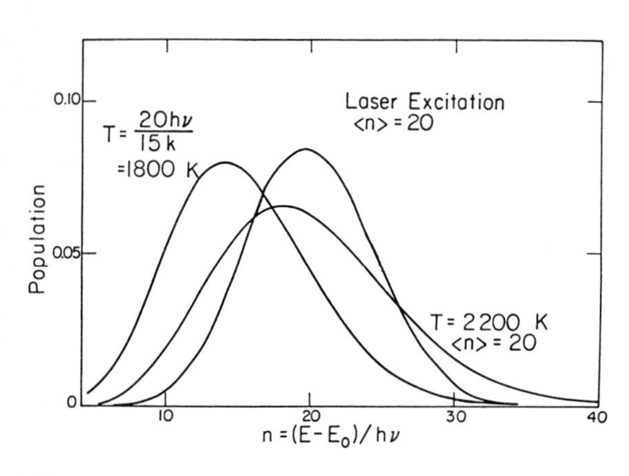

Fig.3.2. Comparison of the laser-excited SF_6 population distribution with $\langle n \rangle = 20$ with two thermal distributions: one with $T = 2200$ K corresponding to $\langle n \rangle = 20$ and the other with $T = 1800$ K obtained from the equipartition theorem $T = 20\ h\nu/15\ K$

that the thermal distribution is broader and has a long high-energy tail. This tail is absent in the distribution obtained from the model calculation which properly takes into account the depletion of population due to dissociation. The fact that the laser-excited distribution is narrower than the thermal distribution follows from the assumption that the absorption cross section decreases with excitation.

A more realistic model calculation should include the intensity-dependent 3-6 photon excitation over the discrete levels of SF_6. The existence of an intensity-dependent "bottleneck" in the excitation into the quasi continuum was originally suggested by *Abartzumian* et al. [3.22] and subsequently shown experimentally [3.14, 23]. Detailed quantum-mechanical models of the discrete levels have been developed [3.22-26] which include the rotational substructure in the vibrational levels. The time development of the excited SF_6 population in the discrete levels has been calculated using such models [3.27-29]. However, these models are too complex for our simple model calculation, and we have used a much simpler phenomenological approach to describe the bottleneck effect.

We assume that the quasi continuum begins at an energy of 4 $h\nu$. The excitation of a population into the quasi continuum requires different laser intensities for different rotational-vibrational states. If a single four-photon process is effective in exciting the population into the quasi continuum, then the net excitation rate is proportional to I^4. We then have

$$\frac{dN_{0,J}}{dt} = -a_J I^4 N_{0,J} \quad , \tag{3.10}$$

$$N_{0,J}(t) = N_{0,J}(0) \exp\left[-a_J \int_0^t I^4(t')dt'\right] \quad , \tag{3.11}$$

where a_J is a constant different for each rotational substate and $N_{0,J}$ is the population in the Jth rotational state of the ground vibrational level. The rate at which the population is coupled into the quasi continuum is then equal to $-d[\Sigma N_{0,J}(t)]/dt$. Once in the quasi continuum, the population evolves according to (3.8). For a bell-shaped laser pulse, the function $N_{0,J}(t)$ is close to a step function and may be approximated by $N_{0,J}(0)$ as long as $I(t) < I_J$, and by 0 after $I(t)$ has reached I_J, where I_J is the coupling intensity for the Jth rotational state.

It is known from two-frequency experiments [3.30,31] that the coupling intensities vary from less than 0.1 MW cm^{-2} to 100 MW cm^{-2}, depending on J. For a bell-shaped laser excitation pulse, the rate of switching the population into the quasi continuum therefore depends on time. In our calculation, we simply assume it takes the form

$$-\frac{d}{dt}\left(\sum_J N_{0,J}\right) = \frac{1}{2}\frac{d}{dt}\left[\mathrm{erf}\left(\alpha \ln \frac{I_{max}(t)}{\beta}\right)\right] \quad . \tag{3.12}$$

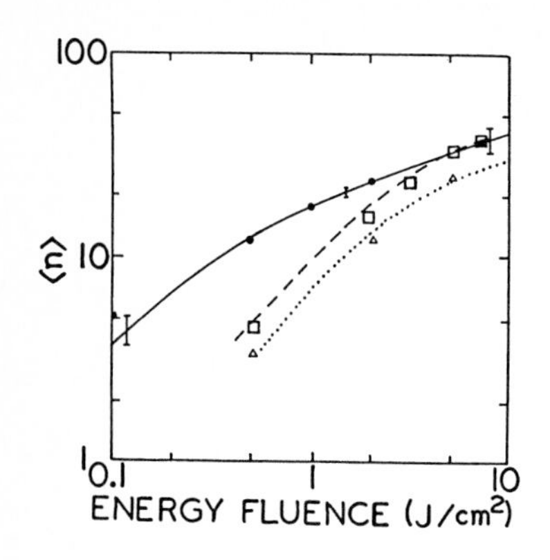

Fig.3.3. Results of the model calculation for the average number of photons absorbed by SF_6 vs energy fluence for three laser pulses. These results are compared with the experimental data of *Black* et al. [3.14]. The experimental and calculated results, respectively, are shown for a 0.6-ns pulse (● ● ● and ——), for a 60-ns multimode pulse (□ and ---), and for a 60-ns single-mode pulse (△ and ...)

This should be included as an additional term in the rate equation for the bottom level of the quasi continuum in (3.8). Here $I_{max}(t)$ is the maximum intensity achieved by the laser pulse as of time t, and α and β are adjustable parameters. For a bell-shaped laser pulse, $I_{max}(t)$ increases monotonically from zero to its maximum value and then "freezes." The intensity β is that at which 50% of the total population is coupled to the quasi continuum, and α is a parameter which reflects the spread in the coupling intensities of the initial rotational states.

As shown in Fig.3.3, our calculation using (3.8) with (3.12) included reproduces the experimental curves of *Black* et al. [3.14] for the absorption of photons by SF_6 obtained with three different laser pulses. The parameters used in the calculation were, for the absorption cross section, $\sigma_0 = 8 \times 10^{-19} cm^2$, $\gamma = 5 \times 10^{-2}$, and for the bottleneck, $\alpha = 0.5$, $\beta = 16$ MW cm^{-2}. In the case of the 0.6-ns laser pulse, the calculation is not at all sensitive to the values of α and β since the peak intensity is much larger than β. For the 60-ns laser pulses, the discrete-state bottleneck has a substantial effect. It is more severe with the single-mode pulses since the peak intensity of a single-mode pulse is less than half that of a multimode pulse. That the model calculation fits the experimental 60-ns pulse results indicates that our treatment of the bottleneck is a fair approximation.

In order to investigate the effect of the bottleneck on the population distribution, the calculation was run with 15-ns and 60-ns laser pulses each at the same energy fluence of 0.3 J cm^{-2} (well below the threshold for dissociation). The results are shown in Fig.3.4. For the 15-ns pulse, 70% of the population is pumped into the quasi continuum, while for the 60-ns pulse only 40% of the molecules overcome the bottleneck. However, the excited population distributions have essentially the same shape and width. While the peak intensity determines the total population in the quasi continuum, it appears that the energy fluence essentially determines the population distribution within the quasi continuum. Thus, the results shown in Fig.3.2 are essentially unchanged when a realistic pulse shape is used and the bottleneck effect is included.

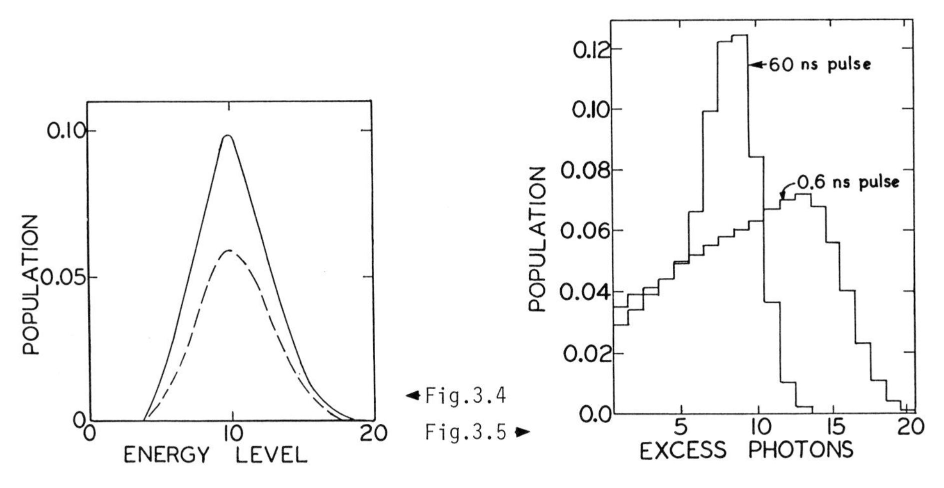

Fig.3.4. SF_6 population distribution for 15-ns (——) and 60-ns (---) laser pulses, each at an energy fluence of 0.3 J cm^{-2}

Fig.3.5. Total dissociation yield of SF_6 vs the number of excess photons for 0.6-ns and 60-ns laser pulses each at an energy fluence of 7.5 J cm^{-2}

Our calculation also answers the question of how the laser intensity affects the distribution of excess energies with which SF_6 dissociates. Figure 3.5 shows the distributions which result from 60-ns and 0.6-ns laser pulses, each at an energy fluence of 7.5 J cm^{-2}. The histograms represent the total dissociation yield from each level, both during and after the laser pulse. (The fraction of dissociation which is completed during the laser pulse is 50% for the 60-ns pulse and 34% for the 0.6-ns pulse.) It is seen that the average excess energy is higher for the short pulse, since the level of excitation is determined by the balance of up-excitation and dissociation, and the up-excitation rate is much greater in this case. This should be true whenever the energy fluence is high enough to saturate the dissociation yield, that is, when there is enough energy in the pulse to pump most of the molecules all the way up through the quasi continuum. At sufficiently low fluences, only a small number of molecules at the high-energy end of the population distribution can dissociate and most of the dissociation will occur from low-lying levels with long lifetimes. Thus, at energy fluences well below saturation, both the dissociation yield and the average level of excitation will depend on energy fluence and not much on the intensity.

The preceding general theoretical considerations on MPE and MPD are quite realistic. The experimental results obtained in molecular-beam experiments which will be described in the following sections are in good agreement with the above description.

3.3 Experimental Arrangement

In order to understand the dynamics of infrared multiphoton dissociation of polyatomic molecules, it is necessary to carry out experiments under collision-free conditions and obtain some information which is directly related to the dissociation products, the measurement of the energy distribution of the fragments and the determination of the lifetime of the excited molecules are important data that need to be obtained in order to make an assessment of the extent of energy randomization and the level of excitation prior to the dissociation of excited molecules.

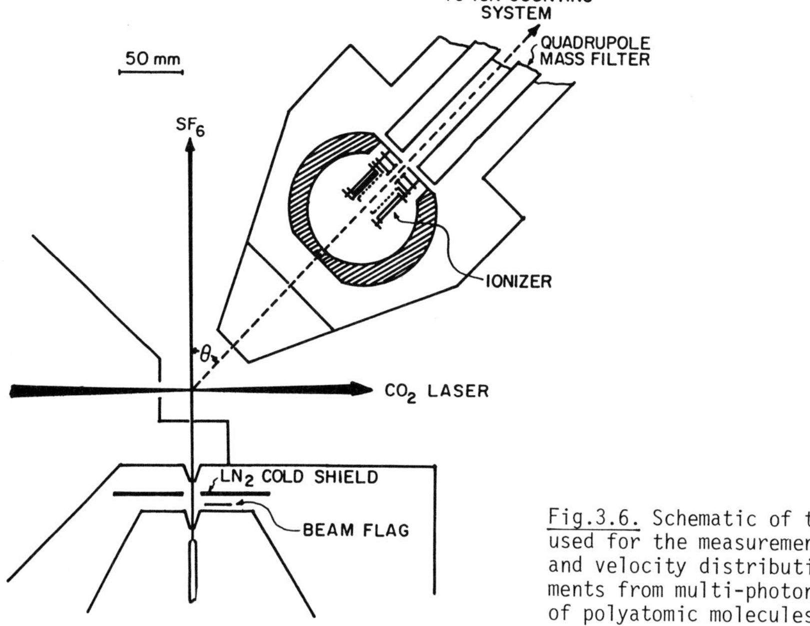

Fig.3.6. Schematic of the apparatus used for the measurements of angular and velocity distributions of fragments from multi-photon dissociation of polyatomic molecules

The arrangement in which the laser and the molecular beam are crossed is most suited for this purpose and is used in our experimental investigations. The molecular beam apparatus used is shown schematically in Fig.3.6. The molecular beam was formed by expansion of the pure gas or a gaseous mixture using a rare gas as carrie- at ~75-200 Torr stagnation pressure from a 0.1 mm-diameter quartz nozzle. Three stages of differential pumping were used along with two conical skimmers and a final collimating slit to produce a well-defined beam ~2 mm in diameter in the laser irradiation region. The molecular beam had a very sharply delineated angular distribution of 1.2° full width at halfmaximum (FWHM). Three stages of differential pumping were found to be necessary for this type of experiment in order to be able to detect the dissociation products near the molecular beam, since the fragmentation of beam molecules in the ionizer of a mass spectrometer produces the same mass peaks as those from the dissociation products. The velocity distribution of the molecular beam typically had a FWHM spread of 25% of the average velocity, or better. The density of molecules in the beam in the irradiation region was $\sim 3 \times 10^{11} cm^3$. The velocity spread and the number density of the molecular beam are both limited by the stagnation pressure which had to be kept low to avoid the formation of Van der Waals dimers and polymers during the expansion.

A Tachisto TAC II grating tuned CO_2 TEA laser (~1.0 J/pulse) was used in our experiments as the excitation source. The laser beam was admitted into the vacuum chamber via a ZnSe lens with a 25.4-cm focal length. The power and the energy fluence of the laser at the molecular beam was adjusted by varying the distance between the focal region of the lens and the molecular beam. The fragments produced by multiphoton dissociation of polyatomic molecules at the small intersection region were detected by a triply differentially pumped quadrupole mass spectrometer utilizing electron bombardment ionization and ion counting. The pressures in the three regions of the detector were maintained at $\sim 10^{-9}$, $\sim 10^{-10}$ and $\sim 10^{-11}$ Torr by a combination of ion pumps, a sublimation pump, a liquid nitrogen trap and a liquid helium cryopump. The partial pressure of the beam molecule in the third region, where the ionizer is located, was usually kept below 10^{-13} Torr. The angular position of the mass spectrometer around the beam intersection point could be varied so that the angular distribution of the fragments could be measured. The mass filter was usually adjusted to provide better than unit mass resolution.

As shown schematically in Fig.3.7, external triggering at 0.5 Hz was used to fire the laser and to enable a dual-channel scaler for recording counts of fragments from the mass spectrometer. Separate adjustments of delay and gate times were made to ensure that one scaler channel recorded only background (i.e., with the laser pulse off) while the other recorded both background and signal. Typically, 100-1000 laser shots were used to measure the fragments produced at each laboratory angle θ (Fig.3.6) for the measurements of angular distributions. The angular resolution of

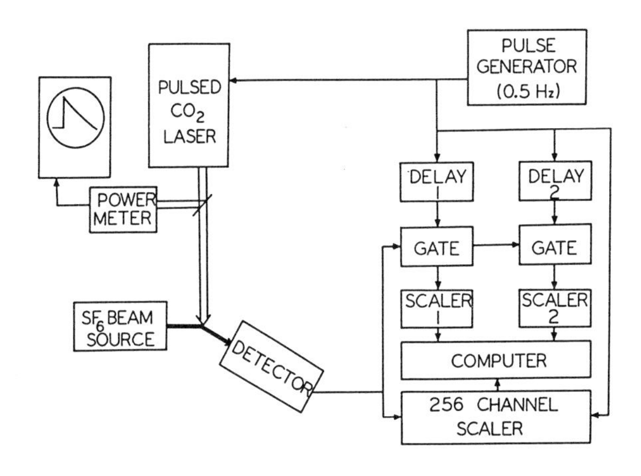

Fig.3.7. Experimental arrangement. The "SF_6 beam source" is the mocular-beam source. The pule generator triggers the laser (which partly dissociates a section of the molecular beam) and the multichannel scaler, opens a gate to scaler 1 a few hundred microseconds later (to count dissociation product signal and background signal) and a gate to scaler 2 a few milliseconds later (to count background signal only)

the detector was 1°. In order to positively identify the dissociation products and to check for possible secondary dissociation of a primary product by the same laser pulse, the angular and velocity distributions were scanned at several mass peaks in the mass spectra of the dissociation products of a molecule under investigation. The fragment velocity distributions at various laboratory scattering angles were obtained by determining the arrival time of each fragment, after a flight path of 21 cm, at the detector, relative to the time origin defined by the laser pulse. This was done by multiscaling the mass spectrometer output signal. Typically, a 10-μs channel width was used in a scan over 2.5 ms.

The dissociation products and their angular and velocity distributions were extensively measured while varying the laser frequency, power and energy fluence, and the vibrational and rotational temperatures of the molecules.

3.4 Experimental Results

The major MPD products identified in our molecular-beam experiments are listed in the first column of Table 3.1, which summarizes our results. The dissociation products observed are typically those from the channel with the lowest activation energy. According to the measurements by several other groups [3.7-12] using laser-induced fluorescence detection, they appear in their ground electronic states, or in some cases [3.32-33], in low-lying electronic states.

For C_2F_5Cl and $CHClCF_2$ two dissociation channels corresponding to the two lowest activation energies have been observed. For CH_3CF_2Cl the HF and HCl molecular eliminations were suggested to have, within experimental uncertainty, the same activation energies in earlier thermal dissociation studies, but the HCl elimination is the only channel observed in our experiments. For SF_6 and $CFCl_3$, secondary dissociation of the primary products is observed at high energy fluence ($SF_5 \rightarrow SF_4 + F$, $CFCl_2 \rightarrow CFCl + Cl$).

In the cases where two competitive dissociation channels are observed, the intensity of the laser pulse was found to influence the branching ratio. Figure 3.8 shows the relative dissociation yield of C_2F_5Cl into $CF_3 + CF_2Cl$ and $C_2F_5 + Cl$ as a function of laser energy. The chlorine atom elimination has a threshold at 0.5 J cm^{-2} and saturates at 1 J cm^{-2}. The channel producing $CF_3 + CF_2Cl$ has approximately the same threshold, but as the intensity is increased, the fraction dissociating by C-C bond rupture continues to increase.

The laboratory angular and velocity distribution for SF_5 in the fluorine atom elimination from SF_6 are shown in Figs.3.9,10. The angular distribution of the SF_5 peaks as close to the SF_6 beam as can be measured (5°) and falls off monotonically with increasing angle. The velocity distributions of SF_5 shown in Fig.3.10 were obtained from the time of flight measurements at three angles. Also shown is the

Table 3.1. Dynamics of multiphoton dissociation

Molecule	Endoergicity [kcal mol^{-1}]	Potential energy barrier [kcal mol^{-1}]	Average translationl energy [kcal mol^{-1}]	Peak of translational energy distribution [kcal mol^{-1}]	Estimated average energy available to products [kcal mol^{-1}]	Estimated lifetime [ns]
$SF_5 \rightarrow SF_5 + F$	93	0	3	0	25	20
$SF_4 + F$	51	0	1	0	7	20
$CF_3Cl \rightarrow CF_3 + Cl$	86	0	1.1	0	4	5
$CF_3Br \rightarrow CF_3 + Br$	71	0	1.2	0	5	2
$CF_3I \rightarrow CF_3 + I$	53	0	1.1	0	4	1
$CF_2Cl_2 \rightarrow CF_2Cl + Cl$	82	0	2	0	10	5
$CF_2Br_2 \rightarrow CF_2Br + Br$	61	0	1.6	0	7	5
$CFCl_3 \rightarrow CFCl_2 + Cl$	75	0	1.2	0	5	12
$CFCl + Cl$	~ 70	~ 0	-	0	-	-
$C_2F_5Cl \rightarrow C_2F_5 + Cl$	83	0	4	0	35	60
$C_2F_5Cl \rightarrow CF_3 + CF_2Cl$	$\geqslant 97$	0	3.3	0.4	21	200
$N_2F_4 \rightarrow 2NF_2$	22	0	0.4	0	2	1
$(NH_3)_2 \rightarrow 2NH_3$	4	0	0.3	0	1.5	-
$CHClCF_2 \rightarrow C_2HF_2 + Cl$	~ 80	0	1	0	-	-
$CHClCF_2 \rightarrow C_2F_2 + HCl$	58	> 0	1	0	-	-
$CHF_2Cl \rightarrow HCl + CF_2$	50	6	8	5	-	-
$CH_3CCl_3 \rightarrow HCl + CH_2CCl_2$	12	42	8	5	-	-
$CH_3CF_2Cl \rightarrow HCl + CH_2CFCl$	14	55	12	6	-	-

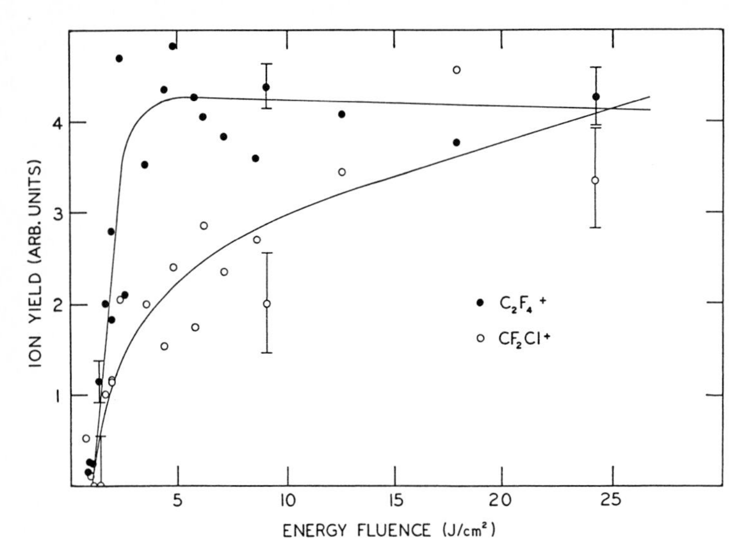

Fig.3.8. Dissociation yields for the products from MPD of C_2F_5Cl: $C_2F_5Cl \rightarrow C_2F_5 + Cl$ ($C_2F_4^+$ detected), $C_2F_5Cl \rightarrow CF_3 + CF_2Cl$ (CF_2Cl^+ detected)

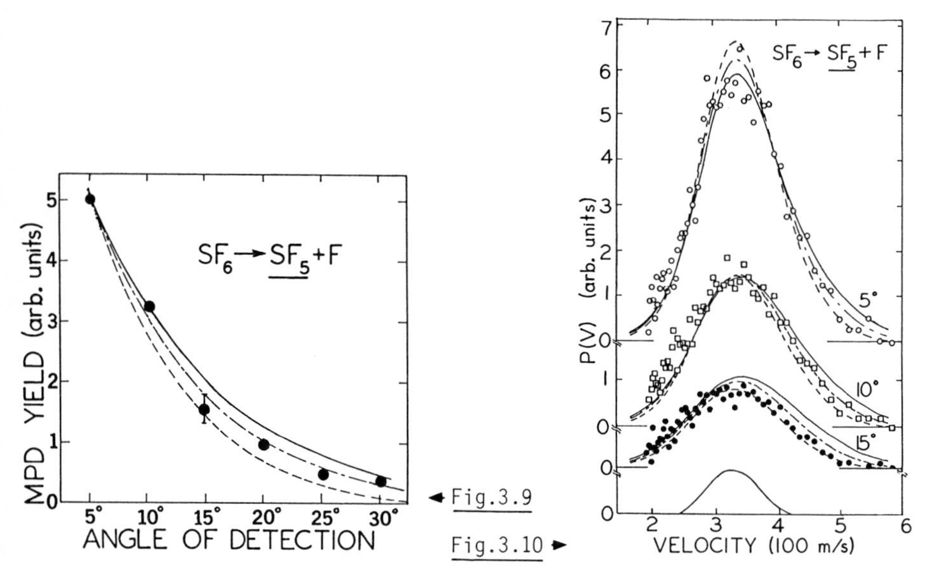

Fig.3.9. Angular distribution of SF_5 fragments from MPD of SF_6: (•) experimental distribution; (----) RRKM theory, 5 kcal mol^{-1} excess energy; (-·-·-) RRKM theory, 8 kcal mol^{-1} excess energy; (——) RRKM theory, 12 kcal mol^{-1} excess energy

Fig.3.10. Speed distribution of SF_5 fragments from MPD of SF_6 at 5°, 10°, and 15° from the SF_6 beam. Symbols as in Fig.3.9. Bottom: SF_6 beam speed distribution

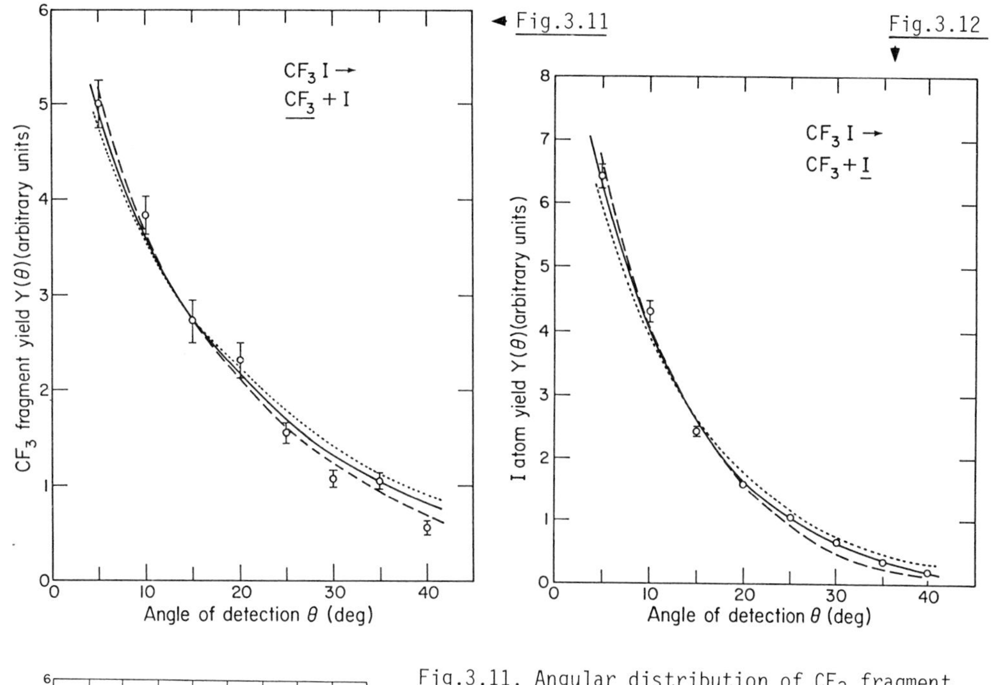

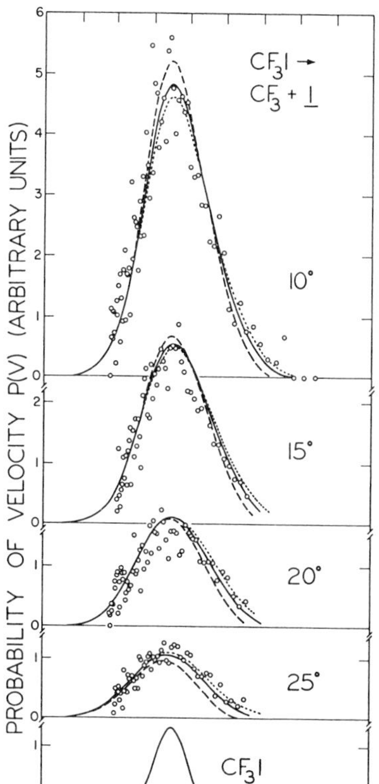

Fig.3.11. Angular distribution of CF_3 fragment from MPD of CF_3I: (○) experimental distribution; (----) RRKM theory, 3 kcal mol^{-1} excess energy; (——) RRKM theory, 4.5 kcal mol^{-1} excess energy; (····) RRKM theory, 6 kcal mol^{-1} excess energy

Fig.3.12. Angular distribution of I atoms from MPD of CF_3I. Symbols as in Fig.3.11

Fig.3.13. Speed distribution of I atoms from MPD of CF_3I at 10°, 15°, 20°, and 25° from the CF_3I beam. Symbols as in Fig.3.11. Bottom: Speed distribution of CF_3I beam

SF_5 beam velocity distribution. The angular and velocity distributions for SF_6 are typical of the other halogen atom elimination reactions. For example, Figs.3.11-13 show the laboratory angular and velocity distributions of CF_3 and I from MPD of CF_3I.

Translational energy distributions of dissociation products are derived from the measured laboratory angular and velocity distributions. First, an assumed center-of-mass translational energy distribution of the fragments is transformed to the laboratory coordinates, including the convolution over the beam velocity distribution and the length of the ionizer in the mass spectrometer. Then, the angular and velocity distributions in the laboratory coordinates can be calculated and fit to the experimental curves. Center-of-mass angular distributions of products are found to be isotropic for all systems studied. This can be concluded from the agreement between experiments and theoretical curves deduced using this assumption, and from the observation that our results were independent of laser polarization. Figure 3.14 shows the translational energy distribution of SF_5 +F derived from the experimental results. The curves drawn in Figs.3.9,10 are the angular and velocity distributions calculated from the translational energy distributions shown in Fig.3.14.

Columns 4 and 5 of Table 3.1 give information on the average translational energy and the peak of the translational energy distribution. It is clearly seen that except for some 3- and 4-center eliminations, which are known to have additional potential energy barriers for dissociation, the translational energy distributions all peak at zero kinetic energy and the average translational energies of the products are generally very low.

The systems with an additional potential energy barrier in the exit channel have characteristically different translational energy distributions, which are reflected in laboratory angular and velocity distributions. An example is 3-center elimination of HCl from CHF_2Cl. The velocity distributions of the HCl in this case is shown in Fig.3.15. The center-of-mass translational energy distribution peaks at 5 kcal mol^{-1} and the average translation energy is as high as 8 kcal mol^{-1}, as shown in Fig.3.16.

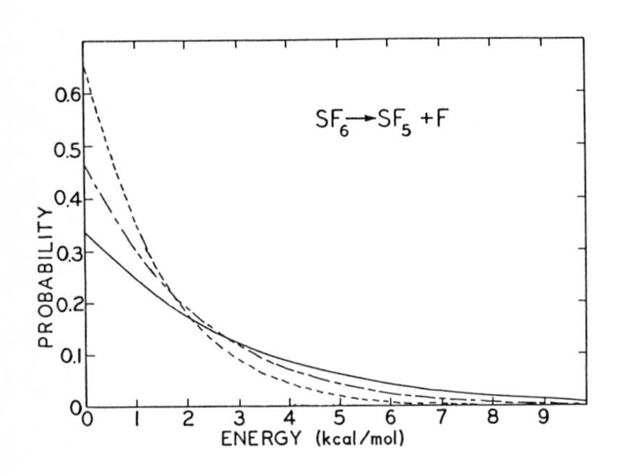

Fig.3.14. Center-of-mass translational energy distribution of the fragments from the MPD of SF_6, calculated from RRKM theory. Symbols as in Fig.3.9

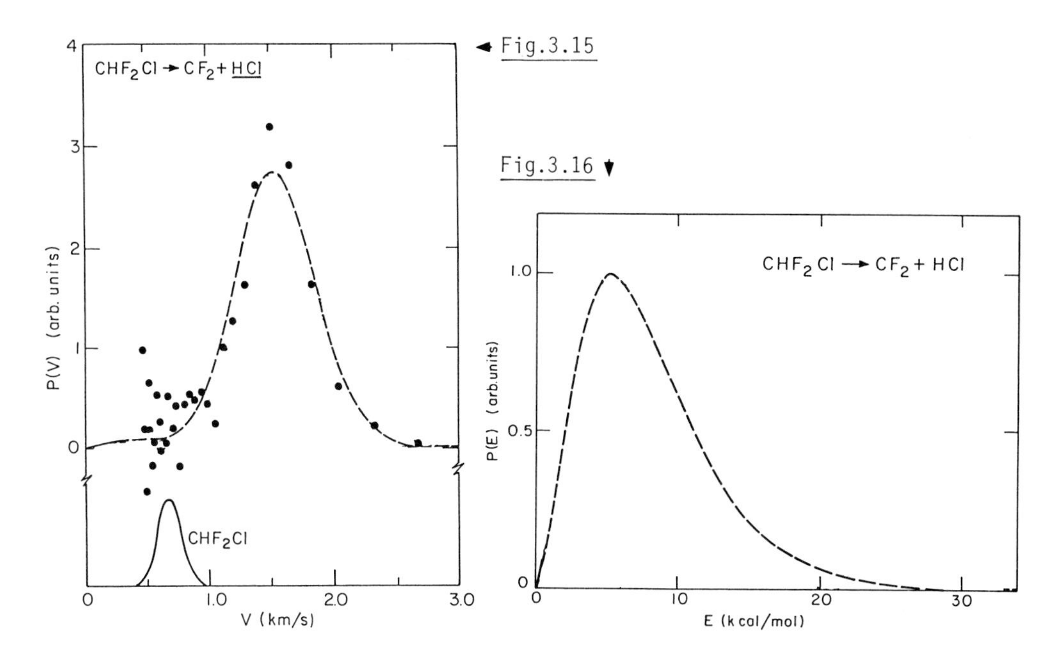

Fig.3.15. Speed distribution of HCl fragments from the MPD of CH_2Cl: experiment (ooo), calculated from Fig.3.16 (---)

Fig.3.16. Center-of-mass translational energy distribution in the fragments from MPD of CHF_2Cl

Both 4-center elimination of HCl from CH_3CCl_3 and Ch_3CF_2Cl and C-C bond rupture in C_2F_5Cl have similar characteristic translational energy distributions.

3.5 Discussion

As already discussed in the theory section, a simple picture can be used to describe MPD: the incoherent stepwise multiphoton process excites the molecules through the quasi continuum to and beyond the dissociation energy level. The level of excitation is eventually limited by molecular dissociation, which can be accurately described by the statistical theory (RRKM theory) of unimolecular reactions. The results from our molecular-beam experiments have provided sufficiently detailed information on important features of the MPD process for drawing such a picture. In the following discussion we will emphasize the collisionless aspects of the process, and only briefly mention a few of the additional effects that have to be taken into account if collisions cannot be neglected.

Let us start by looking at some of our typical experimental results on the translational energy distribution of the dissociation fragments. Shown in Fig.3.10 are the velocity distributions of SF_5 fragments from MPD of SF_6 at various angles with respect to the SF_6 beam [3.18,34]. We see that they are only slightly broader than that of the primary SF_6 beam because the average translational energy imparted to the fragments in the dissociation is quite small. The same conclusion can be drawn from the angular distribution of SF_5 fragments shown in Fig.3.9. It falls off rapidly as the angle from the SF_6 beam increases, again indicating that very little translational energy is released to the fragments. More quantitatively, this can be seen from the translational energy distribution of the fragments as shown in Fig.3.14, where the distribution curves actually yield velocity and angular distributions which fit the measured ones in Figs.3.9,10 very well.

This form of translational energy distribution of the fragments is actually predicted by the RRKM theory. As explained in the theory section, it predicts that as the excitation in a molecule increases above the dissociation energy, the dissociation rate constant increases. This tends to favor dissociation through the lowest-energy dissociation channel. Experiments, in particular those using the molecular beam, have shown that the MPD of most molecules proceeds through the lowest-energy channel [3.6,34-36]. The RRKM theory also predicts how the excess energy (excitation energy minus dissociation energy) is distributed among the various vibrational modes of the molecule in the critical configuration, including the relative motion of the dissociating fragments. Figure 3.14 shows the translational energy distributions that were used to fit the experimental results, calculated from the RRKm theory for excess energies of 5, 8, and 12 CO_2 laser photons. The good fit indicates that the RRKM theory describes MPD quite well.

The MPD results on halogenated methanes show (Table 3.1) that most of the molecules dissociate with excess energies of 1-3 CO_2 laser photons, as compared to around 8 photons for SF_6. On the other hand, C_2F_5Cl dissociates with around 13 photons of excess energy. How can the average excess energy depend so much on molecular structure?

To understand the above results we should consider the laser excitation scheme presented in the theory section and in Fig.3.17. The laser excites the molecule up the ladder of energy levels with a net up-excitation rate proportional to the laser intensity. As the excitation increases above the dissociation level the dissociation rate increases rapidly, and soon starts to compete with the up-excitation. The average level of up-excitation from which most of the molecules will dissociate is then determined by a balance between dissociation and up-excitation. By using this simple picture of competition between dissociation and up-excitation, we can draw a number of important conclusions.

1) In a heavier, more complex molecule that has more degrees of freedom and more low-frequency vibrations (e.g., SF_6 or C_2F_5Cl), the dissociation rate constant should increase more slowly with the increase in energy. This was quite dramatically displayed in Fig.3.1. Consequently, the heavier molecules tend to reach higher levels of excitation before they dissociate. This explains why the C_2F_5Cl molecule

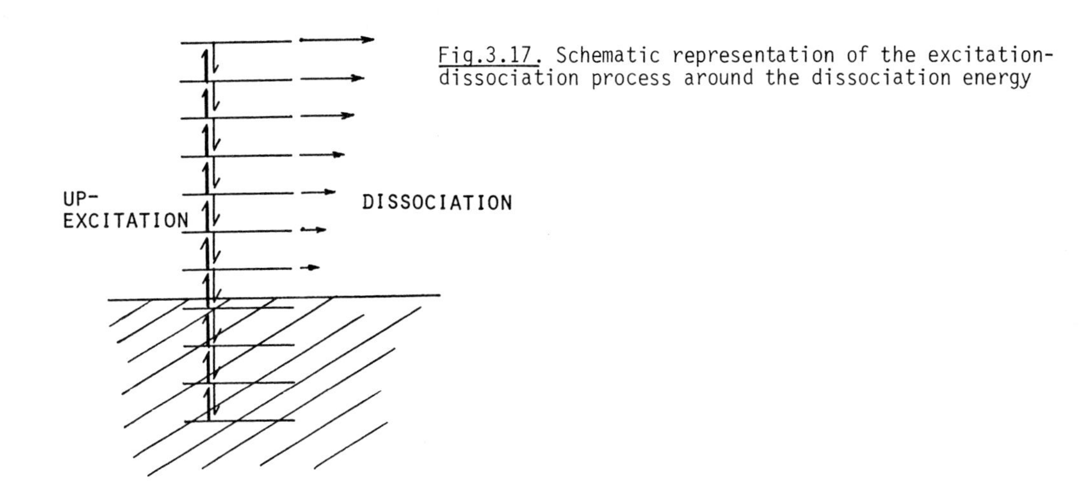

Fig.3.17. Schematic representation of the excitation-dissociation process around the dissociation energy

has a higher excess energy than SF_6, which in turn has more excess energy than the halogenated methanes.

2) If the laser pulse is very short, none of the molecules dissociates before the laser pulse is over. Then the population distribution and the level of excitation from which dissociation occurs is completely determined by the total pulse energy fluence. However, if the laser pulse is sufficiently long, the excitation level reached is limited by the dissociation, and at this level the up-excitation rate and the dissociation rates are about equal. Thus the level of excitation in this case should be higher with higher intensity, or at least at frequencies where the transition rates are higher.

When the dissociation yield is near saturation, the time it takes for a molecule to be pumped up above the dissociation energy is about equal to the pulse duration. The time it takes to make a transition above the dissociation energy is a reasonable fraction of this time (say, 1/10-1/50), since it takes some 10-50 transitions to get above the dissociation energy). Thus, in the case of dissociation rate limited excitation, the lifetime corresponding to the average level of excitation is of the order of 1/10 of the laser pulse duration. Our molecular beam experiments were done with a laser pulse of about 60-ns FWHM. From Table 3.1 we see that the dissociation lifetimes corresponding to the level of excitation calculated from the RRKM theory to fit the observed translational energy distributions are indeed in the 1- to 100-ns range (mostly around 10 ns).

3) The RRKM dissociation rate constant should increase more rapidly with excess energy if the dissociation energy is lowered. The dissociation energy of CF_3I is slightly more than half that of SF_6, and this accounts in part for the difference in their dissociation rate constants shown in Fig.3.1. An even clearer example is N_2F_4, which has a dissociation energy of only about half that of CF_3I. Even though it has one atom more than CF_3I, its dissociation rate grows so rapidly with excess

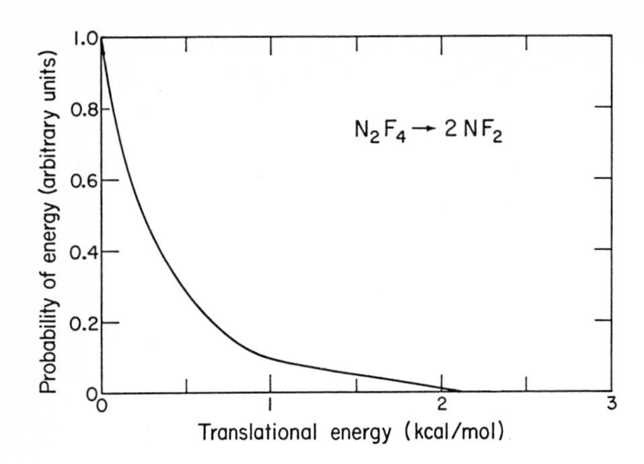

Fig.3.18. Center-of-mass translational energy of a pair of NF_2 fragments from the MPD of N_2F_4

energy that dissociation already dominates over up-excitation at a level one CO_2 laser photon above the dissociation energy. This is shown quite clearly by the translational energy distribution of NF_2 fragments in Fig.3.18, as there are no fragments with more than one photon or 3 kcal mol^{-1} of translational energy.

4) Since only a small fraction of the excess energy is released as translational energy, most of the excess energy should then remain as internal energy in the fragments. For heavy, complex molecules which reach high levels of excitation before dissociating, the framgents emerging from dissociation are already excited to their quasi continuum and can readily absorb more energy from the laser field to go through another MPD process. This process of secondary dissociation is of course more likely to occur if the fragments have a strong absorption band coinciding with the laser frequency. In our experiments, we have observed secondary dissociation in SF_6 and $CFCl_3$, with the fragments SF_5 and $CFCl_2$ dissociating further to form SF_4 and CFCl, respectively. The various products were identified through their different electron impact ionization spectra in the mass spectrometer. The laser frequency used was not in near resonance with any known strong IR absorption lines of $CFCl_2$ or SF_5, so the observed secondary dissociation must result from excitation of SF_5 and $CFCl_2$ already in the quasi continuum. Of course, for this to make sense, the primary dissociation must take place before the laser pulse is over. This is certainly the case — the translational energy distributions of SF_5 and $CFCl_2$ indicate that they are produced from parent molecules with lifetimes of around 10 ns, compared to the laser pulse duration of more than 60 ns.

In MPD of CF_3Cl, CF_3Br, and CF_3I, the CF_3 fragment produced has little internal energy (1-2 CO_2 laser photons), but CF_3 in the ground state is known to absorb close to the laser frequency used. Although the molecular beam experiments were not sensitive enough to detect dissociation of CF_3, in gas cell experiments on the same three molecules [3.37] CF_2 radicals and F atoms have been observed, indicating that a secondary dissocation of CF_3 may have taken place.

Many of the products observed by the extremely sensitive laser-induced fluorescence detection method are probably also produced from sequential dissociations of intermediate products. For example, C_2, CN and CH have been observed [3.10-12,33] in the dissociation of molecules with six or more atoms. Unfortunately, the laser-induced fluorescence detection method are not able to reveal anything about the intermediate steps leading to these small final products. It seems that the secondary or sequential dissociation is an unavoidable effect in the MPD at high energy fluence of all but the lightest, simplest molecules. This is a factor which often complicates the studies of the MPD process, regardless of the method used for detection and analysis of the dissociation products.

5) For the heavier, more complex molecules, competing dissociation channels may also open up, provided their dissociation energies are not too far above that of the lowest-energy channel. If the laser intensity is sufficiently high, the molecule can be excited well above the dissociation energies of several channels before dissociation dominates over up-excitation. Then several dissociation channels may start to compete with the lowest one. We should stress here that this effect is actually expected from the statistical theory of unimolecular dissociation. A system in which such an effect has been observed is C_2F_5Cl. The energetics of the various dissociation pathways are not well known, except for the lowest one, which is the Cl atom elimination, with a dissociation energy of about 83 kcal mol^{-1}. The next lowest channel is probably C-C bond rupture, with a dissociation energy of around 97 kcal mol^{-1}. The RRKM calculations indicate that the rate constant for the C-C bond rupture grows more rapidly with excess energy than that for the Cl atom elimination. As we have already seen, the average level of excitation in C_2F_5Cl pumped by a 1-J TEA laser can be around 13 CO_2 laser photons (40 kcal mol^{-1}) above the C-Cl bond energy, well above the dissociation energy of the C-C bond rupture so that the C-C dissociation rate can be comparable to the C-Cl dissociation rate. In the experiments, competition between the two channels has actually been observed. If we look at low laser intensities, the Cl atom elimination dominates over the C-C bond rupture. As the laser intensity is increased, thus pumping the molecules to higher levels of excitation, the Cl elimination channel very rapidly saturates, whereas the C-C bond rupture becomes increasingly important. This effect is not peculiar to MPD experiments. In pyrolysis of ethane compounds it has long been observed that at low temperatures, atomic elimination reactions dominate the dissociation, but as the temperature is increased, C-C bond rupture becomes progressively more important, making the analysis of such reactions exceedingly complicated.

Now, many of the results discussed under (1-5) above have also been obtained in an explicit model calculation on SF_6, using the simple set of rate equations presented in the theory section [3.17-19]. By fitting the free parameters in the model to experimental results on energy adsorbed as a function of input laser energy fluence and laser pulse duration, we can predict the dissociation yield as a function of energy fluence, the onset of secondary dissociation, the level of excitation from which dissociation occurs, and thus, the translational energy distribution in the fragments. As explained in the theory section and shown in Figs.3.3,5 all the results agree with the experiments, and illustrate quite clearly in a more quantitative way what we have discussed above in qualitative terms.

How will these results be modified if we cannot neglect molecular collisions? Depending on the collision partners, we can have a number of complications:

1) Collisions between excited molecules will lead to a thermalization of the energy deposited by the laser in the molecules via intermolecular vibrational energy transfer. Thus any differences between thermal heating and multiphoton excitation will be washed out. The isotopic selectivity of the process will decrease, due to enery transfer between different isotopic species. Rotational and vibrational intermolecular energy transfer during the laser pulse can increase the number of molecules interacting resonantly with the laser field, thus reducing the bottleneck for excitation out of the discrete levels into the quasi continuum. Collisionally induced dissociation can also occur, even in the absence of multiphoton dissociation.

2) Collisions between excited molecules and cold molecules will lead to a deactivation of the excited molecules. The cold collision partners may be buffer gases, if present, or reaction products from the dissociation. Their presence will increase the energy absorption necessary for a given dissociation yield, and lower the level of excitation. Thus, in cases with competing dissociating channels, the lowest energy channel will be favored. Since the excited products from the dissociation can also be deactivated via collisions, secondary dissociation of the products will be inhibited.

3) Collisions between dissociation products, and between products and other atoms or molecules present, usually lead to chemical reactions. The products from MPD are mostly highly reactive free radicals. Thus recombination or disproportionation of the dissociation products may occur, and complicated chemical reaction chains may follow the primary dissociation. Analysis of the process is complicated, and dependent upon detailed information on the chemical kinetics of the reactions involved. Little information about the dynamics of the primary dissociation can be deduced from the final products.

In the preceding discussion on the translational energy distribution of fragments, we have actually only considered the simple cases where the observed distributions are in agreement with predictions of the RRKM theory. This is usually true for simple bond rupture reactions [3.36]. There is negligible interaction between the fragments once the critical configuration is passed so that the energy distribution in the fragments remains the same as in the critical configuration calculated in the RRKM theory. However, in cases where such interaction cannot be neglected, the simple RRKM theory we have used cannot take this interaction into account, and translational energy distributions very different from the ones we have discussed so far may result. For a number of molecular elimination reactions, such as 3-center elimination reactions from halogenated methanes, and 4-center elimination from halogenated ethanes and ethenes, there is a considerable potential energy barrier between reactant and products. This potential energy will have to be distributed between the various vibrational, rotational, and translational degrees of freedom in the fragments as they move away from the critical configuration on the top of the barrier. The RRKM theory cannot predict anything about how the energy will be distributed. It will depend on the nature of the potential energy surface for the fragments.

As an example, we will discuss the dissociation of CHF_2Cl into CF_2 +HCl, which has been studied in a molecular beam [3.35] as well as with laser-induced fluorescence detection of the CF_2 fragment [3.8]. Thus, translational, as well as rotational and vibrational energy distributions in the CF_2 fragment have been measured. The conclusions that can be drawn from the results on CHF_2Cl are representative for molecules with this kind of dissociation dynamics.

The velocity distribution of HCl fragments at 10° from the CHF_2Cl beam is shown in Fig.3.15, compared to the distribution calculated from the translational energy distribution in Fig.3.16. We see that the products are quite a bit faster than the CHF_2Cl beam, due to the considerable amount of energy gained from the dissociation. Most CF_2 fragments have a translational energy of more than 2 kcal mol^{-1} while only a small percentage have less than 1 kcal mol^{-1}. *Stephenson* and *King* [3.8] found the population distribution in the vibrational modes of CF_2 to be well represented by a thermal distribution of temperature 1160 K. The average rotational energy was also estimated in the experiments, although its value was too high for a detailed measurement of the distribution to be made. However, assuming a thermal distribution, a rotational temperature of about twice the vibrational temperature was obtained. The high translational energy content in the fragments means that there is a strong repulsive interaction between the departing fragments after they pass through the critical configuration. This repulsive interaction is quite asymmetric, giving the fragments considerable rotational energy.

However, we want to emphasize that although RRKM theory alone may be inadequate for predicting the final energy partitioning in the fragments, it still predicts the dissociation rates. If we add up all the energies in the fragments in the HCl elimination from CHF_2Cl, using the results of *King* and *Stephenson*, we get to a level of excitation corresponding to an RRKM lieftime around 1 ns. This is what we should expect from the statistical theory of MPD as in the cases of the other halomethanes. In fact, there exists no evidence in all the cases we have studied that the general statistical picture of the multiphoton excitation and dissociation process does not apply.

3.6 Concluding Remarks

There are still a number of assumptions and theoretical predictions about the dissociation that need to be checked experimentally. The partition of energy between all degrees of freedom in at least one of the two fragments from the dissociation should be measured in a case where the RRKM theory predicts the distributions. The dissociation lifetimes should be measured directly and independently, together with their dependence on laser intensity, under well characterized conditions. The processes of secondary dissociation and competing dissociation channels need better characterization. The methods that so far have revealed the most about the dissociation process are the molecular-beam method and the laser-induced fluorescence method. A natural extension would be to use laser-induced fluorescence as a detection method in a molecular-beam experiment. Studies of this kind are already being prepared in several laboratories. The work is hampered by the low particle densities involved in molecular-beam experiments, insufficient knowledge of the spectroscopy of many of the radicals produced in the dissociation, and lack of tunable lasers in the UV frequency ranges of interest for many compounds.

Although there are some detailed questions which still need to be further investigated, the general physical picture constructed from various experimental and theoretical investigations is quite adequate for understanding and predicting many important features of the MPD process under various conditions. But since MPD is a rather complex process, it is not possible to draw reliable conclusions unless all the factors involved are carefully analyzed. The dependence of the dissociation yield and the dynamics of dissociation on both the laser intensity and energy fluence is an important example.

For a given chemical species, the laser intensity required for a certain fraction of the molecules to overcome the discrete-state bottleneck not only depends on the frequency, but also on the vibrational and rotational temperature of the molecules. Once the molecules are excited to the quasi continuum, the energy fluence, not the power of the laser, was shown to be responsible for driving the molecules through the quasi continuum and beyond the dissociation level. But in most of the gas cell experiments, the dissociation yield of those molecules in the quasi continuum is not simply related to the energy fluence alone. For molecules lying above the dissociation level, there is a complicated competition between unimolecular dissociation, collisional deactivation, and laser up-excitation. Consequently, for a given gas pressure and a given laser energy fluence, a higher laser intensity should result in a higher level of excitation and an increased rate of dissociation. This in turn reduces the effect of collisional deactivation and thus increases the dissociation yield. In general, for smaller molecules, the laser intensity influences the yield by limiting the fraction of the molecules which can be excited to the quasi continuum, but since the lifetime of small molecules becomes very short after only a couple of excess photons are deposited beyond the dissociation threshold, collisional deactivation could be overcome with a rather moderate intensity at low pressure. On the other hand, for larger molecules with many vibrational degrees of freedom, if an appropriate frequency is chosen, a large fraction of the molecules will reach the quasi continuum at a very moderate laser intensity. But since many more excess photons are required before the dissociation lifetime becomes comparable to the mean collision time, the laser intensity is expected to strongly influence both the dissociation yield and the ratio of competitive dissociation channels by controlling the level of excitation beyond the dissociation energy.

In most of the experiments carried out with a CO_2 TEA laser, one often adjusts the laser intensity or energy fluence by either adjusting the focusing condition or attenuating the laser output. Consequently both the laser intensity and energy fluence are often varied simultaneously. If the energy fluence requirement for dissociation is met, the intensity of the laser is already high enough to pump some of the molecules to the quasi continuum and dissociation is observed. However, it is important to keep in mind that both the intensity and energy fluence of the laser can separately affect the experimental results. Once the complicated dependence of the excitation and dissociation dynamics on the initial distribution of molecules over vibrational and rotational states, and on the frequency, intensity, and energy fluence of the laser is properly taken into account, we are indeed in a very good position to understand and predict the general behavior of MPD of the systems of interest.

Since the manuscript for this chapter was completed, there have been many advances in this field. For references, the readers should consult "Multiphoton Bibliography" compiled by J.H. Eberly, N.D. Piltch, and J.W. Gallagher, and the recent series of review articles by *Bagratashvili* et al. [3.38]. Additional results may be found in [3.39,40].

References

3.1 N.R. Isenor, V. Merchant, R.S. Hallsworth, M.C. Richardson: Can. J. Phys. **51**, 1281 (1973)
3.2 V.S. Letokhov, E.A. Ryabov, O.A. Tumanov: Opt. Commun. **5**, 168 (1972)
3.3 V.S. Leotkhov, E.A. Ryabov, O.A. Tumanov: Sov. Phys.-JETP **36**, 1069 (1973)
3.4 R.V. Ambartzumian, V.S. Dolzhikov, V.S. Letokhov, E.A. Ryabov, N.V. Chekalin: Sov. Phys.-JETP **42**, 36 (1976)
3.5 M.J. Coggiola, P.A. Schulz, Y.T. Lee, Y.R. Shen: Phys. Rev. Lett. **38**, 17 (1977)
3.6 Aa.S. Sudbø, P.A. Schulz, E.R. Grant, Y.R. Shen, Y.T. Lee: J. Chem. Phys. **68**, 1306 (1978)
3.7 D.S. King, J.D. Stephenson: Chem. Phys. Lett. **51**, 48 (1977)
3.8 J.C. Stephenson, D.S. King: J. Chem. Phys. **69**, 1485 (1978)
3.9 J.D. Campbell, G. Hanock, J.B. Halpern, K.H. Welge: Opt. Commun. **17**, 38 (1967); Chem. Phys. Lett. **44**, 404 (1976)
3.10 J.D. Campbell, M.H. Yu, M. Mangir, C. Wittig: J. Chem. Phys. **69**, 3854 (1978)
3.11 S.E. Bialkowski, W.A. Guillory: J. Chem. Phys. **68**, 3339 (1978)

3.12 M.L. Lesiecki, W.A. Guillory: J. Chem. Phys. **69**, 4572 (1978)
3.13 P.J. Robinson, K.A. Holbrook: *Unimolecular Reaction* (Wiley, Chichester 1972); W. Forst: *Theory of Unimolecular Reactions* (Academic, New York 1973)
3.14 J.G. Black, E. Yablonovitch, N. Bloembergen, S. Mukamel: Phys. Rev. Lett. **38**, 1131 (1977);
J.G. Black, P. Kolodner, M.J. Shultz, E. Yablonovitch, N. Bloembergen: Phys. Rev. A**19**, 704 (1979)
3.15 R.J. Buss, M.J. Coggiola, Y.T. Lee: Discuss. Faraday Soc. **67**, 162 (1979)
3.16 J.D. Rynbrandt, B.S. Rabinovitch: J. Phys. Chem. **74**, 4175 (1970); J. Chem. Phys. **54**, 2275 (1971)
3.17 E.R. Grant, P.A. Schulz, Aa.S. Sudbø, Y.R. Shen, Y.T. Lee: Phys. Rev. Lett. **40**, 115 (1978)
3.18 P.A. Schulz, Aa.S. Sudbø, E.R. Grant, Y.R. Shen, Y.T. Lee: J. Chem. Phys. **72**, 4985 (1980)
3.19 J.L. Lyman: J. Chem. Phys. **67**, 1868 (1977)
3.20 A.V. Nowak, J.L. Lyman: J. Quant. Spectrosc. Radiat. Transfer **15**, 945 (1975)
3.21 J.F. Bott: Appl. Phys. Lett. **32**, 624 (1978)
3.22 R.V. Ambartzumian, Yu.A. Gorokhov, V.S. Letokhov, G.N. Makarov, A.A. Puretzki: Sov. Phys.-JETP **44**, 231 (1976)
3.23 R.V. Ambartzumian, N.P. Furzikov, Yu.A. Gorokhov, V.S. Letokhov, G.N. Makarov, A.A. Puretzki: JETP Lett. **23**, 217 (1976); Opt. Commun. **18**, 517 (1976)
3.24 D.M. Larsen, N. Bloembergen: Opt. Commun. **17**, 254 (1976)
3.25 R.S. McDowell, H.W. Galbraith, B.J. Krohn, C.D. Cantrell, E.D. Hinkley: Opt. Commun. **17**, 178 (1976)
3.26 C.C. Jensen, W.B. Person, B.J. Krohn, J. Overend: Opt. Commun. **20**, 275 (1977)
3.27 H.W. Galbraith, J.R. Ackerhalt: Opt. Lett. **3**, 109 (1978); Opt. Lett. **3**, 152 (1978)
3.28 C.D. Cantrell, K. Fox: Opt. Lett. **2**, 151 (1978)
3.29 I.N. Knyazev, V.S. Letokhov, V.V. Lobko: Opt. Commun. **25**, 337 (1978)
3.30 R.V. Ambartzumian, Yu.A. Gorokhov, V.S. Letokhov, G.N. Makarov, A.S. Puretski, N.P. Furzikov: JETP Lett. **23**, 194 (1976)
3.31 M.C. Gower, T.K. Gustafson: Opt. Commun. **23**, 69 (1977)
3.32 J. Danon, S.V. Filseth, D. Feldmann, H. Zacharias, C.H. Dugan, K.H. Welpe: Chem. Phys. **29**, 345 (1978)
3.33 J.D. Campbell, M.H. Yu, C. Wittig: Appl. Phys. Lett. **32**, 413 (1978)
3.34 E.R. Grant, M.J. Coggiola, Y.T. Lee, P.A. Schulz, Aa.S. Sudbø, Y.R. Shen: Chem. Phys. Lett. **52**, 595 (1977)
3.35 Aa.S. Sudbø, P.A. Schulz, Y.R. Shen, Y.T. Lee: J. Chem. Phys. **69**, 2312 (1978)
3.36 Aa.S. Sudbø, P.A. Schulz, E.R. Grant, Y.R. Shen, Y.T. Lee: J. Chem. Phys. **70**, 912 (1979)
3.37 E. Wurzberg, L.J. Kovalenko, P. Houston: Chem. Phys. **35**, 317 (1978)
3.38 V.N. Bagratashvili, V.S. Letokhov, A.A. Makarov, E.A. Ryabov: Laser Chemistry **1**, 211 (1983); **4**, 1,171 (1983); **4**, 311 (1984); **5**, 53 (1984)
3.39 V.S. Letokhov: *Nonlinear Laser Chemistry*, Springer Ser. Chem. Phys., Vol.22 (Springer, Berlin, Heidelberg 1983)
3.40 P. Lambropoulos, S.J. Smith (eds.): *Multiphoton Processes*, Springer Ser. Atoms Plasmas, Vol.2 (Springer, Berlin, Heidelberg 1984)

4. Two-Frequency Technique for Multiple-Photon Dissociation and Laser Isotope Separation

R. V. Ambartzumian

With 20 Figures

In this chapter the improvement of selectivity in laser isotope separation by the technique of employing one infrared laser to provide selective excitation and a second to provide the energy required for dissociation is reviewed and illustrated. Since this technique is especially useful for molecules that contain heavy elements and for which the isotopic shift of vibrational frequencies is therefore small (often of the order of 1 cm^{-1}), special emphasis is placed on results obtained with the molecule OsO_4.

4.1 Background

The possibility that an isolated molecule, placed in an intense infrared laser field, can absorb many photons (with enough total energy for dissociation) in a very short time span was suggested in 1973 by *Isenor* et al. [4.1] based on their studies of the prompt visible luminescence that was observed at and near the focal point of a lens used to focus pulsed infrared laser radiation into an absorption cell. Interest in this phenomenon was further generated by observations that under certain conditions the multiple-IR-photon dissociation process is isotopically selective [4.2,3]. In these papers, very large isotopic selectivity factors for boron and sulfur isotopes were reported when BCl_3 and SF_6 molecules were dissociated by CO_2 laser pulses. It is now more or less obvious that the isotopic selectivity of the multiple-photon dissociation process is based on the large isotope shifts of the vibrational absorption bands of the molecules used in these early experiments. For example, the isotope shifts in the ν_3 mode for BCl_3 and SF_6 molecules are of the order 40 and 17 cm^{-1}, respectively, and the fundamental absorption bands of the different isotopic species are well separated.

After the success of experiments on the separation of isotopes of elements of light to intermediate mass such as sulphur or boron, the question was posed whether the same technique might be directly applicable to the isotope separation of elements with much smaller isotope shifts by the dissociation of molecules such as OsO_4 or UF_6, where the isotope shift $\Delta\nu_{isot}$ is small compared to the vibrational bandwidth, or whether a different type of process might be needed for these purposes.

It is clear that the selectivity of the dissociation of a gas consisting of two or more isotopic species is determined by the ratio of the dissociation yield of each isotopic species per laser pulse at a given optical frequency. Therefore the narrower the peak of the curve representing the frequency dependence of the dissociation yield, the more selective the dissociation process. This statement is illustrated in Fig.4.1, where the dissociation yield and the enrichment factor of the reaction products of the SF_6 dissociation fragments with the ambient species are shown as a function of laser frequency. In these experiments the radiation of a CO_2 laser was focused in the center of a cell, so that there was a region in the cell where the intensity of the pumping beam was extremely high, of the order 1 GW cm^{-2}, and where most of the dissociation occurred. The presence of a region with a high-intensity field affects the selectivity of dissociation in two ways. First, there is significant power broadening (determined by the Rabi frequency) so that if the isotope shift is of the same order as the Rabi frequency one cannot expect high enrichment. Second, some molecules containing the undesired isotope are excited by simple vibration-vibration (V-V) energy transfer during the initial part of the laser pulse, and if the energy fluence is sufficiently high they may decompose by absorbing energy from the tail of the pulse.

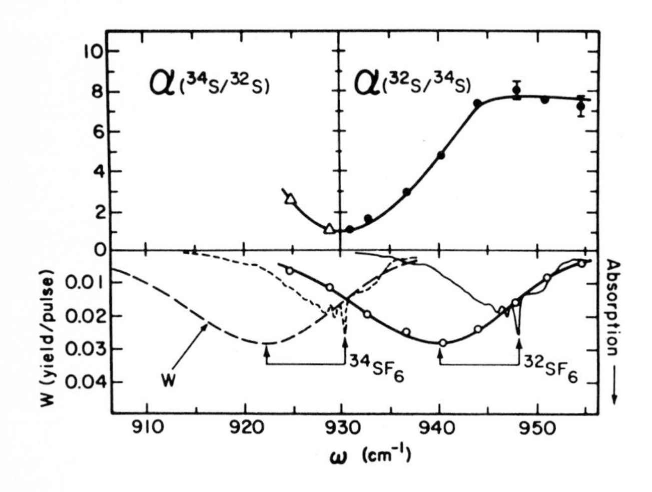

Fig.4.1. Frequency dependence of the dissociation yield per pulse $W(\omega)$ and enrichment coefficient measured in the product SOF_2, which reflects the dissociation selectivity σ. The absorption of the ν_3 mode of SF_6 is shown. It was assumed that $^{32}SF_6$ and $^{34}SF_6$ have identical yield spectra $W(\omega)$

Of course, to obtain the highest possible selectivity in single-frequency experiments one may work exactly at the dissociation threshold. The selectivity of dissociation in this case is enhanced as a result of two factors: (i) The power broadening is reduced to a minimum. (ii) The small isotopic difference in dissociation cross section leads to a maximal difference in yield when one isotope is at the threshold of dissociation. However, in this case the yield per pulse is negligible ($\sim 10^{-5}$-10^{4} of the irradiated molecules). For further details the reader is referred to Chap.2 of this volume.

One possible way to solve this problem is to use laser pulses at two different frequencies to select and then dissociate the desired molecules. This technique permits one to satisfy simultaneously the requirements of selectivity, laser power, and stability.

4.2 Basic Concepts of Two-IR-Frequency Dissociation

The technique of selective dissociation using two infrared frequencies was proposed and the first experiments on SF_6 were conducted in late 1975 [4.4]. The underlying ideas were extremely simple: A polyatomic molecule exhibits multiple-IR-photon absorption in laser fields that are orders of magnitude lower than the fields required for dissociation [4.5]. This property of polyatomic molecules is quite general and has been demonstrated in a large number of molecules of various symmetries. As will be shown later, even in a very low-intensity field a fraction of the molecules experiences strong vibrational excitation. The energy absorbed per vibrationally excited molecule can be of the order of 1 eV or more. As a result of vibrational excitation the frequency of the maximum linear (small-signal) absorption of the excited molecules is shifted with respect to the $v=0 \rightarrow v=1$ ground state absorption. The shift may be to lower or higher frequencies, depending on the sign of the anharmonicity. The spectrum of transitions between highly excited vibrational states is broadened by anharmonic coupling, which at a certain level of excitation may lead to homogeneous absorption. This may be generally expected with vibrational normal modes pumped by a laser. Combination bands show the same behavior (Fig.4.2). The existence of many vibrational modes in a polyatomic molecule leads to the formation of the so-called quasi continuum of vibrational states. In the quasi continuum, resonant transitions are possible for a broad range of laser frequencies.

From these considerations it is clear that if we can preexcite the desired molecules to high vibrational states by a low-intensity pulse at a frequency ω_1 in resonance with the $v=0 \rightarrow v=1$ vibrational transition, we can apply another pulse

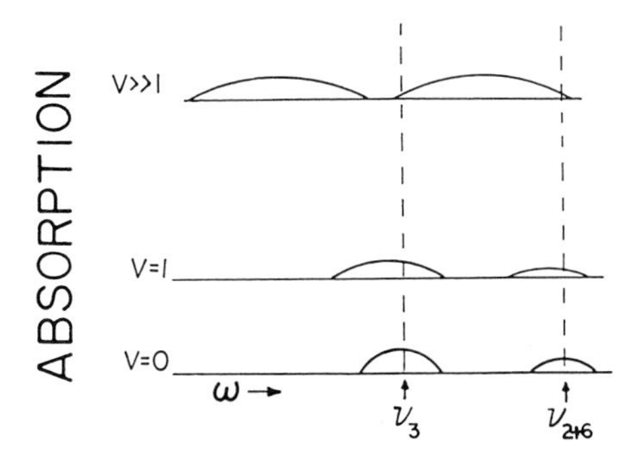

Fig.4.2. Qualitative picture of the linear absorption of SF_6 molecules that experience multiple-photon absorption. Only two bands are considered: ν_3 and $\nu_2+\nu_6$ (not drawn to scale)

at a frequency ω_2 that is nonresonant with the $v=0 \rightarrow v=1$ band. We may choose this pulse to be powerful enough to dissociate only the preexcited molecules by pumping their "quasi continuum" while leaving the nonexcited molecules in the ground state.

This technique of selective dissociation solves the problem of selectivity: the characteristics of the first resonant pulse can be chosen to obtain the maximum selectivity of excitation through the choice of frequency and intensity. The latter can be taken very low to avoid undesirable effects such as power broadening and excitation of molecules through forbidden transitions. The decomposition of molecules excited through undesired V-V transfer processes may be avoided by tuning the frequency of the dissociating pulse ω_2 sufficiently far from the ground-state absorption frequency (to the red, for example). If the detuning is large enough, the pulse at ω_2 will dissociate only those molecules in highly excited states that cannot be populated in a few collisions. We shall assume here that the pressures in the reaction vessel and the laser pulse lengths are such that the laser pulse length is short compared to the collision time of ground-state molecules.

By an appropriate choice of the frequency ω_2 of the dissociating pulse it is possible to reduce the required laser fluence or intensity to the point where it is possible to work at intensities that are easily attainable in unfocused beams of existing lasers and that do not lead to undesired side effects such as window damage. This allows one to use a large fraction of the laser energy for dissociation.

Besides these practical advantages of the two-frequency technique as compared with the single-frequency technique, the two-IR-frequency technique has become a very powerful tool for investigation of the process of multiple-IR-photon excitation. Five parameters can be varied independently in this technique: the frequency ω_1 of the resonant pulse and the energy fluence Φ_1 at ω_1; the frequency ω_2 of the dissociating pulse and its fluence Φ_2; and the time delay τ_d between the pulses at ω_1 and ω_2.

This chapter will review certain experimental results obtained in two-frequency dissociation experiments. First we shall discuss the problem of selectivity enhancement in isotope separation experiments, and in the last section we shall discuss the results of a more fundamental investigation of the multiple-photon excitation process.

4.3 Selectivity of Dissociation

4.3.1 Spectral Measurements

The two-frequency technique was initially proposed to enhance the selectivity of the dissociation process. An enhancement of selectivity was demonstrated in the dissociation of the molecules SF_6 [4.4], OsO_4 [4.6] and SeF_6 [4.7]. In [4.4] the enhancement of selectivity in the dissociation of SF_6 was deduced from a comparison

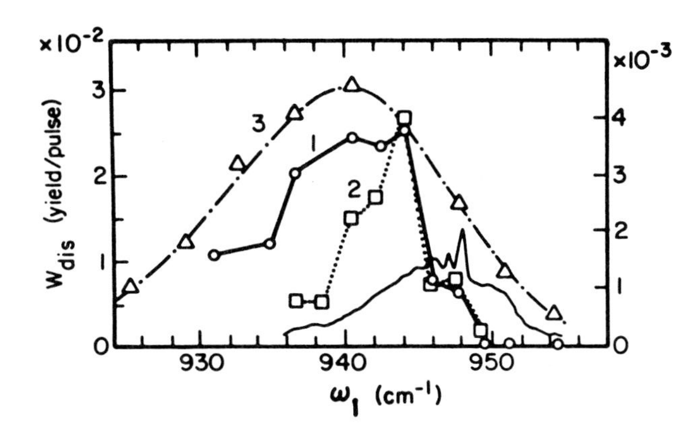

Fig.4.3. Frequency dependence of the dissociation yield in three cases. *1*: two-frequency experiment at 300 K, *2*: two-frequency experiment at 190 K, *3*: single-frequency experiment. The fluence of the $_1$ beam was 0.4 J cm^{-2}, and its intensity was 4 MW cm^{-2}. The intensity of the ω_2 beam averaged over the irradiation volume was 58 MW cm^{-2}. It was focused by an f = 10 cm lens. The delay between the ω_1 and ω_2 pulses was 500 ns

of the widths of the curve giving the frequency dependence $W(\omega)$ of the dissociation yield for single-frequency versus two-frequency dissociation experiments. These results are shown in Fig.4.3, where it is evident that the half-width of $W(\omega)$ decreased roughly by a factor of two in the two-color experiments. In these experiments special care was taken to avoid the dissociation of SF_6 except in the presence of both fields ω_1 or ω_2. The ω_1 beam propagated through the cell unfocused, while the ω_2 beam was focused in the center of the absorption cell. The frequency ω_2 in these experiments was tuned far to the blue ($\omega_2 = 1084$ cm^{-1}) with respect to the ν_3 mode of SF_6. This circumstance played a major role in these observations and will be discussed later in conjunction with other experiments on OsO_4.

One can see that the frequency dependence of $W(\omega_1)$ is quite similar to the spectrum of multiple-photon absorption of SF_6 [4.8,9] obtained at approximately the same fluences as the fluence at ω_1 in these experiments (for details see the other chapters in this volume). Therefore if one finds a narrow multiple-photon absorption contour in a molecule at fluences much smaller than the dissociation threshold, it is possible under certain conditions to reproduce this narrow spectrum in the dissociation yield as well. This was clearly shown in experiments on OsO_4 [4.10,11], where it was found that the multiple-photon absorption spectrum shows several peaks and minima with half-widths of the order of 2-3 cm^{-1}, as shown in Fig.4.4. The quantity plotted is $\langle n \rangle$, the number of infrared photons absorbed in the irradiated volume on the assumption that all molecules are excited, i.e.

$$\langle n \rangle = E_{abs}/\hbar\omega N_0 \qquad (4.1)$$

where E_{abs} is the amount of energy absorbed from the beam and N_0 is the total number of molecules in the beam.

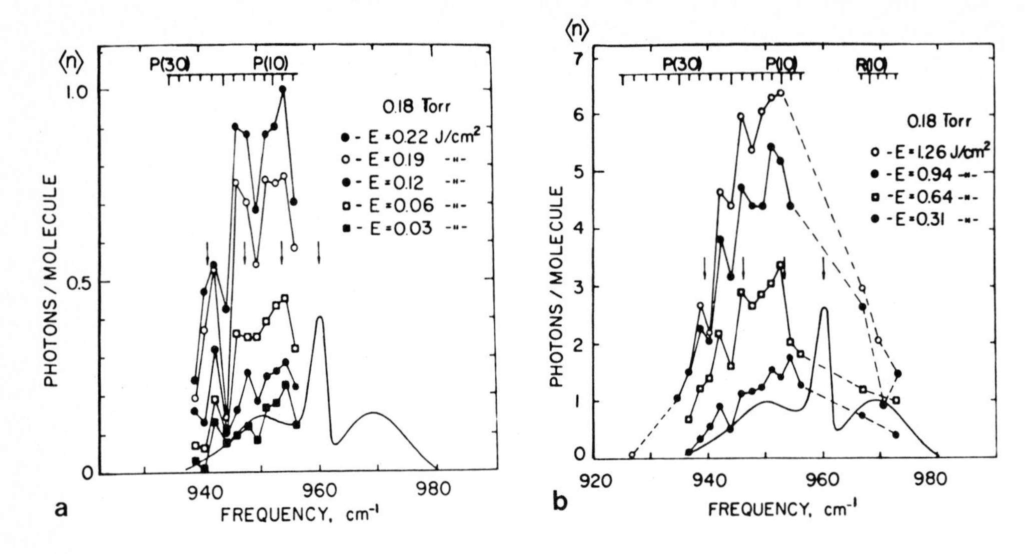

Fig.4.4a,b. Multiple-photon absorption spectrum of OsO_4 at various energy fluences. The OsO_4 pressure was 0.18 Torr. The arrows show the positions of the Q-branches of the transitions in ν_3 mode: $v=1 \rightarrow v=2$, $v=2 \rightarrow v=3$, etc. The laser pulse consisted of a 90-ns spike and a 1-μs tail

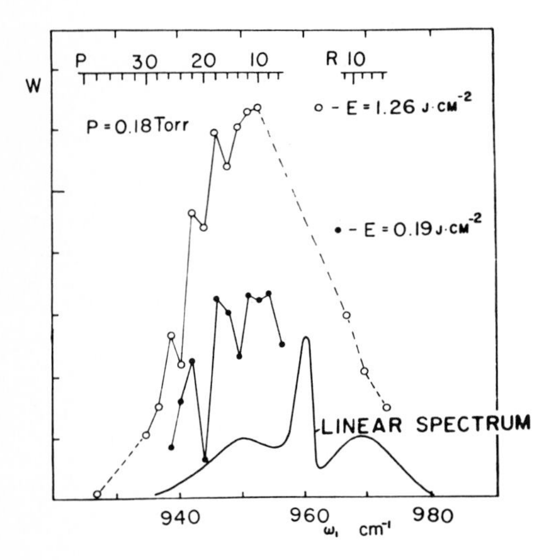

Fig.4.5. Frequency dependence of the dissociation yield in OsO_4 in two-frequency experiments. The curves were obtained at different energy fluences Φ_1 at ω_1; *1*: (●●●) $\Phi_1 = 0.19$ J cm^{-2}, *2*: (○○○) $\Phi_1 = 1.26$ J cm^{-2}. The sample pressure, energy fluence at ω_2, and the time delay between the ω_1 and ω_2 pulses were the same in both cases. The increase of energy fluence in the ω_2 beam when $\Phi_1 = 0.19$ J cm^{-2} led to the same result as shown by curve 2

In two-color dissociation experiments with OsO_4 it was found that when ω_2 is tuned to the red with respect to the Q-branch of the ν_3 mode of OsO_4 and when the energy fluence in the ω_2 beam is far below that at which the yield starts to saturate, then the yield per pulse versus ω_1 resembles the multiple-photon absorption spectrum (Fig.4.5). If we increased the energy fluence at ω_2, which is intended to dissociate only the excited molecules, or increased the energy fluence at ω_1, then we found that the structure in the frequency dependence of the yield was washed out.

This shows that the two-frequency dissociation technique requires delicate adjustment if one wishes to enhance the selectivity of dissociation by this method; i.e., the dissociating frequency ω_2 and the energy fluences at ω_1 and ω_2 should be optimized to obtain the best possible final results.

In the SF_6 experiments, the fact that the dissociating frequency ω_2 was tuned to the blue provided the enhancement of selectivity. Two reasons could account for this fact: the fluence at ω_2 could have been very far from the saturating fluence at the frequency that was used, or the ω_2 beam could have interacted with only a small fraction of the SF_6 molecules that were excited higher than the average level of excitation. The latter explanation seems to be more attractive since (as may be seen from Fig.4.3) when ω_1 coincided with the R branch of the ν_3 mode of SF_6 there was no dissociation. If we look at the multiple-photon absorption spectrum of SF_6, we see that SF_6 exhibits multiple-photon absorption even when the R-branch is pumped. Combining these two observations one can draw the conclusion that when ω_1 coincides with the $v=0 \rightarrow v=1$ R-branch the SF_6 molecules are not pumped high enough to permit the nonresonant frequency $\omega_2 = 1084$ cn^{-1} to interact with the excited molecules. When ω_1 is in the $v=0 \rightarrow v=1$ P-branch region, the SF_6 molecules are pumped to much higher vibrational states and are ready to absorb radiation at $\omega_2 = 1084$ cm^{-1}. In the following section we shall see that this qualitative picture is realistic and is also correct for the OsO_4 molecule.

In the OsO_4 experiments the observation that the frequency dependence of the yield depends also on the energy fluence at ω_2 gives a qualitative idea of the distribution of the excited molecules over the molecular vibrational levels: many of the excited molecules have a low energy of excitation, and a few are excited to very high vibrational levels and can easily be dissociated by the low-intensity radiation at ω_2. Rough estimates show that the small, highly excited fraction is of the order of a few percent of the excited molecules.

4.3.2 Isotope Enrichment Experiments

The enhancement of the selectivity of dissociation by the use of two IR frequencies was demonstrated in two similar experiments: the two-frequency isotopic selectivity was compared with the results of single-frequency dissociation and enrichment experiments on the molecules OsO_4 and SeF_6 [4.6,7].

The isotopic separation of Se was reported in [4.7], where two lasers were used: an NH_3 laser that performed the initial excitation and a small amount of dissociation, and a CO_2 laser that dissociated the preexcited molecules. The CO_2 laser radiation enhanced the dissociation yield in these experiments by a factor of roughly 2.5 measured by the technique of monitoring the IR fluorescence.

In both cases the isotope shifts of the absorption bands pumped by the laser were smaller than the width of the vibration-rotation band. This is an interesting

Table 4.1. Ratios of the dissociation cross section for various Se isotopes: $\sigma(^j\text{Se})/\sigma(^i\text{Se})$. For each value of i,j there are three entries. The upper entry is the experimental single-frequency dissociation cross section using an NH_3 laser; the middle entry (underlined) is the value from the two-frequency experiment; the lower entry (in parentheses) is the calculated value (see text). The SeF_6 pressure was 0.2 Torr; the NH_3 laser output was 35 mJ/pulse at 780.5 cm^{-1}; the CO_2 laser output was 1.57 J/pulse at 944.2 cm^{-1}. The NH_3 and CO_2 laser beams were focused collinearly with lenses of 7.5- and 12.8-cm^2 focal length respectively

i	j: 76	77	78	80
77	1.05 1.10 (1.03)			
78	1.15 1.29 (1.14)	1.10 1.18 (1.11)		
80	1.52 1.73 (1.61)	1.45 1.58 (1.56)	1.32 1.34 (1.41)	
82	2.25 2.17 (2.69)	2.15 1.97 (2.61)	1.95 1.67 (2.36)	1.48 1.25 (1.67)

and useful result since the two-frequency technique does not enhance the selectivity obtainable with the single-frequency technique when the isotope shift is large compared to the total width of the absorption band.

The isotope shift of the laser-pumped mode (ν_3 in both cases) is 0.26 cm^{-1}/amu for OsO_4 ($\nu_3 = 960\ cm^{-1}$) and 1.6 cm^{-1}/amu for SeF_6 ($\nu_3 = 779\ cm^{-1}$).

In both cases mass analyses were conducted on the residual (undissociated) parent molecules at a fixed burnout level, i.e., the measured quantity was β rather than α, which is a direct measure of the selectivity. For a fixed burnout level β is simply related to α. The results of Se isotope enrichment experiments are summarized in Table 4.1 [4.7]. The entries in parentheses were calculated under the following assumptions: (i) The dissociation yield versus frequency has the same dependence for SeF_6 as the dissociation yield of SF_6 in single-frequency experiments (Fig.4.1). (ii) There are no resonances in $W(\omega)$. The experimental conditions were identical for the one- and two-frequency cases; the only change was the presence of CO_2 laser radiation. As seen from Table 4.1, the two-frequency technique led to different results depending on the isotopes considered: for some pairs of isotopes of selenium the selectivity of dissociation was enhanced (i.e., $^{76}\text{Se}/^{77}\text{Se}$, $^{76}\text{Se}/^{78}\text{Se}$, $^{77}\text{Se}/^{78}\text{Se}$) roughly by a factor of two; for $^{76}\text{Se}/^{80}\text{Se}$ up to $^{78}\text{Se}/^{80}\text{Se}$ the enhancement was negligible, and finally for $^{i}\text{Se}/^{82}\text{Se}$ the enhancement was negative, i.e. the selectivity of the dissociation dropped compared with the single-frequency technique.

Although these results are intriguing, the authors of [4.7] made no attempt to explain them. Here we shall propose a tentative explanation based on our OsO_4 experiments.

In single-frequency experiments differences in dissociation cross section for adjacent isotopic species arise primarily because of the frequency dependence of the dissociation cross section. Other contributing factors include small differences in the number of molecules that are excited by the radiation field, and small differences in the average excitation level. In both cases the effect is similar to the effect of the frequency dependence of the dissociation cross section, provided that the laser frequency interacts with approximately the same (adjacent) J values in the rotational distribution and that excitation starts with similar transitions (P, Q or R). If the radiation hits different branches, e.g., R- or Q-branches in different isotopic species, then the situation cannot be so simply understood. When the P-branch is pumped, the excitation level is higher than for Q- or R-branch pumping. But when the Q-branch is pumped, the fraction of molecules that are excited by the radiation is very close to unity. Since the nonresonant field ω_2 dissociates most easily those molecules that are already excited to high vibrational levels, when we pump the P-branch with ω_1 a larger fraction of the excited molecules dissociates than when the Q-branch is prepumped. In the latter case approximately all the molecules are prepumped and a small fraction of them dissociate. The interplay of these two factors in both cases seems to explain the decrease of selectivity in two-frequency experiments when selectivity is measured in the ratio $^{76}Se/^{80,82}Se$. The NH_3 laser radiation in those experiments excited Q branches of the hot bands of $^{76,79}SeF_6$ and in the case of $^{80,82}SeF_6$ the same frequency excited R-branch transitions. The interplay between the easy dissociation of highly excited molecules and the presence of an increased fraction of preexcited molecules had the result that the selectivity decreased in some two-frequency experiments.

Osmium isotope separation experiments were conducted by single- and two-frequency dissociation techniques in [4.6]. In order to detect enrichment in the single-frequency experiments, it was necessary to prepare an equimolar mixture of $^{192/187}OsO_4$. Under conditions where the laser power was held just above the dissociation threshold and the degree of burnout was 95%, the best enrichment factor β was 1.15 giving a selectivity of dissociation α of the order 1.03-1.04 for isotopic compounds of OsO_4 with an isotope shift 1.26 cm^{-1}. In the course of these experiments it was found that some unidentified scrambling chemical reactions occurred in the cell during and after the irradiation resulting in a decrease of the measured selectivity of dissociation α.

Application of the two-frequency technique resulted in a large enhancement of α, making it possible to conduct experiments with OsO_4 with the natural abundance of Os isotopes. In these experiments the radiation at both frequencies propagated unfocused, and the dissociation yield without any one of the two pulses was two orders

Table 4.2. Mass-spectrometer analyses

Laser Frequencies $\omega_1 + \omega_2$	Enrichment Factor (±0.02) $^{192}Os/^{186}Os$	$^{192}Os/^{190}Os$	$^{192}Os/^{189}Os$	$^{191}Os/^{188}Os$	$^{192}Os/^{187}Os$
P(6) +P(20)	1.53	1.11	1.14	1.22	1.62
P(6) +P(12)	1.02	1.04	1.04	1.08	1.04
P(20)+P(20)	0.96	1.02	1.02	1.04	0.98
P(4) +P(20)	1.24	1.10	1.11	1.18	1.29
P(4) (one frequency) 10 J cm^{-2}	1.01	1.01	1.01	1.01	1.01

of magnitude lower than when both pulses were present. The cell was filled with 0.3 Torr of OsO_4 and up to 2.0 Torr of OCS which served as a scavenger. The burnout level was held fixed at 90% of the initial OsO_4. The results of mass-spectrometer analyses of the gas remaining after irradiation are given in Table 4.2.

The last row of Table 4.2 demonstrates that at very high intensity there is no enrichment at all. The highest enrichment factor achieved here was for $^{192}Os/^{186}Os$ and was roughly equal to 1.6. It was found that even if the burnout level was increased β remained the same, showing that an equilibrium was established between the dissociation-enrichment process and the isotope-scrambling reactions. Nevertheless even in this case the two-frequency technique gave an enhancement of selectivity a factor of 4 to 5 larger than the single-frequency technique. An exact comparison is very difficult because of the effects of chemical scrambling reactions. Other scavengers such as C_2H_4 and NO gave approximately the same results. If CS_2 was added to the cell no enrichment was observed.

To measure the initial selectivity of the dissociation the following experiments were done. The enrichment factor β was measured as a function of the number of irradiation pulses, or the burnout level W_n (at $W_n = 1$ the concentration of parent molecules is decreased by a factor of e). The results are shown in Fig.4.6. Instead of growing as a function of the burnout level, β saturates at a certain level. From this graph α can be evaluated by fitting α to the initial slope of β. The value of α measured in this manner is equal to 1.16 ± 0.02. This selectivity is high considering the small isotope shift ~ 1.26 cm^{-1} and that the experiments were conducted at room temperature and other parameters such as OsO_4 pressure and scavenger species were not optimized.

One of the most interesting results in OsO_4 enrichment experiments is the frequency dependence of β obtained under fixed experimental conditions (burnout level, sample pressure, energy fluences in both beams, etc.). The results are shown in Fig.4.7. It is rather surprising that the dependence of β on ω_1 showed no correlation with the multiple-photon absorption spectrum of OsO_4. A detailed interpretation

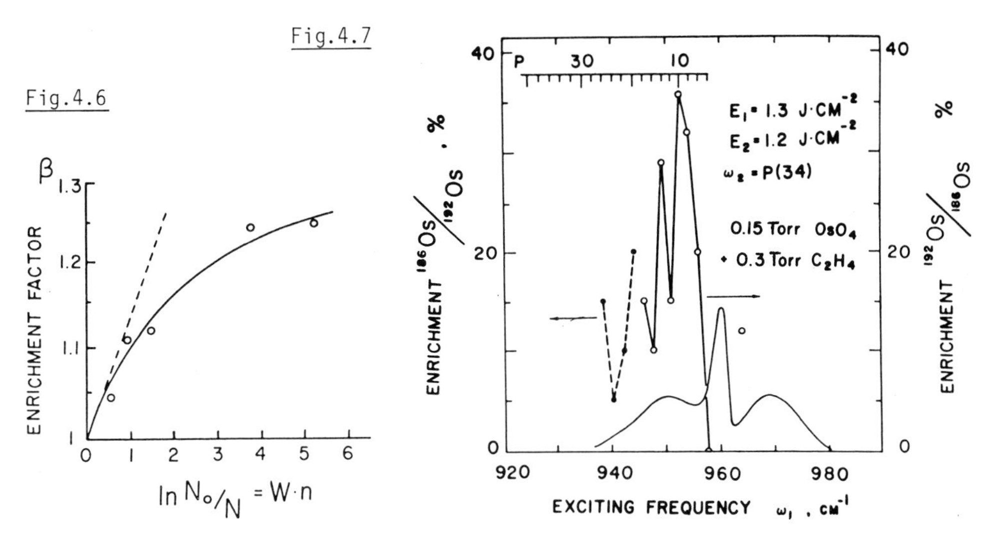

Fig.4.6. Enrichment factor β in OsO_4 as a function of the burnout level W_n (——). The dashed line shows the expected dependence in the absence of scrambling reactions for $\sigma = 1.16$

Fig.4.7. Dependence of enrichment factor in OsO_4 versus frequency ω_1 measured under the same experimental conditions. The energy fluences at ω_1 were $\Phi_1 = 1.3$ J cm^{-2}, and the fluences at ω_2 [=P(34) CO_2 laser line] were $\Phi_2 = 1.2$ J cm^{-2}. The delay between the ω_1 and ω_2 pulses was 250-300 ns

of these results is beyond our present understanding of the processes of multiple-photon excitation and dissociation, but some comments can be made. Unfortunately the multiple-photon absorption spectrum shown in Fig.4.4 is the result of point-to-point measurements with a very limited set of laser frequencies, while a meaningful comparison of $\langle n \rangle$ (ω_1) and β (ω_1) can be made only with the real spectra of these two quantities. It may also be the case that the intensity at ω_2 was so high that the amplitude of the peaks in the multiple-photon absorption spectrum was diminished. It seems that a full answer can be given only on the basis of future experiments.

4.4 Investigation of Multiple-Photon Excitation

The two-frequency technique has been successfully applied to the investigation of the process of multiple-photon excitation [4.12]. Measurements of the energy absorbed per molecule of the type performed in [4.5,8,9] do not give information on the fraction of molecules that interact with the radiation field, the excitation level of the molecules that are excited, or the dependence of these parameters on laser fluence, frequency, etc. Here we describe experiments that provide information of this kind.

The experiments described are of the double-resonance type, but the use of an extremely intense probe pulse at ω_2 gave additional information that cannot be obtained in experiments with a weak probe at ω_2.

It was found experimentally that if ω_2 is tuned to the red with respect to the peak of the ν_3 fundamental band of OsO_4 it is very easy to saturate the dissociation yield, i.e., an increase of the energy fluence in the beam beyond a certain value does not result in a concomitant increase of the dissociation yield. For a frequency $\omega_2 = 927\ cm^{-1}$ the value of the energy fluence at which the dissociation yield saturated was found to be 1.4 J cm^{-2}. Moreover it was found that this value applies to both single-frequency and two-frequency experiments. Within experimental error the saturation fluence Φ_{sat} remains 1.4 J cm^{-2} independent of ω_1 or the laser fluence Φ_1 at ω_1 in two-frequency experiments. The results are shown in Fig.4.8. The saturation of the dissociation yield was interpreted as being due to dissociation of all the molecules that were preexcited by the beam at ω_1. If this assumption is correct it gives a tool for direct measurements of the fraction of molecules q that were excited by the beam ω_1. A knowledge of q and $\langle n \rangle$ (see Fig.4.4) gives the average excitation level ℓ since they are connected by the simple relation

$$\langle n \rangle = q\ell \quad . \tag{4.2}$$

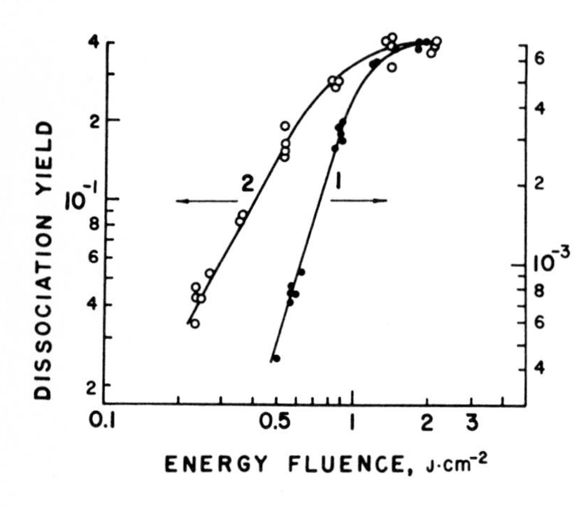

Fig.4.8. Dependence of the dissociation yield per pulse for OsO_4 in two cases. *Curve 1*: single-frequency experiment (right-hand scale) with $\omega = \omega_0 = 927\ cm^{-1}$. *Curve 2*: two-frequency experiments (left-hand scale) with $\omega_1 = 947.7\ cm^{-1}$. Energy fluence in the exiciting beam $\Phi_1 = 1.2$ J cm^{-2}. OsO_4 pressure 0.03 Torr. The pulses used had a 90-ns spike with a 1-μs tail. Note that in both cases the yield saturates at the same energy fluence at 927 cm^{-1}

4.4.1 Evaluation of q and ℓ in OsO_4

The dissociation yield per pulse was measured as a funcition of ω_1 and Φ_1 for an energy fluence at ω_2 equal to 2 J $cm^{-2} > 1.4$ J $cm^{-2} = \Phi_{sat}$. The results of these measurements are given in Fig.4.9 where the frequency dependence of q and ℓ is shown for three fixed energy fluences Φ_1. As can be seen from Fig.4.9 a variation of Φ_1 leads mainly to a variation of the fraction of excited molecules q while the average excitation level ℓ remains appproximately the same. It can also be seen

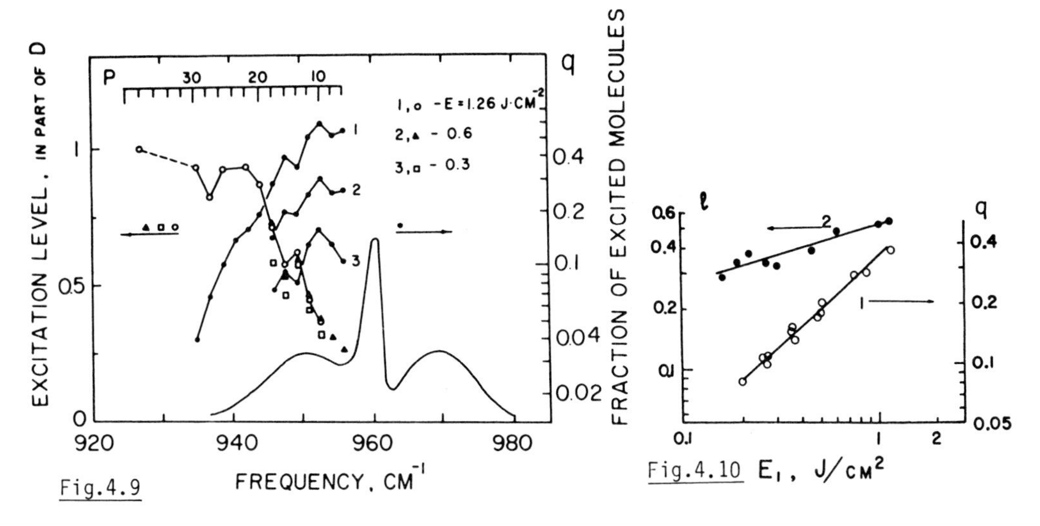

Fig.4.9. Fraction of OsO_4 molecules excited by the pulse at ω_1 (right-hand scale) and average excitation level (left-hand scale) shown in units of the dissociation energy D_0 versus laser frequency for three values of the energy fluence at ω_1, obtained by the yield-saturation technique. The energy fluence at ω_2 was 1.8 J cm^{-2} at $\omega_2 = 927$ cm^{-1}. OsO_4 pressure was 0.03 Torr. The linear absorption of OsO_4 is also shown

Fig.4.10. Dependence of the fraction of molecules excited (q) and average excitation level (ℓ) on energy fluence of the exciting beam at a fixed excitation frequency. All other experimental conditions were the same as in Fig.4.9. The excitation was in the middle of the P-branch, $\omega_1 = 947.7$ cm^{-1}

that when ω_1 is tuned away from the Q-branch of the ν_3 fundamental band the excitation level increases, while the fraction of molecules that interacts with the radiation field decreases. Figure 4.10 shows detailed measurements of q and ℓ as a function of Φ_1 at fixed ω_1.

These data are consistent with other direct and indirect measurements. Among the indirect observations we include investigations of the shape of the visible luminescene pulse emitted by highly excited OsO_4 molecules following IR laser irradiation. When ω_1 is tuned to the red of the ν_3 fundamental Q-branch, the visible luminescence is mainly collision-free and follows the laser pulse shape, indicating that the OsO_4 molecules are prepared in extremely high vibrational states. On the other hand, if ω_1 is in the vicinity of the fundamental Q-branch the visible luminescence is entirely collisional, indicating that the high-lying vibrational levels are reached by V-V energy transfer processes.

The results shown in Figs.4.9,10 are very interesting from the following point of view: even at very low energy fluences the average excitation level ℓ of the OsO_4 molecules corresponds to roughly $9\hbar\omega$ (1/3 of the dissociation energy $= D_0 = 27\,\hbar\omega$) so that by the end of the pulse they should be in the quasi continuum of vibrational states, which is supposed to start in OsO_4 at 3-4 $\hbar\omega$. At this energy the density of vibrational states in OsO_4 is of the order 10^2-10^3 levels per cm^{-1}. Since molecules at $9\hbar\omega$ are already deep in the quasi continuum, they should easily be excited further up to the dissociation states since the quasi continuum is supposed to provide resonances at any IR frequency. However, in reality the OsO_4 molecules practically stop absorbing laser radiation after reaching a certain level of excitation.

From our point of view, taking these experimental results into account, the quasi continuum as formally introduced up to now plays no major role in explaining the multiple-photon excitation of OsO_4. Possibly OsO_4 is unique because of its set of vibrational frequencies; in general there may be deviations in other molecules from the physical situation observed in OsO_4.

Let us consider the SF_6 molecule. There are indications that in this molecule, as in OsO_4, the increase of the amount of absorbed energy with increasing fluence is connected mainly with the enhancement of q when the pumping field is increased. Figure 4.11 shows two curves obtained for SF_6: the average number of IR quanta absorbed per molecule ($\langle n \rangle$) defined by (4.1), and the dissociation yield per pulse as a function of the energy fluence or power when the ω_2 beam was focused in the cell [4.4]. The intensity of the beam at ω_2 was sufficient to produce saturation of the dissociation yield. Of course there were regions in the absorption cell where the ω_2 beam did not saturate the dissociation yield, making the interpretation of the results somewhat ambiguous. As can be seen from Fig.4.11, the curves of $\langle n \rangle$ and W are parallel to each other. From our point of view this shows that the main effect that occurs with an increase of the fluence of the resonant pump at ω_1 is connected with the enhancement of the fraction of molecules q excited by the radiation. Some of the results reported in [4.13] can be interpreted similarly [Ref. 4.13, Figs.1,4].

Naturally, as the energy fluence Φ_1 increases, the fraction of molecules excited tends towards unity. Interpreting the results of [4.13] in our manner, we estimate that in the vicinity of an energy fluence of 0.3-0.5 J cm^{-2} nearly all the SF_6 molecules are already involved in excitation when the laser pumps the Q-branch of the ν_3 mode. The results of [4.14] obtained by the double-IR-resonance technique

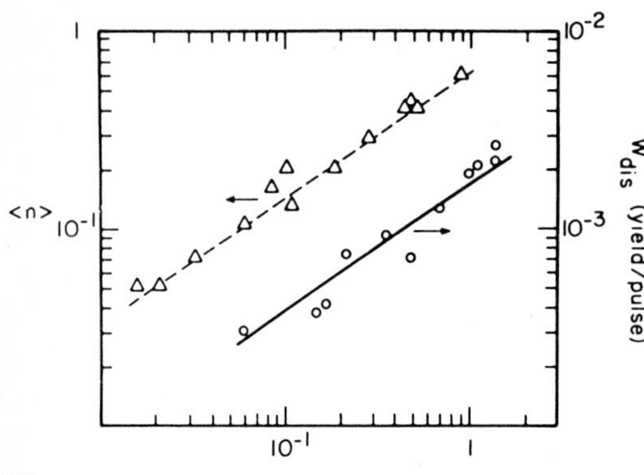

Fig.4.11. Dependence of the number of absorbed IR quanta per molecule ($\langle n_1 \rangle$) and yield per pulse (W) in SF_6 on pump power at ω_1, in two-frequency experiments for $\omega_1 = 924.4$ cm^{-1} and $\omega_2 = 1048$ cm^{-1}. The ω_2 beam was focused. The intensity of the ω_2 beam was 60 MW cm^{-2} averaged over the irradiation volume. The pulse duration was 100 ns (FWHM). Both curves are shown on a logarithmic scale

with a low-intensity probe beam showed that at low laser fluences the enhancement of <n> is connected mainly with an increase of q, while ℓ is a much slower function of Φ_1 than q.

Therefore the physical situation discussed applies generally to the excitation of a large class of molecules.

It is also interesting to compare the frequency dependence of the fraction of excited molecules q with the rotational distribution of the molecules. The results obtained by the saturated-yield technique are given in Fig.4.12, where the intensity of the visible luminescence in OsO_4 was measured as a function of ω_1. Under conditions of saturated yield, the intensity of the visible luminescence signal is proportional to q[1]. In the same figure the linear absorption of OsO_4 is shown on a logarithmic scale. The shape of this spectrum reflects more or less the distribution over rotational levels. It should be noted that the room-temperature linear absorption spectrum of OsO_4 consists mainly of hot bands orginating from the thermally excited vibrational levels at 330 and 660 cm^{-1}.

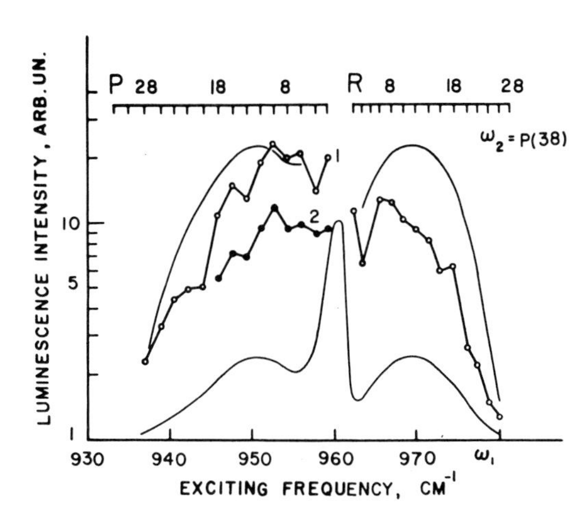

Fig.4.12. Luminescence intensity versus ω_1 in two-frequency experiments on OsO_4. For $\omega_2 = 927$ cm^{-1} the energy fluences were: $\Phi_1 = 1.2$ J cm^{-2} (ooo) and $\Phi_1 = 0.6$ J cm^{-2} (●●●). The fluence at ω_2 was $\Phi_2 = 1.8$ J cm^{-2}; the OsO_4 pressure was 0.03 Torr; the delay between the pulses was 0.5 μs. The IR absorption of OsO_4 is shown on linear (*lower curve*) and on logarithmic scales for comparison

As can be seen from Fig.4.12 the fraction of molecules excited approximately follows the band contours of the P- and R-branches of the linear absorption spectrum of OsO_4, so that q follows the molecular vibration-rotation distribution. This result disagrees with the statements made in [4.15,16], that all molecules are involved in multiple-photon absorption independent of their rotational quantum numbers. The observed correspondence of $q(\omega)$ with the initial molecular vibration-rotation distribution shows that for any initial rotational state there exists a path by which a molecule can reach high-lying vibrational states by the absorption of several quanta of the pump radiation.

1 It was found that for OsO_4 the amplitude of the visible luminescence signal is directly proportional to the dissociation yield per pulse. This proportionality is valid when the yield does not exceed 0.2-0.3.

On the basis of these results some important conclusions can be drawn. First of all a question may be raised concerning the validity of constructing theoretical models of the dissociation process by simply adjusting parameters to fit the curve of dissociation yield versus energy fluence. In all existing theoretical models it is tacitly assumed that all the molecules interact with the laser radiation, but from the results presented above, one can see that an increase of fluence affects not only the energy deposited per excited molecule, but also the number of molecules that interact with the laser radiation. Therefore, adjustments of the parameters of theoretical models should be made using the data provided by two-frequency experiments (the yield versus the fluence at ω_2 as shown in Fig.4.8), since the beam at ω_2 interacts with a fixed number of molecules that were preexcited by the first pulse at ω_1. Even in this case the question arises whether the preexcited molecules interact homogeneously with the ω_2 beam or the ω_2 beam interacts with a variable number of preexcited molecules depending on the energy fluence at ω_2, as is the case for the ω_1 beam.

The same can be said about attempts to describe multiple-photon absorption by a single cross section. Since there are at least two parameters (the fraction of excited molecules q and the excitation level ℓ) that are independent functions of energy fluence in the pump beam, it is not clear what physical meaning can be attached to the multiple-photon absorption cross section that was introduced in [4.8,9] and other papers. From our point of view the cross section introduced by these authors characterizes only the energy deposition in the molecular system. The cross section has the meaning of a real absorption cross section at large energy fluences when all molecules are excited, i.e., $q \cong 1$, and in that region the absorption cross section characterizes energy absorption in the so-called quasi continuum.

4.5 Interaction of the Nonresonant Pulse with Excited Molecules

4.5.1 Absorption Measurements

To understand and describe quantitatively the process of dissociation it is necessary to know how energy is fed into the molecular system not only by the resonant pulse at ω_1, but also by the pulse that is nonresonant and therefore interacts only with excited molecules.

Measurements of this kind were performed in [4.15], where the absorption of preexcited OsO_4 was studied calorimetrically. Quantitative measurements of the absorption of OsO_4 were performed at the frequency $\omega_2 = 927\ cm^{-1}$ which lies below the Q-branch of the ν_3 mode at 960 cm^{-1}.

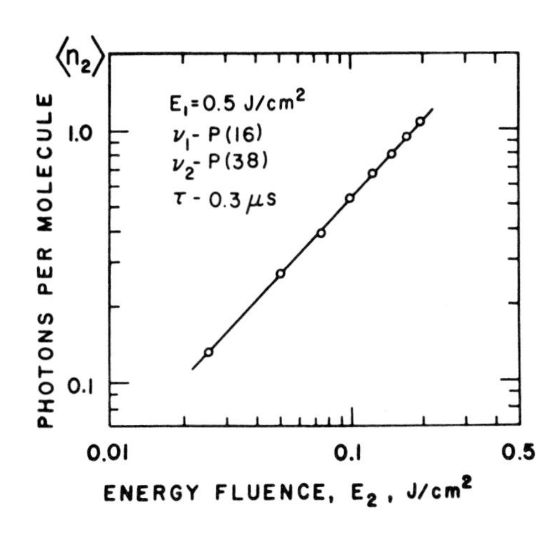

Fig.4.13. Number of IR quanta absorbed per molecule in the irradiation volume from the nonresonant ω_2 beam versus energy fluence in it. The parameters were $\omega_1 = 947.7\ cm^{-1}$, $\Phi_1 = 0.5\ J\ cm^{-2}$, $\omega_2 = 927\ cm^{-1}$, OsO_4 pressure 0.2 Torr, pulse delay 0.3 μs. Note that under these conditions q = 0.18 (Fig.4.10)

In Fig.4.13 we show the number of absorbed quanta per molecule (4.1) versus the energy fluence at ω_2. If we take into account the fact that the beam at ω_1 excited only a fraction q of the molecules out of the ground state and take q from the data presented in Fig.4.10, we estimate that the energy absorbed per excited molecule is at least five times greater than shown in Fig.4.13 since q is equal to 0.18 at the fluence at ω_1 that was used in this experiment. Since the time delay between the two pulses was small and the pressure was low, neither the number of excited molecules nor q was altered significantly as a result of the delay between the pulses.

If the radiation at ω_2 interacts with all the preexcited molecules then the vibrational "temperature" of the excited molecules rises linearly with the laser fluence. Since there is no involvement of new "cold" molecules this "temperature" rise gives the functional dependence of the yield shown in Fig.4.8.

As was previously mentioned, theoretical models of the dissociation process should be compared with the data presented in Figs.4.8,13 rather than with the data provided by single-frequency experiments. From two-frequency measurements it is possible to evaluate the following quantities: first, a measurement of $\langle n_1 \rangle$ and $\langle n_2 \rangle$ when $q = 1$ gives the energy required to dissociate the molecule; second, it is possible to evaluate q by taking the ratio of $\langle n_2 \rangle$ under the desired conditions to $\langle n_2^* \rangle$ when q is equal to unity; finally, it is possible to find the minimal molecular excitation level required to interact with the nonresonant field at ω_2 at a given frequency. All of this supposes that the beam at ω_2 saturates the yield, i.e., that all the preexcited molecules are dissociated through interaction with ω_2.

We now discuss how to obtain a q of unity. If $\langle n_1 \rangle$ is large after V-V equilibrium is established, the molecules have a very high vibrational temperature corresponding to the energy $\langle n_1 \rangle \hbar\omega$ absorbed in the vibrational degrees of freedom. If $\langle n_1 \rangle$ is high enough the fraction of molecules in the ground state and low-lying vibrational states is completely negligible, and it can be assumed that $q \cong 1$.

Very simple equations connect the dissociation energy D_0, the fraction of excited molecules, the average numbers of photons $\langle n_1\rangle$ and $\langle n_2\rangle$ absorbed per molecule at ω_1 and ω_2, and $\langle n_2^*\rangle$ which is defined as $\langle n_2\rangle$ measured with a long delay between the ω_1 and ω_2 pulses, i.e., after vibrational equilibrium is established via V-V process. These relations are:

$$\langle n_1\rangle = q\ell \quad ,$$

$$\langle n_2\rangle = q(D_0-\ell) \quad ,$$

$$\langle n_2^*\rangle = S(D_0-q\ell) = S(D_0-\langle n_1\rangle) \quad ,$$

where S is the saturation parameter (S = 1 when the yield is saturated).

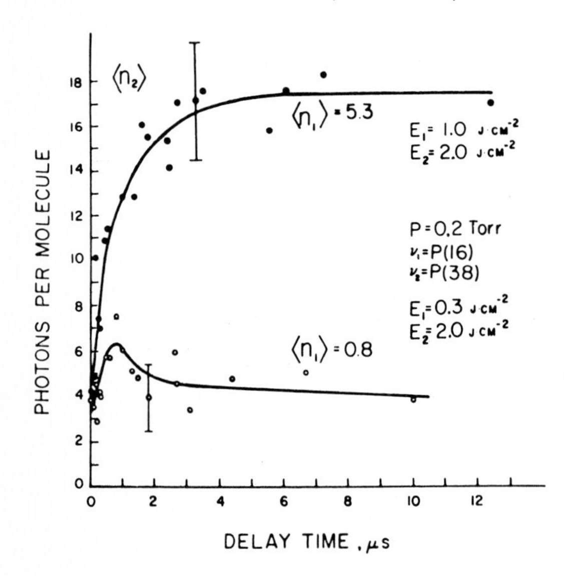

Fig.4.14. The number of absorbed photons per molecule from the nonresonant (ω_2) beam versus the delay between the ω_1 and ω_2 pulses. The data are shown for two fluences at ω_1. The energy fluence at ω_2 was higher than E_{sat}. The OsO_4 pressure was 0.2 Torr; ω_1 = 947.7 cm, ω_2 = 927 cm^{-1}

The results of measurements of $\langle n_2\rangle$ versus the delay between the ω_1 and ω_2 pulses are presented in Fig.4.14. For low fluences at ω_1, immediately after the end of the pulse the number of photons absorbed at ω_2 drops and finally stabilizes at a certain level, namely $\langle n_2^*\rangle \cong 4$. In the case of strong pumping by the ω_1 beam, the number of quanta absorbed at ω_2 after the end of the ω_1 pulse continues to grow with time delay and then after $\tau_d \cong 6$-7 μs reaches the steady-state value $\langle n_2^*\rangle \cong 17$-18.

These data can be explained in a simple way. By introducing a delay between the pulses, we have allowed the establishment of a vibrational temperature corresponding to the amount of energy the system absorbed from the ω_1 beam. In both cases the V-V processes lead to molecular cooling, but in the first case the final temperature is small and only a fraction of the molecules can absorb at ω_2.

If we draw the Boltzmann distribution for equivalent temperatures it is very simple to evaluate the approximate amounts of energy that a molecule needs in order to absorb the ω_2 beam at 927 cm^{-1} by finding the point where the ratio of the areas under the two distributions, counted from the high-energy side, is equal to $\langle n_2^* = 4\rangle / \langle n_2^* = 18\rangle$. With this procedure it was found that an OsO_4 molecule should absorb 0.4 eV to absorb the nonresonant radiation, i.e., the energy needed equals roughly 4 IR photons. In our opinion this explains why the sum $\langle n_1\rangle + \langle n_2^*\rangle = 23$ is less than $D_0 = 27$, since there are molecules that do not absorb ω_2.

From the data given in Fig.4.14 we find that the fraction of excited molecules is roughly 0.6 when $E_1 = 1$ J cm^{-2}. This value coincides with the data provided by Fig.4.10 ($q = 0.5$). As may be seen from Fig.4.10, the excitation level does not depend strongly on the energy fluence in the exciting pulse, while the fraction of molecules changes rather rapidly. Additional evidence for this is provided by experiments where $\langle n_2\rangle$ was measured under conditions of saturated yield versus the amount of absorbed energy $\langle n_1\rangle$ at ω_1. The results are given in Table 4.3.

Table 4.3. Number of quanta per molecule ($\langle n_1\rangle$, $\langle n_2\rangle$) absorbed from the exciting (ω_1) and dissociating (ω_2) fields at different excitation energy fluences Φ_1 at a constant $\Phi_2 = 2.5$ J cm^{-2}. The delay between the two pulses was 0.2 μs. The duration of the exciting pulse was 80 ns FWHM without tail

Φ_1 [J cm^{-2}]	$\langle n_1\rangle$	$\langle n_2\rangle$	$\langle n_2\rangle/\langle n_1\rangle$
0.11	0.45	2.2	4.9
0.20	0.65	3.3	5.1
0.28	0.80	3.8	4.8

It is evident from Table 4.3 that the ratio between $\langle n_1\rangle$ and $\langle n_2\rangle$ at a fixed small delay remains approximately the same, independent of Φ_1. This would not be the case if the number of excited molecules remained constant and the increase of excitation level were responsible for the rise of $\langle n_1\rangle$ versus the energy fluence at ω_1, because at constant q the increase of ℓ should result in a decrease of $\langle n_2\rangle$ under conditions of saturated yield, as was the case in these measurements. As we see, two different types of experiments give the same results for q and ℓ.

4.5.2 Effects of Variation of ω_2

The saturation behavior of the dissociation yield (Fig.4.8) permits on to introduce *formally* a dissociation cross sectin, which may be defined as

$$\sigma = \ell/E_{sat}$$

and which characterizes the whole dissociation process without specifying the nature of the transitions that are saturated. For E_{sat} (measured in photons cm^{-2}) we take the energy fluence at which the yield per pulse becomes nearly insensitive

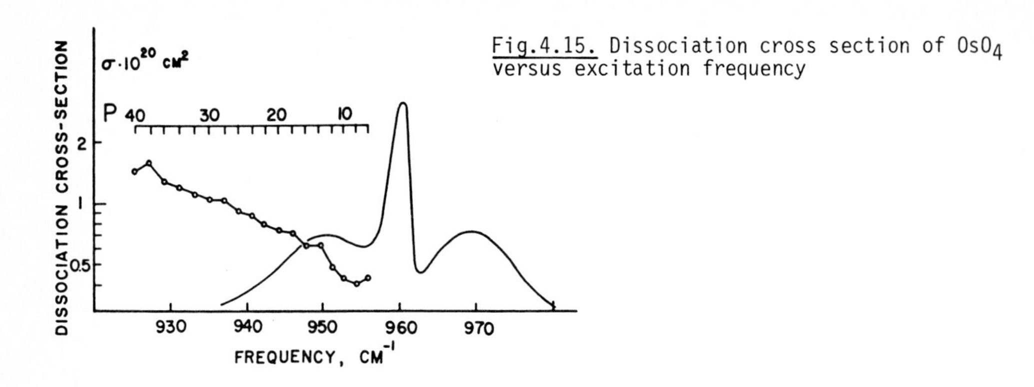

Fig.4.15. Dissociation cross section of OsO_4 versus excitation frequency

to increases in the energy fluence. For each laser frequency the dependence of W on the energy fluence was obtained as shown in Fig.4.8. Once the saturation fluence was obtained, the dissociation cross section was evaluated. These experiments were performed with 100-ns pulses with the tail suppressed to avoid collisional effects on the yield. Essentially the same results were obtained in [4.12] by a somewhat different method. The dissociation cross sections determined in this way are given in Fig.4.15.

As may be seen from Fig.4.15, the dissociation cross section increases exponentially in the frequency region examined when the laser frequency is shifted to the red with respect to the OsO_4 fundamental absorption band as the result of a rather trivial circumstance: the anharmonicity of molecular vibrations. No structure is seen in the frequency dependence of the dissociation cross section. At higher frequencies than the ground-state absorption frequency it was impossible to achieve saturation of the yield in OsO_4 even in focused beam experiments.

The dependence of the dissociation yield on the frequency of the nonresonant field at ω_2 in two-frequency experiments was approixmately the same as that of the dissociation cross section[2]. The shape of the dissociation spectrum $W(\omega_2)$ depends on the delay between the ω_1 and ω_2 pulses, as shown in Fig.4.16. These experiments used a pulse with duration 30 ns (FWHM) with no tail at $\omega_1 = 954.5$ cm^{-1} with energy fluence 30 mJ cm^{-2}.

It can be seen from Fig.4.16 that the vibrational excitation level decreases with increasing delay time, since the maximum of the visible luminescence intensity tends to shift to the blue when the delay between the pulses increases. The set of anharmonically shifted frequencies (estimated using $2X_{33} = 6$ cm^{-1}) shown by the arrows in Fig.4.16 permit an estimate of the minimum excitation level).

If we assume that the maximum of the absorption of the excited molecules at zero time delay corresponds to the frequency ω_2 where the yield is a maximum value, one can deduce that after the pulse at ω_1 the OsO_4 molecules stay in the vibrational

2 In these experiments the quantity measured was the intensity of the visible luminescence in OsO_4, which is proportional to the dissociation yield.

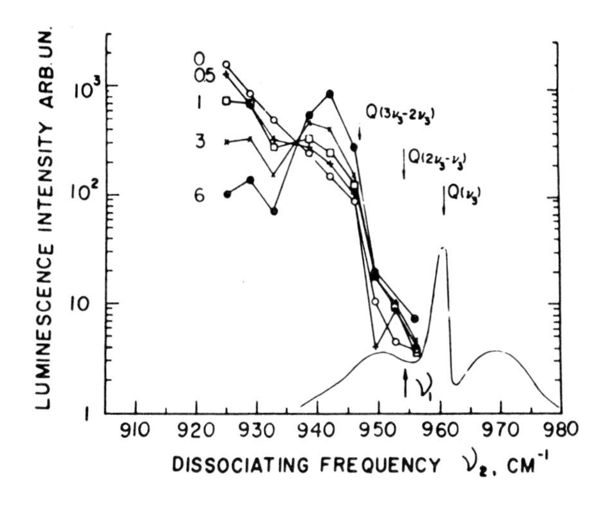

Fig.4.16. Dependence on ω_2 of the dissociation yield in OsO_4 taken with different time delays (in μs) between the ω_1 and ω_2 pulses. The directly measured quantity in these experiments was the intensity of the visible luminescence in OsO_4, which is directly proportional to the dissociation yield. The ω_1 pulse was 30 ns measured at the base; $\Phi_1 = 30$ mJ cm^{-2}

levels $v \cong 7$ of the ν_3 mode, or in other words they have absorbed seven CO_2 laser quanta. After the pulse, due to V-V intermolecular transfer, the excited molecules are cooled vibrationally, and after 6 μs the vibrational excitation has been distributed among all the molecules, with an average excitation of between 1 and 2 CO_2 quanta. An intitial excitation $v \cong 7$ is in rather good agreement with the value ($v = 9$-10) obtained by the technique of yield saturation described in the previous section.

It is more or less obvious that a further shift to the red will cause a departure from resonance with the excited-state absorption, and therefore the yield will drop. The whole resonance curve of the dissociation yield versus ω_2 was measured in [4.16] in experiments with SiF_4 where the ion current was measured versus ω_2 (Fig.4.17)[3]. The conclusion that $W(\omega_2)$ presents a broad maximum shifted to the red from the ground-state absorption was first drawn in [4.17] on the basis of an

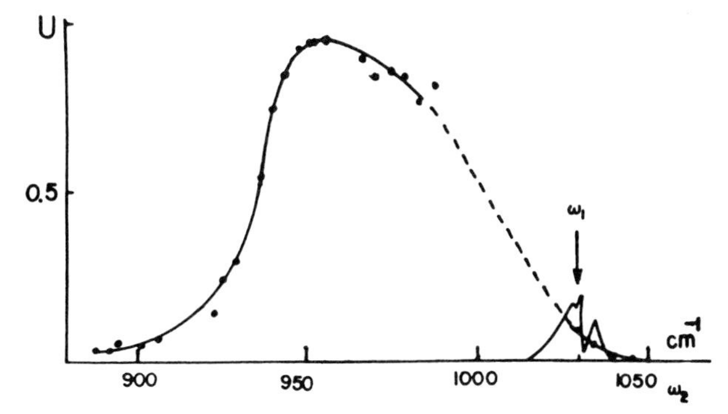

Fig.4.17. Dependence of the dissociation yield in SiF_4 on ω_2. The parameter measured in these experiments was the amplitude of the ion current in the SiF_4 gas that appeared immediately after the laser pulses

3 The formation of the ions that were detected in [4.16] can be explained as the result of secondary reactions of fluorine atoms with traces of hydrocarbons that were present in the cell.

analysis of the excited-state absorption measurements of *Petersen* et al. [4.18] and *Doljikov* [4.19]. In *Doljikov*'s experiments a spinflip laser was employed as a probe beam source to measure the absorption of excited SF_6, as in [4.18]. But even at higher pressures, higher pumping levels and in longer absorption cells than in [4.18] no absorption was found at 850 cm^{-1}.

The comparatively strong red-shifted absorption permits one to think that this excited-state absorption is due to the mode that was prepumped by the laser, namely the ν_3 mode in SF_6 [4.13,18], OsO_4 [4.6,11,12], and SiF_4 [4.16].

Other IR absorption bands should in general show the same effects, as shown in Fig.4.2. This was clearly demonstrated in experiments with SF_6 [4.13] and UF_6 [4.20], the results of which are shown in Figs.4.18,19. The sharp rise of the SF_6 dissociation yield in the frequency region 950-970 cm^{-1} can be explained as the result either of effectively pumping the far blue wing of the highly excited ν_3 mode or of absorption in the $\nu_2 + \nu_6$ combination band, which lies at 970 cm^{-1} in SF_6. The far red wing of the $\nu_2 + \nu_3$ absorption of the excited molecules is responsible for the dissociation of UF_6, as seen in Fig.4.19. The $\nu_2 + \nu_3$ band of UF_6 lies at

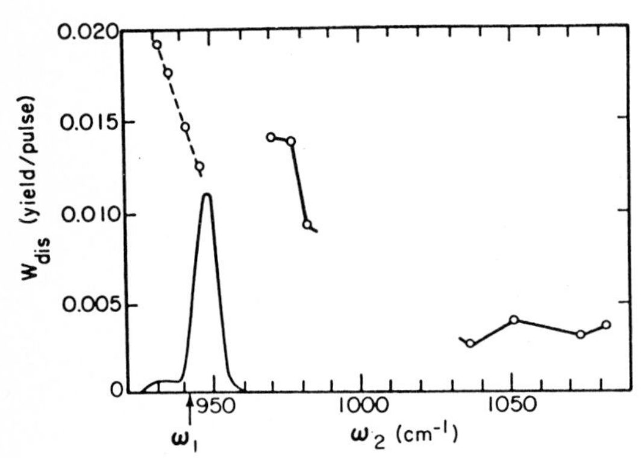

Fig.4.18. Dissociation yield in SF_6 versus the frequency of the nonresonant beam (ω_2); ω_1 was fixed at 942.4 cm^{-1}. The solid line is the experimental curve. The dotted line is the theoretical curve for the frequency region covered by the ν_3 absorption of SF_6

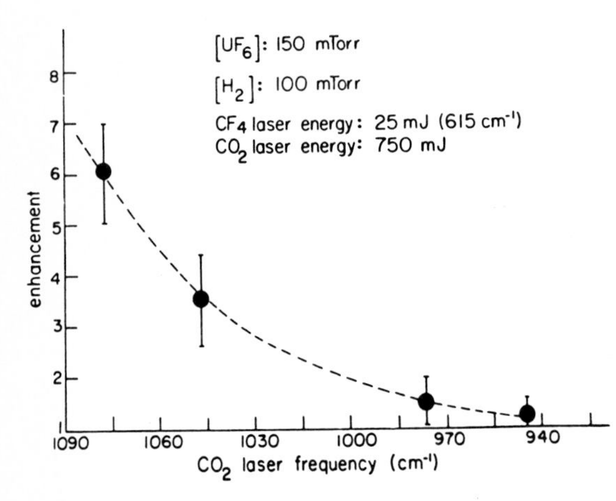

Fig.4.19. Enhancement of the dissociation yield versus frequency ω_2 in UF_6 dissociation experiments. The pulse at ω_1 was provided by a CF_4 laser. The CO_2 laser served as the source of ω_2. In this experiment the intensity of the HF^* fluorescence, which is proportional to the dissociation yield, was measured

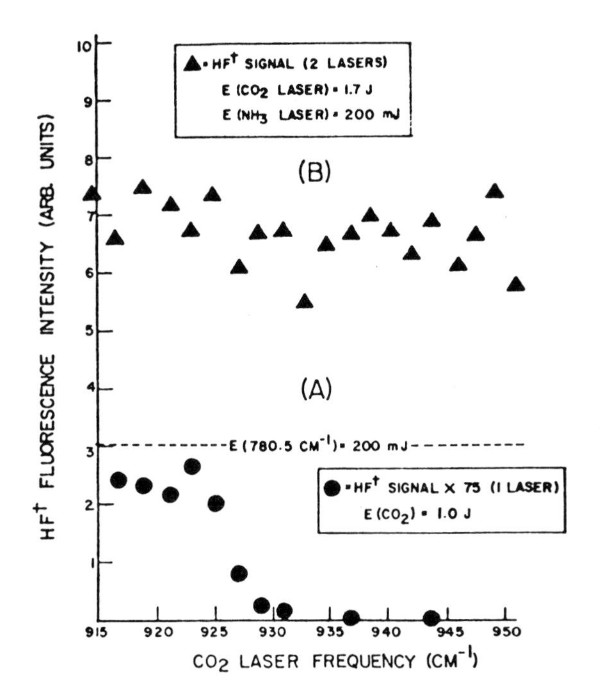

Fig.4.20. Dependence of the dissociation yield in SeF_6 on ω_2. The HF^* fluorescence intensity is a measure of the yield. In the lower part (*A*) the results of single-frequency experiments are given. In the upper part (*B*) the results of two-frequency experiments are given. The NH_3 laser provided the ω_1 pulse, while a CO_2 laser provided ω_2. The NH_3 laser radiation was focused by a 12.8-cm f/1 lens, and the CO_2 laser radiation by a 10-cm f/1 lens. Note that the data for pure CO_2 laser irradiation are amplified (× 75) so that they can be put on the same scale as the other data. The sample in each experiment consisted of 1 Torr of SeF_6 and 0.15 Torr H_2. Note that the enhancement is independent of the CO_2 laser frequency

1156 cm^{-1}. Since $\omega_2 = 1000$ cm^{-1} produces an enhancement of the dissociation yield by a factor of two, one can judge the excitation level of the UF_6 molecules in these experiments.

The absence of any resonance feature in the ω_2 dependence of the dissociation yield in SeF_6 [4.17] when ω_2 was scanned near the $\nu_2+\nu_6$ band is not surprising since ω_2 was tuned from the blue wing of the $\nu_2+\nu_6$ ground-state absorption to frequencies still further to the blue. In prepumped molecules the $\nu_2+\nu_6$ absorption is shifted to the red and ω_2 could not pump this mode in the excited molecules.

As can be seen from Figs.4.18,20, ω_2 caused some dissociation with very low efficiency even where no absorption is seen in the linear absorption spectrum (in SF_6 and SeF_6) even at high pressures. In our opinion this is due to pumping of the so-called vibrational quasi continuum, which is formed by extremely weak combination bands. In this case the process is very inefficient. For example, laser fluences at ω_2 up to 100 J cm^{-2} did not cause saturation of the yield in SeF_6 when this background was pumped. The same situation exists with OsO_4. When ω_2 was tuned to the blue with respect to the ν_3 absorption ($\omega_2 = \nu_3 + 20$ cm^{-1}), the dissociation yield was a factor of 10^7 smaller than when $_2$ was shifted to the red by the same amount ($\omega_2 = \nu_3 - 20$ cm^{-1}), giving a dissociation cross section of the order of 10^{-24}-10^{-25} cm^2.

From the results presented here one can conclude that molecules reach the dissociative states through absorption in a vibrational mode or possibly a combination band which is close to the pumped mode, but shifted to the blue. In most single-frequency dissociation experiments the pumping of the real quasi continuum that has been introduced formally does not play any major role, especially in experiments with SF_6, which has been called "the hydrogen atom of the multiple-photon dissociation process".

4.6 Concluding Remarks

The technique of two-frequency dissociation has been shown to be a very powerful tool for the investigation of the processes of multiple-photon excitation and dissociation. It has permitted the observation of many features of excitation and dissociation that could not be obtained in single-frequency experiments, or by any other technique.

The use of the two-frequency method has also permitted an increase in the selectivity and yield of the dissociation process at moderate intensities.

Finally, I would like to comment that most dissociation and excitation experiments were conducted at pressures in the region 0.1-0.3 Torr with 100-ns pulses. There is some strong evidence [4.11,21] that this pressure is not sufficiently low to ensure that all processes occur in the collision-free regime, since the collisional cross sections for highly excited molecules can be quite different from those in the ground state. If this is established more definitely then it will be necessary to reinterpret some of the results discussed here.

Acknowledgements. The author is indebted to all his colleagues at the Institute of Spectroscopy and especially to Drs. A.A. Puretzky, G.N. Makarov, V.S. Letokhov, as well as to C.D. Cantrell, N. Bloembergen, C.P. Robinson, C. Wittig, and K. Kompa for valuable discussions.

References

4.1 N.R. Isenor, W. Merchant, M.S. Richardson: Can. J. Phys. **51**, 1281 (1973)
4.2 R.V. Ambartzumian, V.S. Letokhov, E.A. Ryabov, N.V. Chekalin: Zh. Eksp. Teor. Fiz., Pis'ma Red. **20**, 273 (1974)
4.3 R.V. Ambartzumian, Yu.A. Gorokhov, V.S. Letokhov, G.N. Makarov: Zh. Eksp. Teor. Fiz., Pis'ma Red. **21**, 171 (1975)
4.4 R.V. Ambartzumian, Yu.A. Gorokhov, V.S. Letokhov, G.N. Makarov, A.A. Puretzky, N.P. Furzikov: Zh. Eksp. Teor. Fiz., Pis'ma Red. **23**, 194 (1976); Opt. Commun. **18**, 517 (1976)
4.5 R.V. Ambartzumian, Yu.A. Gorokhov, V.S. Letokhov, G.N. Makarov, A.A. Puretzky: Sov. Phys.-JETP **42**, 993 (1975); Zh. Eksp. Teor. Fiz. **71**, 440 (1976)
4.6 R.V. Ambartzumian, Yu.A. Gorokhov, G.N. Makarov, A.A. Puretzky, N.P. Furzikov: Kvantovaya Elektron. (Moscow) **4**, 1590 (1977)
4.7 J.J. Tiee, C. Wittig: J. Chem. Phys. **69**, 4756 (1978)
4.8 T.F. Deutsch: Opt. Lett. **1**, 25 (1977)
4.9 D.O. Ham, M. Rothschild: Opt. Lett. **1**, 28 (1977)
4.10 R.V. Ambartzumian, I.N. Knyazev, V.V. Lobko, G.N. Makarov, A.A. Puretzky: Appl. Phys. **19**, 75 (1978)

4.11 R.V. Ambartzumian, V.S. Letokhov, G.N. Makarov, A.A. Puretzky: Opt. Commun. **25**, 69 (1978)
4.12 R.V. Ambartzumian, G.N. Makarov, A.A. Puretzky: Opt. Commun. **27**, 29 (1978)
4.13 G.P. Quigley: Opt. Lett. **3**, 106 (1978)
4.14 V.N. Bagratashvili, V.S. Doljikov, V.S. Letokhov, E.A. Ryabov, V.V. Tyakht: Zh. Eksp. Teor. Fiz. **77**, 2238 (1979)
4.15 R.V. Ambartzumian, G.N. Makarov, A.A. Puretzky: Opt. Lett. **3**, 103 (1978)
4.16 V.M. Akulin, S.A. Alimpiev, N.V. Karlov, A.M. Prokhorov, B.G. Sartakov, E.M. Khokhlov: Zh. Eksp. Teor. Fiz., Pis'ma Red. **25**, 428 (1977)
4.17 R.V. Ambartzumian, V.S. Letokhov: In *Chemical and Biochemical Applications of Lasers*, Vol.3 (Academic, New York 1977)
4.18 A.B. Petersen, J. Tiee, C. Wittig: Opt. Commun. **17**, 259 (1976)
4.19 V.S. Doljikov: Private communication
4.20 J.J. Tiee, C. Wittig: Opt. Commun. **27**, 337 (1978)
4.21 J.D. Campbell, M.H. Yu, C. Wittig: Appl. Phys. Lett. **32**, 413 (1978)

5. Excitation Spectrum of SF_6 Irradiated by an Intense IR Laser Field

S. S. Alimpiev, N. V. Karlov, E. M. Khokhlov, S. M. Nikiforov, A. M. Prokhorov, B. G. Sartakov, and A. L. Shtarkov

With 7 Figures

The fine structure of the spectrum of excitation originating at the vibrational ground state of SF_6 molecules that have been gas-dynamically cooled to a temperature of approximately 30 K has been investigated using two-frequency infrared laser excitation. The absorption feature with highest contrast corresponds to the two-photon resonance between the vibrational ground state and the A_{1g} sublevel of $v = 2$, and occurs at the laser frequency 944.39 cm^{-1} with a FWHM not exceeding 0.1 cm^{-1}.

5.1 Background

The division of the multiphoton dissociation (MPD) of polyatomic molecules into three sequential stages [5.1,2] is well established. The first stage corresponds to the resonant excitation of the lower levels; the second, to the deposition of energy in the quasi continuum; and the third, to the dissociation of molecules excited above the dissociation threshold. The third stage is traditionally described by the Rice, Ramsperger, Kassel, and Marcus (RRKM) theory, while excitation in the quasi continuum is usually described as a sequence of single-quantum transitions with phenomenologically introduced absorption cross sections. Such a description of the last two stages is generally considered to be valid in the limit of a high density of excited vibrational states, and is certainly appropriate to large polyatomic molecules for which the quasi continuum begins at the first or second vibrational level. But the quasi continuum, i.e., the range of excitation energies where the levels of a resonantly excited mode can be mixed with combination levels due to occasional Fermi resonances, may begin at a relatively high degree of excitation in the case of molecules having fewer than 10 atoms, especially in the case of symmetric molecules. For this reason the first stage, i.e., the resonant excitation of the first three or four vibrational levels of the laser-excited mode, is very important in the MPD of small molecules.

The radiation pumping of the lower vibrational levels of polyatomics, as has been shown in SF_6 in [5.3], is very efficient, beginning at a relatively low laser intensity ≅100 kW cm^{-2}, where molecules in many different rotational states take part in the process of excitation [5.4]. To explain the effectiveness of excitation, many different models of the excitation process have been proposed, such as rotational compensation of anharmonicity [5.5], strong anharmonic splitting of the excited vibrational levels of the ν_3 mode of SF_6 [5.6,7], and the possibility of exciting low vibrational levels by "weak" vibration-rotation transitions [5.8]. All of these models base their explanation of multiphoton excitation (MPE) on the resonant coincidence of the laser frequency with different vibration-rotation transitions. The differences among the models arise from different assumptions concerning the vibration-rotation structure of the molecule. Accordingly different predictions have been made concerning resonances in the excitation spectrum starting at the ground vibrational level, and different explanations of the character of the excitation have been given.

The first experimental observation of resonant structure due to multiphoton transitions in SF_6 in an intense laser field was reported in [5.9], where it was shown that the main channel of excitation out of the lower level is provided by two-photon transitions. These results give evidence in favor of the model of excitation discussed in [5.6]. Experimental investigation of the excitation spectrum opens up the possibility of determining the spectral structure of the lower vibrational states of the ν_3 mode of SF_6. Analysis of the spectrum obtained in [5.9] showed that models assuming a weak anharmonic splitting of the ν_3 vibrational levels are not relevant to SF_6, and allowed a measurement of the position of the A_{1g} $v=2$ level [5.10]. An independent measurement of the $3\nu_3$ overtone spectrum of SF_6 with a resolution of $10^{-4}cm^{-1}$ [5.11,12] permitted the determination of the three fundamental anharmonicity constants of the ν_3 mode. The values of these constants imply a strong anharmonic splitting of the excited vibrational levels.

Narrow resonances in the excitation spectrum of polyatomic molecules are very important from a practical point of view; they open up the possibility of isotopically selective excitation of heavy molecules having very small vibrational isotopic shifts. The width of one of these resonances, which in essence is the Q-branch of a two-photon transition, depends strongly on the rotational temperature of the gas [5.10] and can be significantly decreased by cooling. The aim of this paper is to report our investigations of the spectral characteristics of the lower vibrational levels of the ν_3 mode of SF_6.

5.2 The Experiment

The first observation of the multiphoton resonance structure of the spectrum of SF_6 vapor was made possible by static cooling of the gas to a temperature $T=140$ K (thereby removing hot bands from the spectrum) and by the use of a high-pressure continuously tuned CO_2 laser instead of the usual atmospheric-pressure line-tuned

CO_2 laser. As has been shown in [5.10], the Coriolis vibration-rotation interaction in the ν_3 mode broadens some of the resonances. At a gas temperature of 140 K the broadening of the two-photon resonances E_g and F_{2g} is large enough to prevent their resolution from the background absorption in a strong field. A significant narrowing of these resonances and their observation in the absorption spectrum can be expected only at SF_6 rotational temperatures below 30 K.

To achieve a temperature of about 30 K we cooled the SF_6 gas by a pulsed gas-dynamic expansion. But here one encounters a difficulty in measuring the absorption spectrum because the traditional techniques of optoacoustic and transmission spectroscopy do not work, for reasons that are easy to comprehend. To avoid this difficulty we dissociated the SF_6 gas in a pulsed gasdynamic stream by two IR laser pulses with different frequencies, as described in Chap.4. The first field was used for resonant excitation of the lower vibrational levels and served to investigate the excitation spectrum starting at the ground state. The second, auxiliary field dissociated the molecules excited by the first field. The fluorine atoms originating from the dissociation of SF_6 were detected using the luminescence of HF^* [5.13]. For this purpose, and to enhance the gas-dynamic cooling, the SF_6 gas was diluted with hydrogen. Due to collisions in the gas-dynamic stream between F atoms and H_2 molecules, vibrationally excited HF^* molecules were formed.

The experimental setup for investigating the absorption spectrum of gas-dynamically cooled SF_6 gas under two-frequency irradiation is shown schematically in Fig.5.1, which shows the laser sources (L_1 and L_2), the control system for the frequency of the first laser, the gas-dynamic cooling, and the detector system.

The source of the first field was a continuously tuned, electron-beam-sustained, high-pressure ($p = 6$ atm) (CO_2: $N_2 = 1:1$) CO_2 laser. A detailed description of this laser has been given elsewhere [5.14]. The laser beam is characterized by a pulse energy of 0.1 J, a pulse duration $\tau_{p1} \cong 40$ ns (at the base), and a beam diamter of 0.5 cm.

The source of the second field was a line-tuned TEA CO_2 laser with an output aperture of 2×2 cm^2. The duration of the L_2 laser pulse was $\tau_{p2} \cong 100$ ns (FWHM) and the maximum pulse energy was 2 J. The delay time of the L_2 laser pulse relative to the L_1 laser pulse was $\tau_d = 500 \pm 30$ ns and was measured by a photon-drag detector.

The radiation of the L_2 laser was focused into cell C by a KBr lens ℓ_1, where it propagated in the opposite direction to the beam from L_1. A grating G (100 grooves/mm) provided spatial separation of the L_1 and L_2 laser beams in the L_2 laser cavity, thereby excluding any influence of the first laser beam on the active medium of the second laser. The energy of both lasers was measured by pyroelectric detectors. Calibrated CaF_2 discs were used as the attenuators At_1 and At_2.

The frequency of the first laser was measured by reference to N_2O absorption lines (the $00^\circ1$-$10^\circ0$ transition) detected optoacoustically. The gas pressure in the optoacoustic detector cell was 150 Torr. Frequencies were interpolated between N_2O

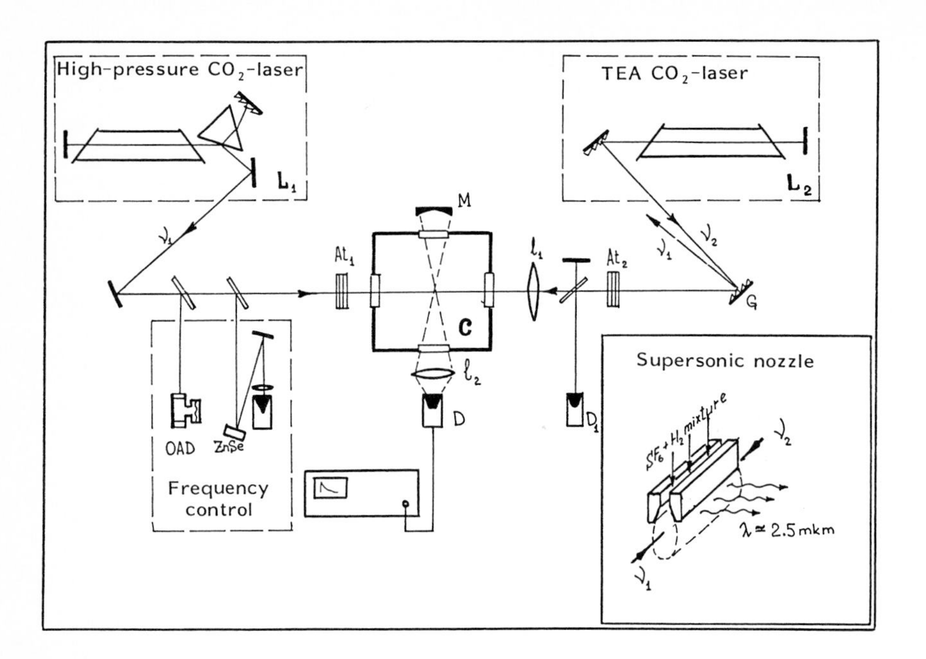

Fig.5.1. Schematic diagram of the experimental setup. (L_1 and L_2: lasers; C: gas-dynamically cooled cell; G: grating; M: spherical mirror; D: detector for 2.5-μm HF^* fluorescence; At_1 and At_2: attenuators; OAD: optoacoustic detector)

lines with the help of a Fabry-Perot interferometer made of plane-parallel ZnSe plates 0.58 cm thick.

For gas-dynamic cooling the cell C was provided with a slit nozzle having a rectangular aperture of 0.5 mm by 3 cm at the throat, and an outlet angle of 11°. To control the nozzle operation a fast-pulsed valve was used. The cell was connected to a pumped vacuum chamber. During operation of the valve at a repetition rate of 0.1 Hz, the gas pressure in the pumped chamber was less than 10^{-3}Torr. Behind the valve there was a mixture of SF_6 and H_2 in the ratio 1:20 at a pressure of 300 Torr. The L_1 and L_2 laser beams were directed coaxially and oppositely into the cut-off area of the nozzle. The rotational temperature of the gas in the irradiated area of the gas-dynamic stream at the above parameters of the nozzle and gas mixture was no more than 30 K [5.15]. The magnetic control of the valve kept it completely open during an integral of 10 ms. The L_1 laser pulse delay relative to the moment of full opening of the valve was 2 ms.

The HF^* luminescence was observed by a room-temperature PbS photodetector followed by an amplifier with gain $K_a = 500$ and a bandwidth of 200 kHz. The image of the irradiated volume of the gas-dynamic stream was focused on the aperture of detector D by lens ℓ_2 and mirror M.

5.3 Experimental Results and Discussion

Since the HF^* luminescence signal was proportional to the number of F atoms resulting from the two-frequency dissociation, it was used as a measure of the number of dissociated SF_6 molecules. The magnitude of the signal depends on the peculiarities of the lower levels and especially on the excitation of the quasi continuum, and thus on the parameters of the first and second fields. The dissociation yield produced by the second field depends on its frequency and fluence, because different vibrational levels can be dissociated with different probabilities. The dissociation by the second field is most effective when its frequency is shifted to longer wavelengths relative to the fundamental vibrational transition of the excited mode (red shift) [5.16,17].

However, the bigger the red shift of the frequency of the second field, the higher the excitation produced by the first field should be in order to get efficient dissociation by the second field at a moderate intensity. In principle, a strongly red-shifted second field of very high intensity could dissociate molecules from low vibrational levels via multiphoton transitions, but to increase the intensity of the second field significantly one would need sharp focusing, which would decrease the irradiated volume and make the determination of the experimental parameters difficult. Thus, it is not advisible to shift the second field frequency too far to longer wavelengths, in order to provide for efficient dissociation by the second field of the molecules excited by the first field.

Figures 5.2-4 show the luminescence signal S_L versus the frequency ν_1 of the first field taken when laser L_2 operated on the lines P(26), P(30), and P(40) of the 10-μm band of CO_2. When L_2 operated on the P(26) CO_2 laser line, its radiation was focused by lens ℓ_2 with a focal length $f = 11$ cm; for P(30) and P(40), $f = 50$ cm. At frequency intervals of 0.06 cm^{-1}, the luminescence signal $\nu_1 S_L$ was measured as a function of the fluence Φ_1 of the first laser. The spectra of Figs.5.2-4 are drawn for $\Phi_1 = 135$ mJ cm^{-2}. The upper parts of the figures show the N_2O reference lines and the Fabry-Perot interferometer fringes. The interferometer fringes are slightly nonequidistant because of a small frequency pulling of the L_1 laser in vicinity of the CO_2 laser lines. The uncertainty of the L_1 frequency measurement by the N_2O reference lines and interferometer fringes is 0.03 cm^{-1}. The values of the N_2O reference frequencies have been taken from handbook [5.18]: R(6)-944.54 cm^{-1}, R(7)-945.35 cm^{-1}, R(8)-946.15 cm^{-1}, R(9)-946.95 cm^{-1}, R(10)-947.75 cm^{-1}, R(11)-948.54 cm^{-1}, R(12)-949.33 cm^{-1}, R(13)-950.12 cm^{-1}. To compare the recent result with the previous one, Fig.5.5 shows the intense laser field absorption spectrum of the SF_6 molecules statically cooled to 140 K [5.9]. The arrows A, B, and C show the positions of the two-photon resonances A_{1g}, E_g, and F_{2g} calculated using the anharmonicity constants of [5.12].

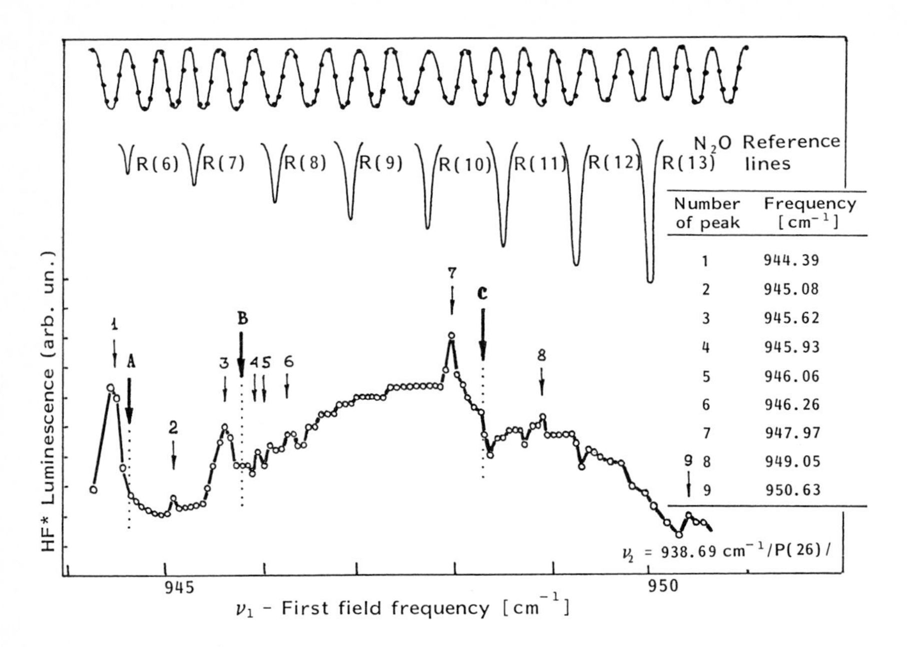

Fig.5.2. The HF* luminescence signal vs the first field frequency at a fluence $\Phi_1 = 135$ mJ cm^{-2}. The second field frequency was 938.69 cm^{-1}, energy fluence $\Phi_2 = 40$ J cm^{-2}

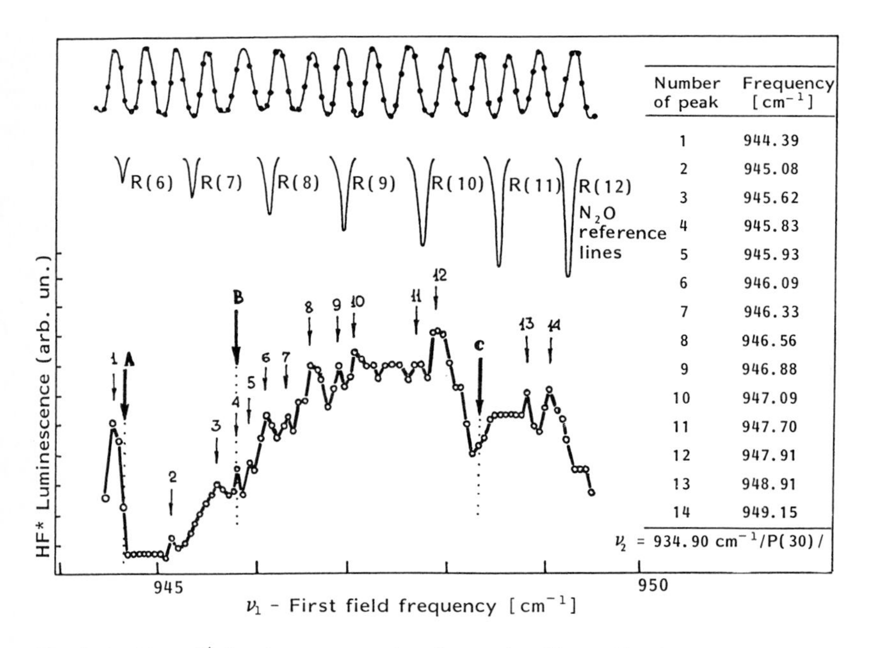

Fig.5.3. The HF* luminescence signal vs the first field frequency at a fluence $\Phi_1 = 135$ mJ cm^{-2}. The second field frequency was 934.90 cm^{-1}, energy fluence $\Phi_2 = 10$ J cm^{-2}

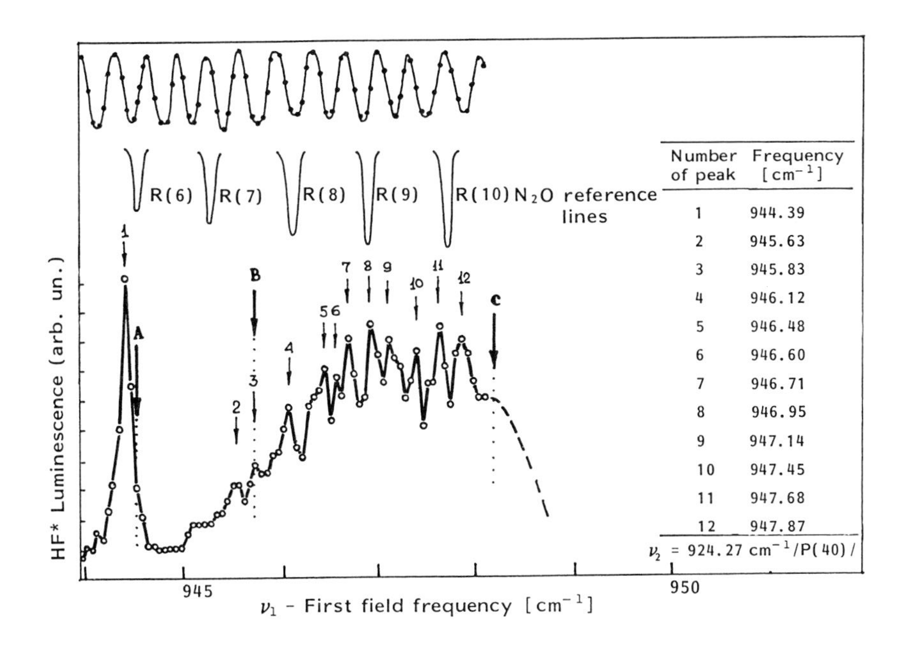

Number of peak	Frequency [cm^{-1}]
1	944.39
2	945.63
3	945.83
4	946.12
5	946.48
6	946.60
7	946.71
8	946.95
9	947.14
10	947.45
11	947.68
12	947.87

Fig.5.4. The HF* luminescence signal vs the first field frequency at a fluence $\Phi = 135$ mJ cm^{-2}. The second field frequency was 924.27 cm^{-1}, energy fluence $\Phi_2 = 10$ J cm^{-2}

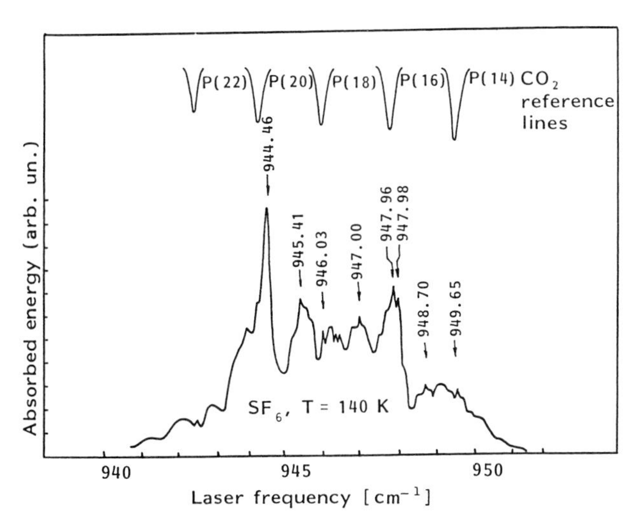

Fig.5.5. The intense laser field absorption spectrum of SF_6 gas cooled to a temperature of 140 K [5.9]

Figures 5.6,7 show the luminescence signal S_L versus Φ_1, measured at the first field frequencies 947.97 am^{-1} (Q-branch of the fundamental transition) and 944.39 cm^{-1} (feature 1, corresponding to the two-photon resonance of the A_{1g} sublevel of $v_3 = 2$). These dependences have been taken for L_2 laser operation on lines P(26), P(30), and P(40). It can be seen that the slope of the curves increases with an increase of the wavelength of the second field. This is in agreement with the state-

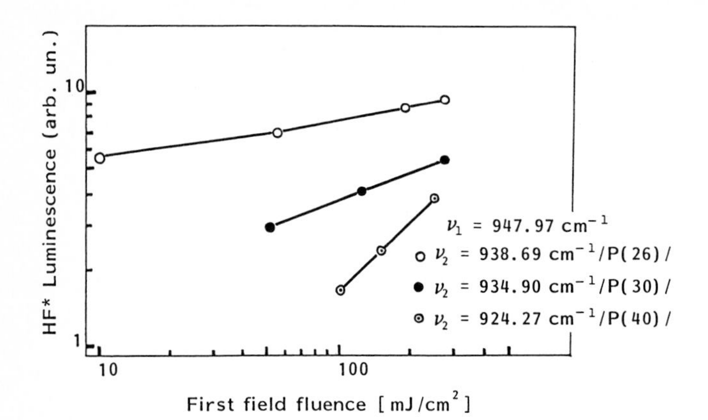

Fig.5.6. The HF^* luminescence signal vs the first field fluence at a frequency $\nu_1 = 947.97\ cm^{-1}$

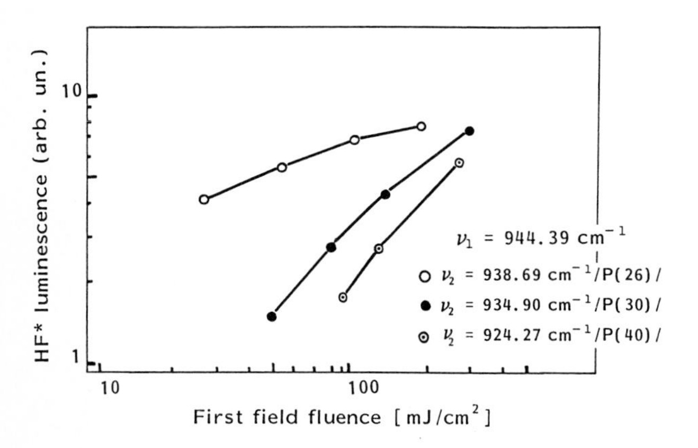

Fig.5.7. The HF^* luminescence signal vs the first field energy fluence Φ_1 at a frequency $\nu_1 = 944.39\ cm^{-1}$

ment that an increase of the second field frequency red shift beyond the optimal value makes this field unable to excite the molecules starting at lower vibrational levels ($v = 1,2$). A strongly red-shifted second field can dissociate molecules only from rather high vibrational states ($v \geqq 3$) populated by the first field via multiphoton transitions, which explains the increase of slope in Figs.5.6,7.

The Q-branch of the fundamental transition was observed only when the L_2 laser was operating on the P(26) line at an energy fluence of 40 J cm^{-2}. It appears that the molecules are no longer resonant with the frequency 947.97 cm^{-1} (the Q-branch of the fundamental transition) after the absorption of one quantum, so that further excitation becomes difficult. At the same time according to [5.6] at a fluence $\Phi_1 \sim$10-100 mJ cm^{-2} saturation of the two- and three-photon vibrational-rotational transitions should occur even if the laser radiation is detuned from resonance with the intermediate states by a few wave numbers. For this reason one might expect the appearance of resonances corresponding to two- or three-photon transitions, which could mask the fundamental transition in the spectra shown in Figs.5.2-5.

The spectral feature of 944.39 cm^{-1} (1 in Figs.5.2-4) and the corresponding feature at 944.46 cm^{-1} (Fig.5.5) are the two-photon resonance from the ground state to the state A_{1g} $v = 2$ [5.10]. For this transition $\Delta J = 0$ and all molecules are resonant with the field apart from a very small change of rotational energy caused by the change of rotational constant with vibrational excitation. The spectral width of this line is equal in order of magnitude to that of the Q-branch of the fundamental transition and is significantly decreased by lowering the rotational temperature. As shown in Figs.5.2-5, the spectral width of the A_{1g} $v=2$ resonance is about 0.2 cm^{-1} at 140 K and less than 0.1 cm^{-1} at 30 K.

One should expect to observe resonances associated with the two-photon transitions to the E_g and F_{2g} states of the level $v=2$. These resonances, unlike the A_{1g} resonance, are subject to additional broadening due to the Coriolis interaction [5.10]. The value of this broadening can be estimated as $1/2\ (2B\zeta\ J_T)^2/\Delta E$. Here ζ is the Coriolis constant, $J_T=\sqrt{kT/B}$, T is the rotation temperature, and ΔE is the energy gap between the vibrational states E_g and F_{2g}. For the case of SF_6 $B \cong 0.09$ cm^{-1}, $\zeta \cong 0.7$ [5.12,19]. These data give the estimate 0.37 cm^{-1} for the E_g and F_{2g} resonances. The expected positions of these resonances calculated using the data of [5.12] are shown in Figs.5.2-4 by the arrows B and C, but the resonances are absent. More than that, the expected position of the A_{1g} resonance shown by arrow A differs by 0.15 cm^{-1} from the measured one. This difference is well above our experimental uncertainty. Apparently the values of the anharmonicity constants of the ν_3 mode of SF_6, which determine the positions of the resonances A_{1g}, E_g and F_{2g}, differ slightly from those given in [5.12].

In contrast to the A_{1g} $v=2$ resonance, all the other features of the spectra of Figs.5.2-4 have a significantly smaller magnitude and look rather like random modulation of the spectral envelope. An adequate assignment of these features is difficult at present. They could result from random coincidence of the frequencies of many vibration-rotation transitions. It is also possible that some of these resonances are related to multiphoton transitions to high vibrational states.

5.4 Conclusion

In conclusion, a high-contrast feature has been observed in the absorption spectrum of the lower vibrational levels of SF_6 under conditions of deep cooling. This feature corresponds to a two-photon transition from the vibrational ground state to the A_{1g} sublevel of $v=2$ and manifests itself most prominently for a moderate but significant red shift of the frequency of the second field.

References

5.1 N.V. Karlov, A.M. Prokhorov: Usp. Fiz. Nauk **118**, 583 (1976)
5.2 V.S. Letokhov, C.B. Moore: Kvantovaya Elektron. **3**, 248 (1976)
5.3 R.V. Ambartzumian, Yu.A. Gorokhov, V.S. Letokhov, G.N. Makarov, A.A. Puretzky, N.P. Furzikov: Opt. Commun. **18**, 517 (1976)

5.4 S.S. Alimpiev, V.N. Bagratashvili, N.G. Karlov, V.S. Letokhov, V.V. Lobko, A.A. Makarov, B.G. Sartakov, E.M. Khokhlov: Zh. Eksp. Teor. Fiz., Pis'ma Red. **25**, 582 (1977)
5.5 P.B. Ambartsumian, Yu.A. Gorokhov, V.S. Letokhov, G.N. Makarov, A.A. Puretzky: Zh. Eksp. Teor. Fiz. **71**, 440 (1976)
5.6 V.M. Akulin, S.S. Alimpiev, N.V. Karlov, B.G. Sartakov: Zh. Eksp. Teor. Fiz. **74**, 490 (1978)
5.7 C.D. Cantrell, H.W. Galbraith: Opt. Commun. **18**, 513 (1976); **21**, 374 (1977)
5.8 I.N. Knyazev, V.S. Letokhov, V.V. Lobko: Opt. Commun. **25**, 337 (1978)
5.9 S.S. Alimpiev, N.V. Karlov, E.M. Markarov, S.M. Nikiforov, A.M. Prokhovrov, B.G. Sartakov, A.Z. Shtarkov: Opt. Commun. **31**, 309 (1979)
5.10 S.S. Alimpiev, N.V. Karlov, B.G. Sartakov: Izv. Akad. Nauk SSSR, Ser. Fiz. **45**, 1078 (1981)
5.11 A.S. Pine, A.G. Robiette: J. Mol. Spectrosc. **80**, 388 (1980)
5.12 C.W. Patterson, B.G. Krohn, A.S. Pine: Opt. Lett. **6**, 39 (1981)
5.13 C.R. Quick, C. Wittig: Chem. Phys. Lett. **48**, 420 (1977)
5.14 S.S. Alimpiev, Yu.I. Bychkov, N.V. Karlov, G.A. Mesyats, S.N. Nikiforov, V.M. Orlovsky, A.M. Prokhorov, E.M. Khokhlov: Pis'ma Zh. Eksp. Teor. Fiz. **5**, 814 (1979)
5.15 G.S. Baronov, A.D. Britov, S.M. Karavaev, S.Yu. Kulikov, A.V. Marzlyakov, S.D. Sivachenko, Yu.I. Scherbina: Kvantovaya Elektron. **8**, 1573 (1981)
5.16 V.M. Akulin, S.S. Alimpiev, N.V. Karlov, A.M. Prokhorov, B.G. Sartakov, E.M. Khokhlov: Pis'ma Zh. Eksp. Teor. Fiz. **25**, 428 (1977)
5.17 C.D. Cantrell: "Coherent versus Stochastic Theories of Collisionless Multiple-Photon Excitation of Polyatomic Molecules" in *Laser Spectroscopy III*, ed. by J.L. Hall, J.L. Carlsten, Springer Ser. Opt. Sci., Vol.7 (Springer, Berlin, Heidelberg 1977) pp.109-115
5.18 A.M. Prokhorov (ed.): *Laser Handbook*, Vol.1, Sov. Radio (Moscow 1978) (in Russian)
5.19 Ch.J. Bordé, M. Ouhayoun, A. Van Lerberghe, C. Salomon, S. Avrillier, C.D. Cantrell, J. Bordé: "High Resolution Saturation Spectroscopy with CO_2 Lasers. Application to the ν_3 Bands of SF_6 and OsO_4", in *Laser Spectroscopy IV*, ed. by H. Walther, K.W. Rothe, Springer Ser. Opt. Sci., Vol.21 (Springer, Berlin, Heidelberg 1979) p.142

6. Laser-Induced Decomposition of Polyatomic Molecules: A Comparison of Theory with Experiment

M. F. Goodman, J. Stone, and E. Thiele

With 23 Figures

A theory of laser-induced multiple-photon dissociation is presented which gives a unified description of discrete and quasi-continuum level excitation and unimolecular decay. Comparisons are made with experiments on SF_6 and CF_2HCl, providing tests of predicted absorption properties over a broad range of molecular energies. Models for a restricted intramolecular flow of energy that would lead to nonrandomization of energy and non-RRKM behavior in the dissociation are analyzed. We suggest that for most systems, current experiments are insufficiently sensitive to rule out this possibility. The relationship of theory involving phenomenological T_1 and T_2 relaxation rates to the microscopic energy level picture is brought out. Theoretical techniques using a diagrammatic treatment of coherence effects are discussed. Finally, we review recent developments relating restricted coupling (restricted quantum exchange) to the energy dependence of linewidths and recent classical calculations which have awakened new interest in the possibility of laser-induced mode-selective chemistry.

6.1 Brief Historical Review

The possibility that a complex polyatomic molecule could be vibrationally excited to the point of dissociation in the presence of an intense electromagnetic field was first investigated as a theoretical problem by *Bunkin* et al. [6.1] who studied vibrationally excited harmonic oscillators and by *Askar'yan* [6.2] who looked at Morse oscillator approximations. Shortly thereafter, we modified *Slater*'s [6.3] classical theory to predict the effect of infrared laser radiation on the high-pressure rate constant for unimolecular decay [6.4]. In 1972-1973 we introduced a quantum-mechanical Boltzmann equation [6.5-7] to describe phase coherent multiple-photon absorption in a gas phase polyatomic molecule undergoing unimolecular decay and included the effects of phase-loss and energy-changing collisions with a surrounding bath of inert solvent molecules.

This theoretical work predated by several years the bulk of experimental effort to observe the occurrence of laser-induced decomposition. Early evidence for laser-induced vibrational excitation leading to decomposition of biphenyl was obtained by *Eisenthal* et al. [6.8] using a ruby laser. The advent of high-power lasers in the infrared led to studies by *Karlov* et al. of decomposition in BCl_3 [6.9], by *Richardson* and *Isenor* of decomposition in SiF_4, CF_4, and CCl_2F_2 [6.10], and by *Basov* et al. of both unimolecular and biomolecular laser-induced reactions in a variety of species [6.11]. The need to distinguish carefully between reactions induced directly by the absorption of radiation in the reacting molecule and reactions brought about indirectly by generalized heating effects prompted *Bauer* and his colleagues to carry out a series of careful kinetic studies of laser-induced reactions under very low-pressure and low-intensity conditions [6.12]. One of the first clear-cut examples of a direct laser-induced reaction was documented by *Isenor* et al. [6.13] by following the time dependence of luminescence from SiF_4 decomposition products. More recently a large number of groups around the world have been extremely active in studying multiple-photon dissociation of complex polyatomic molecules in the gas phase [6.14-27], and experimental techniques have evolved [6.28-32] to provide time-resolved probes for several important systems.

Widespread application of the absorption of infrared radiation in isotope separation [6.33-39] followed the pioneering efforts of *Ambartzumian* et al. [6.14] and of *Lyman* et al. [6.15] and led to attempts to formulate detailed models [6.40-46] for the laser-molecule interaction which leads to unimolecular decay. Our interests and the subject of this article lie in developing a microscopic formulation of multiple-photon absorption in polyatomics and applying this formulation to the construction of models to be used in analyzing data currently available. We also discuss the distribution of vibrational energy in a laser-driven polyatomic and introduce a formalism that focuses on intra- and intermolecular relaxation as well as laser selectivity.

6.2 Models of a Complex Polyatomic Molecule

6.2.1 Laser Excitation Mechanisms

Some of the earliest theoretical models for laser-induced dissociation of molecules by vibrational excitation utilized a Morse oscillator potential [6.2] to represent the molecule or, in the case of a polyatomic molecule, that vibrational motion of the molecule which couples with the laser field. Enormous laser intensities were required to pump through the resultant manifold of unequally spaced levels, and dissociation thresholds of the order of GW cm^{-2} were predicted for reactions which had typical activation energy requirements corresponding to about 30 quanta.

The discovery that polyatomic molecules are efficiently dissociated at far lower intensities, corresponding to about 1 MW cm^{-2} or less, shows that radiative transitions between vibrational states of a polyatomic molecule are far more resonant than between levels of an isolated Morse oscillator. Since the density of energy

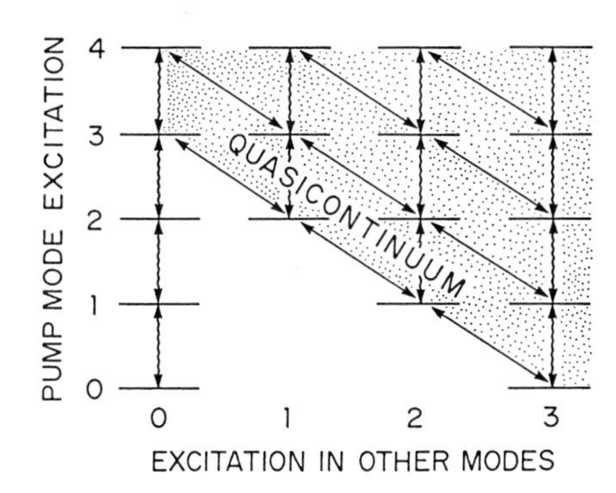

Fig.6.1. States in heat bath feedback model. Excitation is specified for both pump mode and the other vibrational modes which are treated as a finite heat bath. Wavy lines indicate up or down radiative transitions and sloping arrows indicate intramolecular T_1 transitions between quasi-continuum states of the same total energy. In the figure it is assumed that the quasi-continuum begins at a total energy of 3 quanta

states in a polyatomic molecule becomes quite large at energies just a few quanta above the ground-state energy, there is in a polyatomic the possibility of numerous excitation paths not present in an isolated anharmonic oscillator. The description of radiative pumping through this dense set of upper levels, or quasi continuum, presents the theoretician with some new and interesting problems. One fruitful approach focuses on a single infrared-active mode (the pump mode), which plays a special role among the vibrational degrees of freedom as a focus for input of energy from the laser field and allows for a transfer of energy from the pumped mode to remaining degrees of freedom by intramolecular relaxation [6.40,41,46,47].

A diagram that illustrates laser pumping through the lowest lying discrete levels as well as pumping through the quasi continuum is shown in Fig.6.1. The "wavy" vertical transitions represent the absorption of one quantum of laser energy, and the "solid" diagonal transitions represent isoenergetic intramolecular transitions in which energy is scrambled between the pumped mode and the heat bath of the remaining vibrational degrees of freedom. In polyatomic molecules it is unlikely that the 30 or so quanta required for reaction will ever be fed directly into the pump mode because of anharmonic shifts in the energy level spacing. Rather, a continual feeding of energy into the heat bath modes must accompany the pump mode excitation once a sufficient amount of energy to reach the quasi continuum has been acquired. In SF_6, the quasi continuum probably begins at 3 or 4 quanta of CO_2 laser energy above the ground state [6.48]. In very complex molecules such as S_2F_{10} [6.49] and certain uranyl compounds [6.50], the quasi continuum may actually be almost 100% populated (because of high degeneracy factors) at temperatures of just a few hundred Kelvin. As a general rule, laser pumping is more efficient in more complex molecules because feeding of energy from the pump mode to the heat bath modes is more efficient. Also, in the quasi continuum, resonance requirements between the laser frequency and energy level difference will be less stringent because of line broadening due to intramolecular relaxation. We expect that typical excitation mechanisms will lead to a distribution of energy between the pump mode and the heat bath modes of the type shown in Fig.6.2, where more complex molecules have more energy in the heat bath modes. In this discussion we have implicitly assumed that

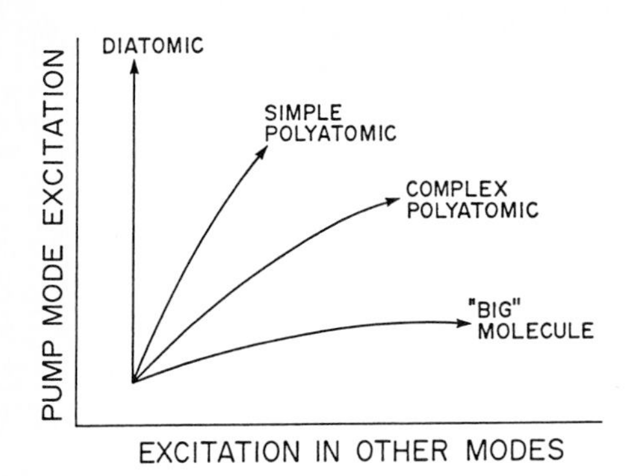

Fig.6.2. Schematic representation of the energy distribution between the pump mode and the other vibrational modes of a laser-excited polyatomic. It is assumed that intramolecular relaxation is rapid so that statistical factors favor "other mode" excitation in more complex molecules

the intramolecular relaxation processes in the quasi continuum have a high overall efficiency and also that the energy shifts caused by increasing heat bath energy, hot band shifts, are not so large as to destroy near resonances with the laser frequency. A complete theory must address these questions on a microscopic level; in Sect.6.2.3 a microscopic point of view is adopted to treat some general aspects of intramolecular relaxation in the quasi continuum.

The details of the intramolecular relaxation processes are also of central importance in discussing selective chemistry and are taken up in Sect.6.4.2. Intermolecular relaxation (collisions) can be expected to be quite efficient in scrambling energy between the pump and heat bath mode. Perhaps it has not been fully appreciated that at some pressures collisions may actually increase the efficiency of pumping in the quasi continuum.

When intramolecular relaxation is rapid compared to radiative pumping, one may describe the laser excitation in the quasi continuum in a simplified way by using a set of rate equations for the time evolution of total energy shell populations, N_k [6.32,47,51-54],

$$\frac{dN_k}{dt} = T_{k\leftarrow k-1}N_{k-1} + T_{k\leftarrow k+1}N_{k+1} - T_{k+1\leftarrow k}N_k - T_{k-1\leftarrow k}N_k \quad , \tag{6.1}$$

where $T_{k\pm 1\leftarrow k}$ is the radiative rate for a transition from a total energy state with k laser quanta to a total energy state with $k\pm 1$ laser quanta. In the next section we develop (6.1) from equations describing more generally the heat bath excitation process and provide an explicit expression for the radiative transition rates. Since the $T_{k\pm 1\leftarrow k}$ are proportional to the laser intensity I, it follows that N_k's determined from (6.1) depend only on the fluence $\int I\,dt$. The notion that the rate or extent of a laser-induced reaction should depend only on the total fluence has proved very useful [6.55], but it is not as general as it may have seemed [6.56-58]. In Sect.6.3.2 we discuss theoretical restrictions on this conclusion and give a specific experimental example of intensity dependence.

Black et al. [6.51] have recently pointed out that when the $T_{k\pm1\leftarrow k}$ are calculated for a hypothetical molecule with constant radiative cross section and degeneracy factors equal to those of an s-fold degenerate oscillator, then a thermal distribution of the N_k's with time-dependent temperature solves (6.1). We elaborate on this interesting result in Sect.6.4.1.

One easily accounts for chemical reactions by including chemical decay terms of the form

$$\frac{-2\Gamma_k}{\hbar} N_k \quad , \qquad k > k_c \tag{6.2}$$

in (6.1) for all energy shells above some critical total energy equal to k_c laser photons. Because (6.1) describes the time evolution of the total energy shell population rather than the pump mode populations, one can consistently calculate the decay constants $2\Gamma_k/\hbar$ from one of the statistical theories (RRK or RRKM) of unimolecular decay.

We complete this general sketch of the theory of laser-induced chemical reactions by mentioning that the pumping of molecules from the ground state to the beginning of the quasi continuum is, in general, at least partially coherent, requiring for its quantitative description something more than rate equations. Some form of the optical *Bloch* [6.59]or *Boltzmann* [6.5] equations, properly interfaced to the rate equations for the quasi continuum, is sufficient to describe coherent effects in the discrete level region [6.6,32,60-62] and account for isotope selectivity and details of the laser frequency tuning curve [6.63-68]. Before taking up the comparison of theory and experiment, we want to look carefully in Sect.6.2.2 at the assumptions going into the derivation of (6.1) and in Sect.6.2.3 at the microscopic calculation of the line broadening factors which enter into the definition of the radiative transition rates $T_{k\pm1\leftarrow k}$.

6.2.2 The Heat Bath Feedback Model

Consider the laser excitation mechanism illustrated in Fig.6.1. Letting $n_{i,j}$ be the population of molecules having i quanta in the pump mode and j quanta in the heat bath modes, the following rate equation [6.47]

$$\begin{aligned}\frac{dn_{i,j}}{dt} &= W_{i,i-1}(j)(n_{i-1,j} - n_{i,j}) + W_{i+1,i}(j)(n_{i+1,j} - n_{i,j})\\ &+ \sum_{\substack{j'=0\\ j'\neq j}}^{i+j} \left(\frac{1}{T_1}\right)_{i,j\leftarrow i+j-j',j'} n_{i+j-j',j'} - \left(\frac{1}{T_1}\right)_{i+j-j',j'\leftarrow i,j} n_{i,j}\end{aligned} \tag{6.3}$$

describes the excitation mechanism in a more fundamental way than (6.1) because it includes intramolecular relaxation represented by the transition rates $1/T_1$. In all terms in (6.3) the index i refers to pump mode excitation and the index j refers to heat bath excitation. The level of heat bath excitation affects the radiative pumping most directly through the j dependence of the radiative transition rates $W_{i\pm1\leftarrow i}(j)$, hence the name heat bath feedback model. In writing down this set of rate equations to describe laser pumping in the quasi continuum we are implicitly assuming that coherence effects are negligible in the quasi continuum. This is a useful assumption to make, but it must be kept in mind that high-intensity short-pulse experiments are likely to reveal some degree of coherence in the quasi continuum. In Sect.6.5 we shall return to this question of coherence, pointing out that (6.3) is valid only if the so-called T_2 or dephasing relaxation is rapid compared to the radiative pumping rate, and we shall also derive the following explicit expressions for $W_{i\pm1,i}(j)$:

$$W_{i\pm1,i}(j) = \frac{\alpha^2_{i\pm1,i}(j)A^2}{2\hbar^2} \frac{(1/T_2)_{i\pm1,i,j}}{(1/T_2^2)_{i\pm1,i,j} + (\chi - |\omega_{i\pm1,j} - \omega_{i,j}|)^2} . \tag{6.4}$$

Here A is the amplitude of the laser field, χ is the laser frequency, $\hbar\omega_{i,j}$ is the energy of the state with i quanta in the pumped mode and j quanta in the heat bath modes, and $\alpha_{i+1,i}(j)$ is the dipole coupling element between states with i and i+1 pump mode quanta, with j heat bath quanta.

An essential property of the $1/T_1$ intramolecular relaxation terms explicit in (6.3) follows from the "ergodic-type" assumption that intramolecular relaxation alone must eventually bring about a random, or microcanonical, distribution of states within a given energy shell. In particular, one can assume microscopic reversibility, that is, the rate of transitions between any pair of states on the energy shell exactly balance when the state populations are distributed randomly. Therefore

$$\left(\frac{1}{T_1}\right)_{i+j-j',j'\leftarrow i,j} \zeta_j = \left(\frac{1}{T_1}\right)_{i,j\leftarrow i+j-j',j'} \zeta_{j'} , \tag{6.5}$$

where ζ_j is a statistical density-of-states factor proportional to the number of heat bath states per unit energy with j quanta of heat bath excitation.

When the relaxation rates T_1 in (6.3) are very rapid compared with radiative transition rates, a statistical distribution of energy is quickly achieved within each energy shell before significant changes in the total energy shell population

$$N_k = \sum_{i=0}^{k} n_{i,k-i} , \tag{6.6}$$

can occur. In this case all relaxation terms T_1 are eliminated and one is led back to (6.1) for the energy shell populations. To arrive at (6.1) from (6.3), one first derives the following exact equations for the total energy shell populations:

$$\frac{dN_k}{dt} = \sum_{i=1}^{k} W_{i,i-1}(k-i)(n_{i-1,k-i} - n_{i,k-i})$$

$$+ \sum_{i=0}^{k} W_{i+1,i}(k-i)(n_{i+1,k-i} - n_{i,k-i}) \quad , \tag{6.7}$$

where the relaxation terms T_1 present in (6.3) have cancelled in the addition of the contributions from each $dn_{i,k-i}/dt$, because the intramolecular relaxation process conserves the total population within each energy shell [6.47].

The microscopic reversibility condition (6.5) guarantees that the intramolecular relaxation matrix has at least one zero eigenvalue corresponding to the equilibrium distribution

$$n_{i,j} = s_{i+j}(i)N_{i+j} \quad , \tag{6.8}$$

$$s_k(i) = \zeta_{k-i} \Big/ \sum_{\ell=0}^{k} \zeta_1 \quad . \tag{6.9}$$

If we now assume that intramolecular relaxation is rapid or, more precisely, that all the nonequilibrium eigenvalues of the $1/T_1$ relaxation matrix correspond to rates that are large compared to the radiative transition rates, one can substitute the equilibrium value of $n_{i,j}$ given by (6.8) into (6.7). In this way one obtains the *closed* set of rate equations relating only the total energy shell populations

$$\frac{dN_k}{dt} = T_{k\leftarrow k-1}N_{k-1} - T_{k-1\leftarrow k}N_k + T_{k\leftarrow k+1}N_{k+1} - T_{k+1\leftarrow k}N_k \quad , \tag{6.1}$$

where the radiative transition rates are

$$T_{k\leftarrow k-1} = \left[\sum_{i=1}^{k} W_{i,i-1}(k-i)\zeta_{k-i}\right] \Big/ \sum_{\ell=0}^{k-1} \zeta_\ell \quad ,$$

$$T_{k-1\leftarrow k} = \left[\sum_{i=1}^{k} W_{i,i-1}(k-i)\zeta_{k-i}\right] \Big/ \sum_{\ell=0}^{k} \zeta_\ell \quad . \tag{6.10}$$

Since the transition rates $T_{k\leftarrow k-1}$ and $T_{k-1\leftarrow k}$ which appear in (6.10) are, like the $W_{i,i+1}$'s, linear in the intensity, the total populations N_k in an energy shell described by (6.1) are fluence dependent in agreement with the Golden Rule argument. Equation (6.1) is thus similar in principle to equations which have been postulated [6.55,69] to explain the fluence dependence of many laser-induced reactions. The important difference is that in deriving (6.1) from a set of rate equations (6.3) which includes heat bath feedback, we obtain explicit expressions (6.10) for induced emission and absorption coefficients in terms of densities of states ζ, intramolecular dephasing rates $1/T_2$ (6.4), and the dipole coupling strength α_{01} of the pump mode. Note that the absorption and emission coefficients calculated in (6.10) satisfy a microscopic reversibility condition of the form

$$\eta_{k-1} T_{k \leftarrow k-1} = \eta_k T_{k-1 \leftarrow k} \quad , \tag{6.11}$$

where

$$\eta_k = \sum_{\ell=0}^{k} \zeta_\ell \tag{6.12}$$

is a density-of-states factor proportional to the total number of states with k laser quanta of energy. If the $W_{i,i-1}(k-i)$ which enter the sum in (6.10) have widths $1/T_2$ and energy differences that do not depend on the distribution of quanta between the pump mode and heat bath modes, but depend only on the total energy of the pair of adjacent energy shells, then (6.10) simplifies to a single Lorentzian for the effective radiative transition rate between energy shells:

$$T^e_{k-1 \leftarrow k} = \frac{A^2}{2\hbar^2} \frac{\langle \alpha^2_{k-1,k} \rangle^e \, (1/T^e_2)_{k-1,k}}{(1/T^e_2)^2_{k-1,k} + (\chi - \omega_k + \omega_{k-1})^2} \tag{6.13}$$

where $(1/T^e_2)_{k-1,k}$ can be regarded as an effective dephasing width for transitions between the k-1 and k energy shells, and

$$\langle \alpha^2_{k-1,k} \rangle^e = \sum_{i=1}^{k} \zeta_{k-i} \alpha^2_{i,i-1}(k-i)/\eta_k \tag{6.14}$$

is an averaged dipole coupling element between the k-1 and k energy shells. When one sets $\alpha_{i,i-1} = i\alpha_{0,1}$ and applies Fermi's Golden Rule arguments to a set of "equivalent state" oscillator strengths, (6.13) is identical to a result derived earlier by *Stephenson* et al. [6.32].

In the RRK or RRKM theory of unimolecular decay, it is assumed that redistribution of energy on any given energy shell is rapid compared to the chemical decay from the energy shell. This fundamental assumption is clearly consistent and, in fact, almost equivalent to the assumption of rapid T_1 relaxation that was used to derive (6.1) for the time evolution of the total energy shell populations N_k. Thus the self-consistent way in which RRK or RRKM theory can be applied to laser-induced reactions is to add chemical reaction rate terms

$$-(2\Gamma_k/\hbar)N_k$$

to (6.1) and to equate the microscopic rate constant $2\Gamma_k/\hbar$ with the expression from RRK or RRKM theory for molecules with a randomly distributed total energy E_k. The total decay rate

$$-\frac{d}{dt} \sum N_k = \sum N_k \left(\frac{2\Gamma_k}{\hbar}\right) \tag{6.15}$$

depends on the total energy distribution as defined by the N_k's. The initial decay rate as calculated from (6.14) is dependent only on the total fluence if the N_k's are independent of intensity. This is true of N_k's calculated from (6.1) without chemical decay, but is true more generally only if the "perturbation" caused by introducing decay terms into (6.1) does not significantly alter the decay-free energy distribution. There are practical cases in which (6.1) is itself invalid, e.g., for radiative pumping rates on the same time scale as intramolecular relaxation, or radiative pumping bottlenecks in discrete levels. Under these conditions an intensity dependence rather than merely a fluence dependence is expected for the initial rate.

6.2.3 Some Microscopic Aspects of the Theory

To bring the theory for pumping in the quasi continuum to a point where calculations can be made, one must be able to estimate the dephasing times T_2 in (6.4) or (6.13). A useful way of looking at the origin of the T_2's and their relationship to microscopic quantities is illustrated in Fig.6.3 for a transition between vibrational levels in the quasi continuum of an excited polyatomic molecule. Exact energy eigenstates can be represented by a sum of many 0th-order normal mode states (the distribution in Fig.6.3 with anharmonic interaction bandwidth w_{k-1}^{int}) and similarly 0th-order states can be represented as a sum of exact energy states (the distribution in Fig.6.3 with anharmonic interaction bandwidth w_k^{int}). The anharmonic interactions which couple the infrared-active normal mode to the other vibrational modes in the molecule cause a spread in the dipole oscillator strength, $w_{k,k-1}^{o.s.}$, in Fig.6.3 that is ultimately responsible for the T_2 line broadenings in (6.4) and (6.13). *Mukamel* [6.70] has taken an important step by formally relating the T_2's to certain dipole coupling correlation functions. For our purposes we are interested in obtaining generally applicable and qualitatively reasonable estimates of $w_{k,k-1}^{o.s}$. To this end one notes that according to the physical picture of Fig.6.3, in which the energy level spacing is constant and the dipole oscillator strength is not spread in the 0th-order representation, the oscillator strength distribution is something like a convolution of the two distributions for the anharmonic interactions of the k-1 and k levels. If the anharmonic interaction distributions are assumed to be Lorentzian,

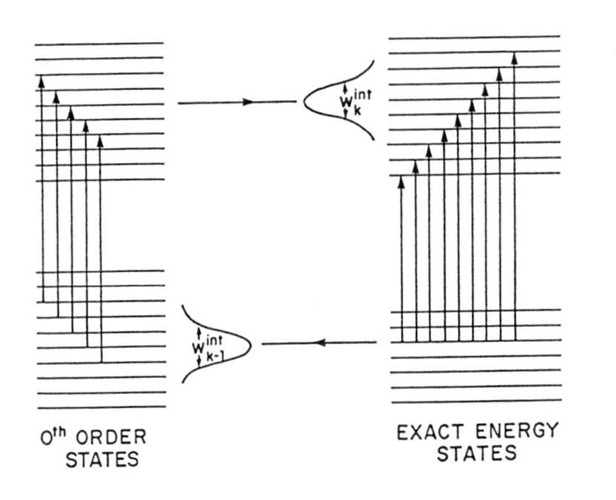

Fig.6.3. Schematic representation of the origin of the spread in oscillator strength for a transition, between two quasi-continuum states. The width in oscillator strength, $w_{k,k-1}^{o.s.}$, is the sum of the anharmonic interaction widths of the two sets of states involved in the transition

then the distribution of the oscillator strength is also a Lorentzian with bandwidth

$$w_{k,k-1}^{o.s.} = w_k^{int} + w_{k-1}^{int} \quad . \tag{6.16}$$

Equation (6.16) will hold approximately for any reasonably smooth anharmonic interaction distribution. We can also identify the linewidth $1/T_2^e$ in the Lorentzian given in (6.13) as

$$\left(\frac{1}{T_2^e}\right)_{k,k-1} = \frac{w_{k,k-1}^{o.s.}}{h} \quad . \tag{6.17}$$

Model calculations to determine the anharmonic interaction bandwidth w_k^{int} in closed form have necessarily assumed rather specialized models for the interaction coupling the zero-order states. Lineshapes developed by *Bixon* and *Jortner* [6.71] to study vibronic coupling were based on the assumption that the matrix elements coupling an optically accessed state to a set of background states were equal in magnitude and sign. The bandwidth interaction in their model, taking the coupling to be H_{anh}, was proportional to $N_k H_{anh}$, where N_k is the number of states in width H_{anh}. More recently, *Tric* [6.72] investigated a model which would appear to be very similar in physical content to the *Bixon-Jortner* model [6.71], but which shows strikingly different dynamic behavior. By including coupling among the background states themselves as well as between background states and the optically selected state, one obtains a system in which the population of the initial state exhibits periodic behavior rather than decay. Furthermore, both the bandwidth and the period of oscillation were shown to be independent of H_{anh} in the limit of high densities of states. On the other hand, a model investigated by *Heller* and *Rice* [6.73] and by *Tric* [6.72], which also included coupling among the background states, but assumed randomly distributed rather than equal coupling elements, did show irreversible decay. In this case the decay rate, as in the *Bixon-Jortner* model [6.71], was dependent on the interaction matrix elements, but only through the variance in their distribution, not through their average magnitude. Finally, in the absence of detailed knowledge concerning interactions in a molecule, we have recently adapted an intuitively based model employing a random walk in energy space to determine a level-dependent anharmonic interaction width [6.40]

$$w_k^{int} = N_k^{1/2} H_{anh} = \eta_k^{1/2} H_{anh}^{3/2} \quad . \tag{6.18}$$

This width depends, as in the *Bixon-Jortner* model [6.71], on the average coupling strength H_{anh} but changes less rapidly with the density of states.

6.3 Comparisons of Theory with Experiment

6.3.1 Laser-Induced Decomposition of CF_2HCl

In a collaborative effort with *Stephenson* et al. [6.32], the theory developed in the previous section has been applied to the CO_2 laser-induced reaction

$$CF_2HCl \xrightarrow{10.6\ \mu m} CF_2 + HCl$$

in the presence of an Ar buffer gas.

A time-resolved probe of the decomposition rate for this reaction is available. Ultraviolet irradiation from a frequency-doubled N_2-pumped dye laser induces fluorescence in the CF_2 radical; product concentration is monitored with a time resolution of the order of 20 ns. From an analytical viewpoint, there is considerable advantage to measuring the rate of product formation as a function of time during the CO_2 laser pulse in preference to measuring a total product yield. This is especially true when collisions are present which redistribute energy and cause further reaction after the laser pulse has ended.

The initial rate of CF_2 production measured as a function of Ar pressure is shown in Fig.6.4 [6.32]. At a laser intensity of 15 MW cm^{-2}, increasing the frequency of collisions increases the rate of CF_2HCl dissociation up to a pressure of about 50 Torr, after which the rate is constant. In Fig.6.5, the CF_2 production rate is plotted as a function of laser intensity at a fixed Ar pressure of 300 Torr. We see that there exists a narrow "linear" range in product formation above threshold between 5-25 MW cm^{-2} and a smooth leveling off between 25-150 MW cm^{-2}. Theoretical predictions are included on Figs.6.4,5 (solid curves), but a full discussion of the comparison of theory and data is postponed until after we summarize some additional aspects of the theoretical treatment.

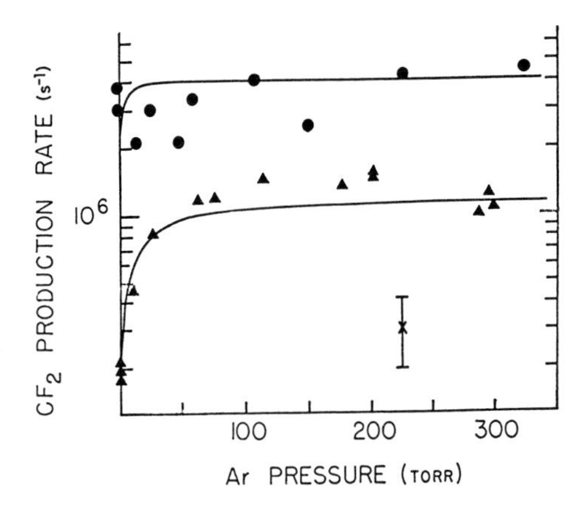

Fig.6.4. CF_2 production rate as a function of buffer gas pressure for the laser-induced dissociation of CF_2HCl, from [6.32]. The pressure dependence was measured for two experimental conditions: (●●●) CO_2 laser operating on the R(32) transition at 1086 cm^{-1} which is resonant with the CF_2HCl one-photon absorption, peak laser intensity of 150 MW cm^{-2}; (▲▲▲) R(32) laser energy reduced by a factor of 10. The curved lines drawn through the data points are the theoretically calculated results, based on the model described in the text. The isolated point shows the magnitude of the error

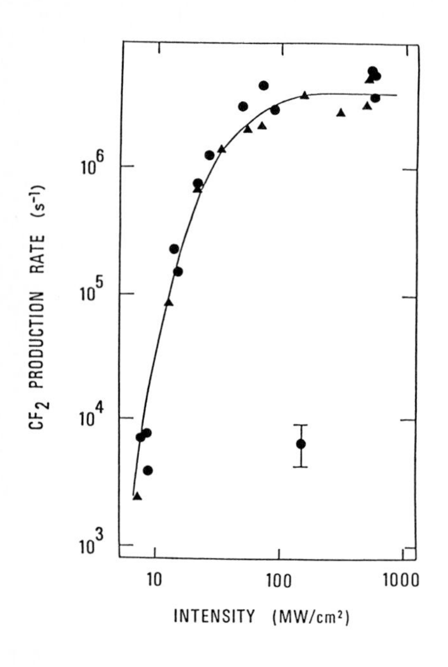

Fig.6.5. Dependence of CF_2 production rate on CO_2 laser intensity for CF_2HCl reactant very dilute (approximately 4 mTorr) in 300 Torr of argon from [6.32]. For the data points indicated by (●●●), the CO_2 laser pulses were partially mode-locked; for the data points indicated by (▲▲▲), the CO_2 laser pulses were not mode-locked. The isolated point shows the magnitude of the error. The solid line shows the model prediction for these experimental conditions

a) Coherent Discrete Level Pumping and Interface with Rate Equation

As described in Sect.6.2.2, the polyatomic molecule is characterized by an infrared-active pump mode in intimate contact with the remaining modes acting as a finite heat bath. The process of driving the molecule from the ground state through sets of discrete and quasi continuum levels to dissociation can be explicitly decomposed into two energy regions. In the low-energy region pumping is essentially coherent, while in the high-energy region pumping is incoherent and rate equations apply. The energy level at which the interface to a set of rate equations is made is not arbitrary but is determined by the condition that the dephasing rate ($1/T_2$) exceeds the dynamic radiative transition rate (Rabi frequency). For the first dynamic region dominated by coherent pumping, the Boltzmann equation discussed in Sect.6.5,

$$\frac{d\rho_{jj}}{dt} = \frac{i}{\hbar}\,[\rho,H]_{jj} \quad , \qquad j = 1,m-2 \quad , \tag{6.19}$$

$$\frac{d\rho_{m-1,m-1}}{dt} = \frac{i}{\hbar}\,[\rho,H]_{m-1,m-1} + T'_{m-1,m}N_m - T'_{m,m-1},\rho_{m-1} \quad , \tag{6.20}$$

$$\frac{d\rho_{jk}}{dt} = \frac{i}{\hbar}\,[\rho,H]_{jk} - \left[\left(\frac{1}{T_2}\right)_{jk} + \frac{1}{\tau}\right]\rho_{jk} \quad , \quad j \neq k \quad , \quad j,k = 0 \ , \quad m-1 \quad , \tag{6.21}$$

is used to include both the effects of coherent pumping and of dephasing processes. The last two terms in (6.20) allow for transitions entering and leaving the uppermost discrete level m-1 coupled to the lowest band of the high-energy region quasi continuum states with population N_m. The Hamiltonian for the system is given by

$$H_{nn} = \varepsilon_n \quad , \tag{6.22}$$

$$H_{n-1,n} = H_{n,n-1} = \alpha_n A \sin\chi t \quad ,$$

where the dipole coupling matrix element $\alpha_n = \alpha_{01}\sqrt{n}$.

Dephasing rates $1/\tau$ and $1/T_2$ associated with collisions and intramolecular relaxation appear explicitly in (6.21) for the off-diagonal elements. The uppermost discrete level, m-1 quanta above the ground state, is the level which provides an effective interface between the discrete levels and the quasi continuum and as such contains both a term $(1/\hbar)\,\rho,H_{m-1,m-1}$, giving the rates of transitions into and out of the discrete levels below, and terms $T'_{m\leftarrow m-1\leftarrow m-1}\rho_{,m-1}$ and $T'_{m-1\leftarrow m}N_m$ describing the rates of transitions into and returning from the quasi continuum, respectively. The rate $T'_{m\leftarrow m-1}$ is given directly by the pumped mode transition rate of (6.4), with an effective T_2:

$$T'_{m\leftarrow m-1} = W_{m\leftarrow m-1}(j=0) = \frac{(\alpha^2_{m,m-1}/2\hbar^2)A^2(1/T^e_2)_{m,m-1}}{(1/T^e_2)_{m,m-1}+(\chi-\omega_m+\omega_{m-1})^2} \quad . \tag{6.23}$$

Since the mth state is actually a band of states of width w_m^{int}, microscopic reversibility implies

$$T'_{m\leftarrow m-1} = T'_{m-1\leftarrow m}\zeta_m w_m^{int} \quad . \tag{6.24}$$

Equations (6.23,24) are crucial in interfacing the discrete coherent pumping to the quasi continuum incoherent pumping without the need to bring in any new adjustable parameters.

The second dynamic region treated in the calculation is governed by a set of rate equations with radiative transitions connecting adjacent energy levels only. The appropriate equations are the energy shell population equations given by (6.1), with unimolecular decay terms included as described at the end of Sect.6.2.2:

$$\frac{dN_j}{dt} = T_{j\leftarrow j-1}N_{j-1} + T_{j\leftarrow j+1}N_{j+1} - T_{j-1\leftarrow j}N_j - T_{j+1\leftarrow j}N_j - \frac{2\Gamma_j}{\hbar}N_j \quad , \tag{6.25}$$

where N_j is the total population of vibrational states where the sum of energy in all modes is equal to j quanta. That the T_1 relaxation terms do not appear explicitly in (6.25) has been discussed in Sect.6.2.2 and is a consequence of rapid equilibration of energy within each vibrational energy shell [6.47]. The terms $2\Gamma_j/\hbar$ are energy-level-dependent microscopic unimolecular decay rates that were obtained through an RRKM analysis of CF_2HCl. The averaged radiative transition rates $T_{j\leftarrow i}$ are given in (6.9,10).

b) Rotational Hole Filling by Collisions

To study the effect of collisions on the rate of laser-induced decomposition, *Stephenson* et al. [6.32] varied the buffer gas pressure in dilute mixtures of CF_2HCl in Ar. Because the reactant molecule concentration was low enough to ensure that the effects of collisions between a pair of reactant molecules could be ignored, the experiments were kept within the domain of applicability of the linear Boltzmann equation formalism, (6.19-21). This is an important consideration in designing experiments to study the effects of collisions. The other situation in which collisions between pairs of reactant molecules cannot be ignored presents severe theoretical difficulties, since some type of nonlinear collision term would have to be included in the Boltzmann equation. We recognize that the study of thermal unimolecular reactions has revealed no recognizable new factors in the case where reactant molecule collisions dominate. However, unlike the thermal reaction, in the laser-driven reaction one produces high transient fractions of very excited reactant molecules during the time interval when the reaction is occurring; this situation is clearly unfavorable to any theory that ignores the collisional exchange of energy between reactant molecules.

The effects of phase randomization are included explicitly in (6.21). This collisionally induced phase randomization, occurring with the gas kinetic collision frequency $1/\tau$, is relatively unimportant for the analysis of *Stephenson* et al. [6.32] since it is small compared to the $1/T_2$ intramolecular dephasing. In the model calculations [6.32], V-T energy transfer does not have to be included, but it is important in the experiment discussed in Sect.6.3.3.

A third effect, "rotational hole filling" [6.74], is most important in the CF_2HCl reaction. The laser selectively depletes the population of those molecules in rotational states whose allowed radiative transition is most resonant with the laser frequency while collisions simultaneously act to restore the distribution of rotational level populations to thermal equilibrium.

The discrete energy level structure of CF_2HCl is not known in detail. If we make the assumption that rotational states for a molecule with unknown level structure can be divided into two classes according to whether or not they are efficiently pumped, then rotational hole filling can easily be incorporated into (6.19-21). *Stephenson* et al. [6.32] considered two classes of vibrational ground states: those which are efficiently pumped by the laser and those which, in the absence of collisions, remain bottlenecked in the discrete levels. Taking ρ_{00} and n_h to be the ground-state population of laser-pumpable and nonpumable states, respectively,

$$\frac{dn_h}{dt} = (1 - f_p)\frac{1}{\tau}\rho_{00} - f_p\frac{1}{\tau}n_h \quad , \tag{6.26}$$

$$\frac{d\rho_{00}}{dt} = f_p\frac{1}{\tau}n_h - (1 - f_p)\frac{1}{\tau}\rho_{00} + \frac{i}{\hbar}[\rho,H]_{00} \quad , \tag{6.27}$$

where $1/\tau$ is the gas-kinetic frequency of collisions between Ar and CF_2HCl molecules, and f_p is the fraction of rotational states capable of interacting with the laser field. Equations (6.26,27) account for the action of collisions in replacing states depleted by the laser. The last term in (6.27) represents the rate of change of the ground state due to laser pumping.

c) A Comparison of Theory with Experiment for CF_2HCl

A fit of the model to the data for the CF_2 production rate as a function of Ar pressure and CO_2 laser intensity gave excellent quantitative agreement (Fig.6.4). The calculation was carried out with 20 vibrational levels of bound CF_2HCl with the computed RRKM decomposition rates, $2\Gamma_i/\hbar$, of 10^7, 10^8, 10^9, 3×10^9 out of the top four levels. Transitions both in the discrete levels and the quasi continuum were taken on resonance (or near band center). There are two adjustable parameters in the model: the fraction f_p of laser-pumpable states (6.26), and the coupling constant H_{anh} that determines the effective width of the interaction between the pump mode and background states. The coupling constant H_{anh} determines the energy-level-dependent dephasing (T_2) relaxation time which in the actual calculation varies between 1 ps at $v=5$ and 0.03 ps at $v=15$ (6.17,18).

A meaningful comparison of model predictions and data can be made because the two adjustable parameters in the model can only be used to fix (1) the CF_2 production rate at two laser intensities in the absence of Ar and (2) the number of infrared pumpable states at a single Ar pressure. The shapes of the curves for the pressure dependence and the decomposition dependence on the laser intensity in the presence and absence of Ar are then strictly model-based predictions.

For each of the curves in Fig.6.4, the parameter f_p was chosen to fit the difference in CF_2 production rates at two values for the Ar pressure, 0 Torr and 300 Torr. The fraction f_p can be determined empirically by measuring the ratio of CF_2 product formation at zero Ar to high Ar pressures (> 100 Torr) at constant laser intensity; given a sufficient number of collisions with Ar, all CF_2HCl molecules can be brought into pumpable states by the mechanism of rotational hole filling on the time scale of radiative pumping. In the results shown in Fig.6.4, accurate estimates for f_p can be made since there is no further increase in CF_2 production at Ar pressures exceeding 100 Torr. The other adjustable parameter, H_{anh} (2.1 cm^{-1}), is chosen to fix the difference between the rates of CF_2 production at the two intensities (15 and 150 MW cm^{-2}) at one fixed value of the Ar pressure (300 Torr).

The match between theory and data in Fig.6.5 involves no further parameter adjustments. For Fig.6.5 the model curve was generated *prior* to obtaining the experimental data. It is worth noting that the calculation of the intensity-dependent falloff in rate (Fig.6.5) depends on only one adjustable parameter, H_{anh}. Thus the shapes of each curve as a function of Ar pressure and the intensity dependence at high Ar pressures are strictly predictive results of the model. Thus, by varying f_p and H_{anh} one cannot arbitrarily reproduce either the shape of each curve as a function of pressure (Fig.6.4) or the intensity dependence of CF_2 production for a fixed Ar pressure (Fig.6.5).

d) Additional Theoretical Predictions

The primary purpose of providing a theoretical model to fit the pressure dependence of CF_2HCl decomposition is to attempt to gain new insight into the dynamics of competing processes simultaneously going on within a polyatomic molecule—absorption of intense phase coherent radiation, intra- or intermolecular relaxation, and unimolecular decay. We view the agreement between model and data (Figs.6.4,5) as but a starting point in a broader effort to predict the response of a polyatomic molecule to laser driving. Beyond the initial data fit discussed in Sect.6.3.1c, the model has been used to predict additional features of the multiphoton absorption process which should be amenable to experimental verification.

Using the model, we computed [6.32] a cross section σ for absorption of laser radiation by CF_2HCl,

$$\sigma = \frac{\hbar\chi}{\text{Intensity}}\left[\sum_{j=m-1}^{N-2} T_{j+1\leftarrow j}N_j - \sum_{j=m}^{N-1} T_{j-1\leftarrow j}N_j + \sum_{j=1}^{m-1} \frac{\alpha_{j-1,j}A}{\hbar}(\rho_{jj-1} + \rho_{j-1,j})\right] , \qquad (6.28)$$

where the density matrix elements N_j, ρ_{ij} are obtained by solving (6.19-21,25-27), and model-based expressions for transition rates $T_{i,j}$ are given by (6.9,10,23,24). For very early times, the absorption rate is dominated by the coherent evolution terms in (6.28) and shows negative values (emission), characteristic of Rabi cycling. At later times, when pumping in the quasi continuum dominates, σ was found to decrease from 10^{-18} to $10^{-20}cm^2$ for absorption of between 5 and 15 photons.

The distribution of internal energy among molecules for the laser-induced excitation process plays an important role in determining chemical decay rates. For the laser-induced decomposition of CF_2HCl, we plot in Fig.6.6 the energy distribution at times during the 150 MW cm^{-2} excitation pulse when the molecules have on average absorbed 5, 10, and 15 photons. These are compared with a Boltzmann distribution corresponding to the same average absorbed energy. Distributions are obtained which are distinctly non-Boltzmann. The irradiation times required to achieve excitation of $\langle v \rangle = 5, 10$, and 15 are 0.5 ns, 2 ns, and 30-100 ns, respectively, according to the model using a square pulse shape. However, the times required experimentally are probably longer for $\langle v \rangle = 5$ and 10 because of the finite pulse rise time.

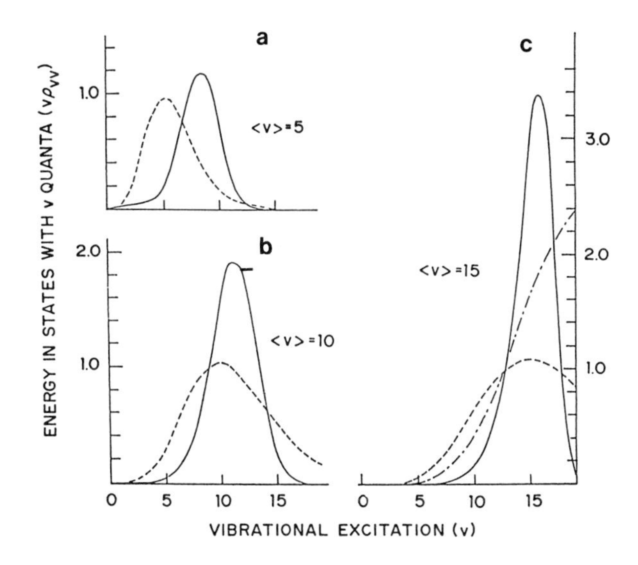

Fig.6.6a-c. Theoretical energy distributions (——) for CF_2HCl dissociation compared with thermal distributions (---), from [6.32]: (**a**) Average energy <v> = 5, (**b**) average energy <v> = 10. The extra curve (— - —) in (**c**) for <v> = 15 is the thermal distribution for a molecule arbitrarily truncated at 20 levels. These model predictions are for a CO_2 laser intensity of 150 MW cm^{-2} and 300 Torr Ar

6.3.2 Laser Intensity Versus Fluence Effects

When rate equations of the form given in (6.25) apply, one expects the initial reaction rate to depend only on fluence, $\int I dt$, and not on intensity or pulse shape. In further experiments on the decomposition of CF_2HCl, *King* and *Stephenson* [6.58] have demonstrated a "fluence-breaking" dependence on the intensity. One of their results, reproduced in Fig.6.7, shows that initial CF_2 production rates are 5-10 times larger for mode-locked pulses; the effective intensity is much higher for the

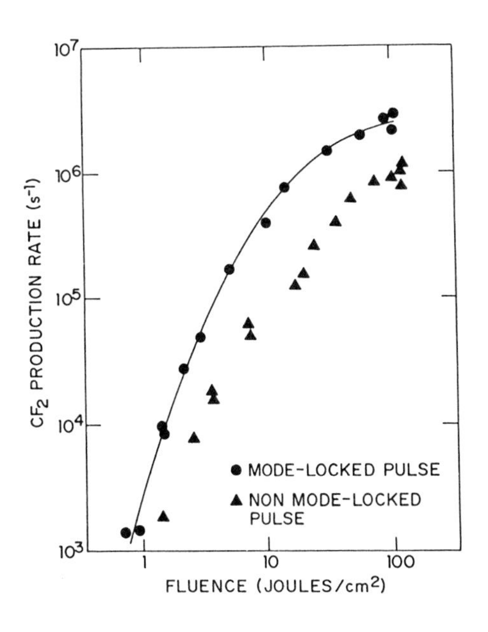

Fig.6.7. Effect of laser mode-locking on CF_2CHl dissociation, from [6.58]. First-order rate constants are given for the dissociation of very low-pressure neat CF_2HCl by either mode-locked or non-mode-locked CO_2 laser pulses (1086 cm^{-1}) of various total energy fluences. Dissociation rates are 5-10 times faster for the mode-locked pulses than for the non-mode-locked pulses, implying that the rates of photon absorption are 300-500 times faster during the intense spikes of the mode-locked pulse train than during the non-mode-locked pulse excitation for pulses of the same fluence

mode-locked pulses. *Ashfold* et al. [6.57] have also concluded that intensity effects are important in the decomposition of CH_3NH_2.

In Sects.6.3.1b,c we showed that the observed increase in CF_2 production rate with Ar pressure is quantitatively accounted for by a model that includes a bottleneck in the discrete level pumping, which is overcome by collisions that cause rotational hole filling. *King* and *Stephenson* [6.58] have suggested that their intensity results are consistent with the idea that the bottleneck in the discrete levels can also be overcome by the dynamic power broadening associated with the laser pulse intensity. The increased rate at high laser intensities results from the increased rate of pumping through the discrete level bottleneck. A secondary effect is also possible in which the shape of the energy distribution curve at fixed average energy is dependent on the rate of pumping through the bottleneck [6.51].

There are conditions under which one also expects to find that the product yield is dependent only on fluence. For example, under steady-state conditions,

$$\text{product yield } |_{s.s.} \sim 1 - e^{-k_{s.s.}t} \quad ,$$

where $k_{s.s.}$ is the steady-state rate constant, As *Quack* [6.75,76] has pointed out, one expects in the weak-field limit that

$$k_{s.s.} \sim \text{Intensity} \quad .$$

In the weak-field limit then the *steady-state* product yield is clearly a function of fluence, intensity × time. Only *Quack* [6.75] has also carried out some numerical calculations indicating that deviations from an intensity-proportional steady-state rate constant are not too large, i.e., constant to within 10% when the laser intensity is changed by a factor of 10-30. Thus one can expect the steady-state product yield to be primarily dependent only on fluence over a usable range of intensities when (6.25) applies.

It was the discovery of *Kolodner* et al. [6.77] and *Lyman* et al. [6.78] that the product yield for the decomposition of SF_6 is only fluence dependent that initially focused attention on this question. This result for the product yield, along with *Quigley*'s [6.79] results for the effects of inert gases on the amount of energy absorbed by SF_6, strongly indicates that above the reaction threshold there is at most only a weak discrete level bottleneck in SF_6 (see also Sect.6.3.4a).

It is important to realize that there are a number of factors, besides bottlenecks in discrete level pumping, that may in principle give rise to fluence-breaking intensity dependence. Since we expect this to be an area of expanding interest, especially as more resolved determinations of decomposition rates are made, we will briefly discuss two other possible sources for this type of intensity dependence.

In Sect.6.2.1 we pointed out that the expectation of a rate dependent only on laser fluence rests on the applicability of a rate equation of the form of (6.1) to pumping through the quasi continuum. In deriving (6.1) from (6.3) for the heat bath feedback model (Sect.6.2.2), a sufficiently rapid exchange of energy between the pump model and heat bath modes of vibration is explicitly assumed. If there is slow intramolecular relaxation among the molecule's vibrational degrees of freedom, i.e., on the same time scale as radiative pumping, then an equation of the form

$$\frac{dN_k}{dt} = T_{k\leftarrow k-1}N_{k-1} + T_{k\leftarrow k+1}N_{k+1} - T_{k+1\leftarrow k}N_k + T_{k-1\leftarrow k}N_k$$

$$+ \text{ intramolecular relaxation terms that are not simply proportional to intensity} \qquad (6.29)$$

replaces (6.1) for the time evolution of the energy shell populations, N_k; see, for example [Ref.6.47, Eq.A-3].

The presence of chemical decay terms in (6.1) can also bring about a fluence-breaking effect. As mentioned in Sect.6.2.2, the solution to the full rate equations including decay terms (6.25) may be significantly dependent on intensity as well as fluence because of the perturbation caused by the decay terms.

6.3.3 High-Pressure Fallof Reaction Rate

It seems somewhat surprising that the CF_2 production rate, plotted as a function of Ar pressure in Fig.6.4, remains constant over the full experimental range of pressure upto 300 Torr after an initial rise due to rotational hole filling. Since it must be true that at sufficiently high pressures deactivation by collisions with buffer gas molecules will begin to compete effectively with radiative pumping, thereby depleting the population of highly excited molecules, one expects a decreasing rate of reaction at high pressures. For this reason we have included collision terms, based on a generalized Landau-Teller mechanism for V-T collisional relaxation, in (6.19-21,25-27) and repeated the 15 MW cm^{-2} calculations for CF_2HCl decomposition. The results of this calculation are plotted in Fig.6.8.

In the Landau-Teller mechanism all collision terms are fixed once $P_{0\leftarrow 1}$ (the probability of a transition from initial level 1 to final level 0 upon collision) is fixed. The larger the value of $P_{0\leftarrow 1}$, the stronger the collisional deactiviation.

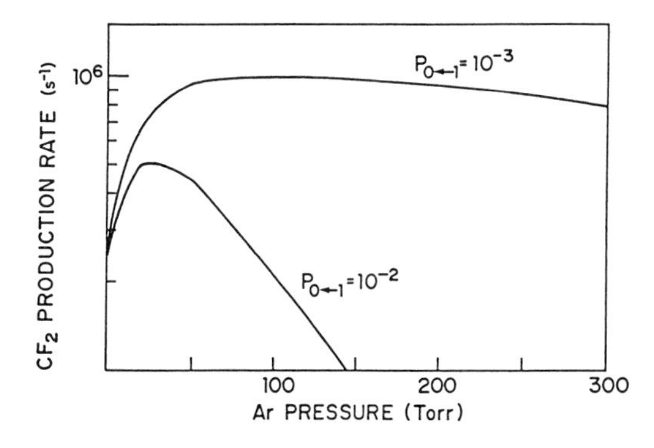

Fig.6.8. Theoretical curves showing a pressure falloff in rate due to deactivation of excited molecules by collisional V-T energy transfer. $P_{0\leftarrow 1}$ is the probability per collision of a V = 2 to V = 0 transition in the generalized Landau-Teller collision mechanism. Laser intensity is 15 MW cm^{-2}. Other parameters are the same as in Fig.6.4

The theoretical curves in Fig.6.8 show a downturn in production rate due to V-T collisional deactivation. As expected, the effect is more pronounced for the larger $P_{0\leftarrow1}$. For $P_{0\leftarrow1} < 10^{-4}$, the production rate curve is essentially identical over an Ar pressure range of 0-300 Torr to the theoretical curve calculated without collisional deactivation. At higher laser intensities the downturn in the production rate is, as one would expect, shifted to higher pressures. There is in fact just a hint of a downturn in the lower intensity experimental data for CF_2HCl decomposition plotted in Fig.6.4.

There is, however, no doubt about the downturn in production rate for the experimental data plotted in Fig.6.9. This data obtained by *King* and *Stephenson* [6.80] for the decomposition of $F_2C{=}CFCl$ shows an immediate decrease in production rate with increasing Ar pressure. *Quick* [6.81] has observed a similar drop-off in production rate for the decomposition of SF_6 buffered with He. There seems to be no reason to doubt that in both these cases the decrease in rate is caused by V-T collisional deactivation.

In the King and Stephenson rate data presented in Fig.6.9, there is no indication of an initial increase in rate at low pressures as was observed for CF_2HCl. In fact, in both those cases for which there is a very pronounced drop-off in rate with increasing noble gas pressure, there is no indication of increasing reaction rate or product yield at low noble gas pressures. Other reactions which show an increase in rate at low buffer gas pressure but no pronounced falloff in rate at higher pressures have been observed. They include the decompositions of $H_2C{=}CF_2$, $H_2C{=}CHF$ and $H_3C{-}CHF_2$ buffered with He [6.82]. We seem to be presented with the following two classes of reactions:

i) Reactions such as the SF_6 decomposition for which there is at most only a weak discrete level bottleneck and for which a pronounced downturn in reaction rate with increasing buffer gas pressure is observed.
ii) Reactions such as the CF_2HCl decomposition, for which there is evidence of a strong discrete level bottleneck but no noticeable downturn in reaction rate with increasing buffer gas pressure.

Since low laser intensity should favor the observation of both a discrete level bottleneck and a high pressure falloff, we expect that

iii) Reactions such as the theoretical decomposition plotted in Fig.6.8, for which there is evidence of a discrete level bottleneck and a pronounced downturn in rate with increasing buffer gas pressure, will be observed in experiments performed at relatively low laser intensity and long pulse times. The isomerization of cyclopropane [6.83] may be an example (see Fig.6.14).

The following argument may explain why class (iii) reactions are harder to observe. The high-pressure falloff is favored when the time required for excitation through the quasi continuum is long (e.g., reactions having high activation energy), while

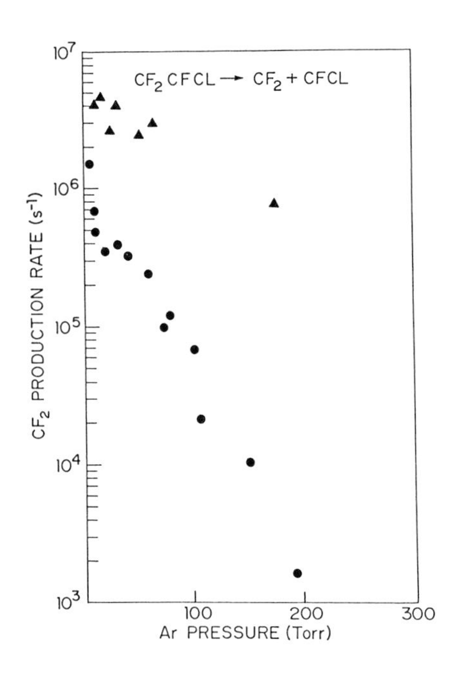

Fig.6.9. Experimental data points of King and Stephenson showing a pronounced pressure fall-off in rate for the laser-induced dissociation of CF_2CFCl, from [6.80]. Data points are given for two laser intensities: (▲▲▲), 101 MW cm^{-2}; (●●●), 57 MW cm^{-2}

discrete level bottlenecks cause the time for pumping through the discrete levels to be long. In order to observe a laser-induced reaction at all one must use a sufficiently high laser intensity so that the sum of these two times is not longer than the laser pulse time. It seems possible that class (iii) reactions have not been easily observed simply because inert buffer gas studies have to date all been carried out with pulsed CO_2 lasers whose pulse duration is not sufficiently long.

6.3.4 Laser-Induced Decomposition of SF_6

a) Discrete Level Spectrum and Tuning Curve

Over the past several years, a large experimental [6.69,77,84-95] and theoretical [6.40-44,54,54,65-67,96-100] effort has been devoted to a study of SF_6 decomposition in the presence of CO_2 laser light. In contrast with the paucity of spectroscopic data for CF_2HCl, the discrete energy level region in SF_6 has been subject to a close scrutiny, no doubt "fueled" by its applications to isotope separation.

Cantrell et al. [6.65,98,99] and *Ackerhalt* et al. [6.100] have postulated on the basis of *Hecht*'s spherical top Hamiltonian [6.101] a model for the discrete energy levels comprising the triply degenerate ν_3 mode in SF_6. Recently, *Horsley* et al. [6.68] have been investigating the resonant pathways through these levels as a first step in including them in a larger calculation embracing the full range of vibrational energies from ground state to dissociative states.

Recent spectroscopic studies [6.88,100] of the $3\nu_3$ overtone band in SF_6 have indicated that this molecule is best represented by a spherically symmetric Hamiltonian with different components of each overtone corresponding closely to different eigenvalues of the vibrational angular momentum quantum, ℓ. Further, the splitting between the two components of $3\nu_3$ ($\ell = 3$ and $\ell = 1$) is small (~ 2 cm^{-1}) compared to the anharmonic shift (~ 15 cm^{-1}). This means that the anharmonic splitting of the overtone levels is not particularly important in providing a resonant pathway to high vibrational levels, contrary to previous suggestions [6.98]. Indeed, *Ackerhalt* and *Galbraith* [6.66] have shown that the anharmonic splitting may be neglected in calculating the time-averaged populations of the vibrational levels up to $3\nu_3$, provided rotational structure is included. It is therefore possible to ignore the anharmonic splitting and represent all the components of a given overtone by a single degenerate level as in [6.40]. To the same degree of accuracy, the rotational structure of the low-lying vibrational levels is described with a single effective rotational constant.

The vibrational levels of the discrete region of the spectrum are always unequally spaced because of anharmonicity. *Ambartzumian* et al. [6.64,102] proposed that with a suitable sequence of vibrational-rotational transitions it should be possible to reduce the differences in level spacing, thereby minimizing the lack of resonance that accompanies pumping through a sequence of unequally spaced levels. This process, known as rotational compensation, is illustrated in Fig.6.10 by the P, Q, R sequence appropriate to depositing three photons in SF_6. More generally, for an excitation of three photons there are 27 different vibrational-rotational pathways, corresponding to the three choices $\Delta J = 0, \pm 1$ for each of the three vibrational-rotational transitions. In studying the relative efficiency of these various excitation pathways, *Horsley* et al. [6.68] found that for coherent pumping the P, Q, R pathway is sometimes optimal, but that more generally the optimal pathway is defined

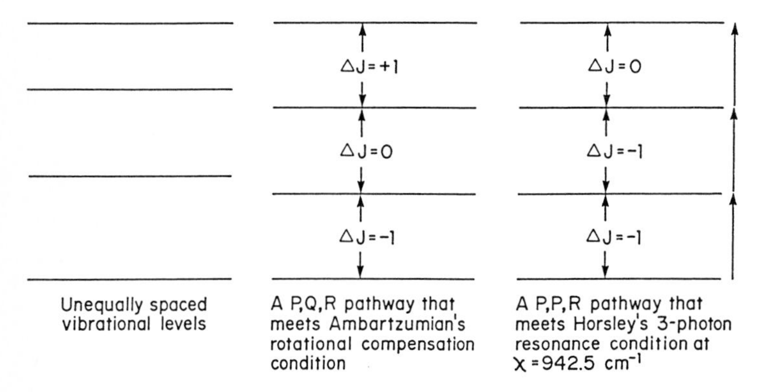

Fig.6.10. Schematic representation of two proposed schemes by which rotational structure helps to overcome restrictions in pumping efficiency caused by unequal spacing of vibrational levels. Small differences in the actual location of the rotational levels for the two schemes are ignored in the diagram

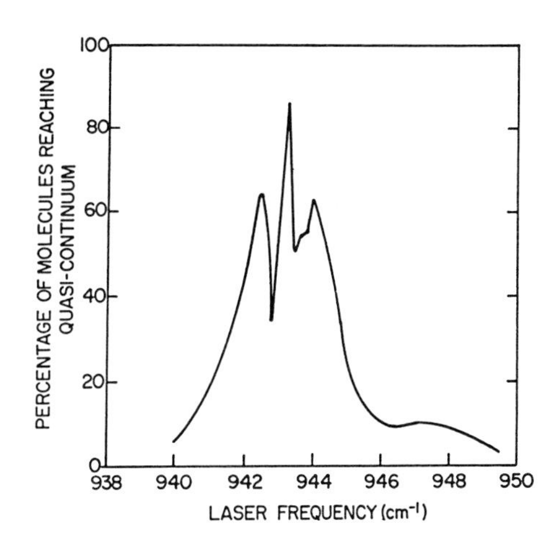

Fig.6.11. Predicted fraction of all SF_6 molecules in the rotational levels J = 1, 100 driven from the vibrational ground state into the quasi continuum during a 250 ns pulse, from [6.68]. Laser frequency was varied. Intensity = 4 MW cm^{-2}

by the requirement that the energy difference between initial and final vibrational-rotational states should be closest to exactly three laser quanta of energy. This three-photon resonance pathway is of course subject to the selection rule requirement that the magnitude of the overall ΔJ change for the three transitions be no more than 3. The P→P→R pathway, which *Horsley* et al. found to be an optimal three-photon pathway at laser frequency χ = 942.5 cm^{-1}, is illustrated in Fig.6.10.

Figure 6.11 shows the percentage of ground-vibrational-state molecules that reach the quasi continuum by the three-photon resonance pathway, calculated as a function of laser frequency at a laser intensity of 4 MW cm^{-2}. This curve, computed for a Boltzmann distribution of rotational states T = 198 K, is fairly broad with a structure obviously reminiscent of the P, Q, R branches of a rotational spectrum. The determination of which pathway best satisfies the three-photon resonance condition depends on both the laser frequency and the initial rotation state; nevertheless the usual situation is that at any given laser frequency one particular pathway applies for most of those molecules that are pumped up to the quasi continuum. For the "P-branch" peak at about 942.5 cm^{-1} over 85% of the excited population uses the P→P→R pathway; for the "Q-branch" peak at about 943 cm^{-1} over 95% of the excited population uses the P→Q→R pathway; for the "R-branch" peak at about 944 cm^{-1} over 90% of the excited population uses the P→R→R pathway. These three pathways are not by themselves sufficient to account for the width of the calculated curve. A large part of the width must be attributed to other pathways, which begin to contribute as the contribution from the above three pathways falls off. For example, at frequencies to the red of the "P-branch" peak the pathways P→P→Q and P→P→P start to contribute to some extent, although less efficiently than the three principal pathways.

The curve shown in Fig.6.11 can be compared with the curve of *Ambartzumian* et al. [6.102] showing the dependence of the dissociation yield on the frequency of the low-power laser when the SF_6 was cooled to 190 K. The experimental curve peaks at about 944 cm^{-1} and is about 6 cm^{-1} wide at half maximum. The calculated three-photon resonance curve agrees rather well with this curve, with a peak at 943.2 cm^{-1} and a width of $\sim$4 cm^{-1} at half maximum.

Figure 6.11 implies that at an intensity of 4 MW cm^{-2} there is no significant "bottleneck" in the pathway to the quasi continuum over a fairly wide frequency range. Between 941 and 945 cm^{-1} more than 30% of the molecules reach the quasi continuum at a given laser frequency. This agrees with the results of *Alimpiev* et al. [6.94] and *Akhmanov* et al. [6.95] who found that a large fraction of the rotational levels were depleted by pumping at moderate laser powers ($\sim$10 MW cm^{-2}). How does the relatively low-intensity laser pump such a large fraction of the molecules? Part of the answer lies in the large number of nearly resonant three-photon transitions that can be obtained in this frequency range. Although the single-photon transition from $v = 0$ to $v = 1$ for a given J state may be quite far off resonance, the three-photon transition for the same J state may be very close to resonance because the final state can be chosen to be any state within the range $J \pm 3$. There is a fairly high probability that for one of the seven states within this range the change in rotational energy will compensate for the anharmonic shift almost exactly. For example, for $J = 57$, pumping at 942 cm^{-1}, the single-photon (P-branch) transition is 2.88 cm^{-1} off resonance. However, the three-photon pathway is only 0.36 cm^{-1} off resonance.

b) Dissociation Yield and Cross Section

We have made preliminary calculations [6.68] for the laser-induced dissociation of SF_6 under collisionless conditions using the heat bath feedback equations to model pumping through the quasi continuum $v > 3$. In applying the specific rate expressions in (6.10) to SF_6, straightforward generalization of the techniques used to derive (6.1) to the case of a triply degenerate infrared-active mode results in the rate equations

$$
\begin{aligned}
T_{k-1 \leftarrow k} &= \sum_{j=1}^{k} R(k-j)_{j,j-1} \zeta_{k-j} / g_k \ , \\
T_{k \leftarrow k-1} &= \sum_{j=1}^{k} R(k-j)_{j,j-1} \zeta_{k-j} / g_{k-1} \ ,
\end{aligned}
\tag{6.30}
$$

where

$$
R(\ell)_{j,j-1} = \frac{\alpha_{01}^2 A^2}{2\hbar^2} \ \frac{(j+1)(j+2)}{2} \ \frac{(1/T_2)_{j,j-1;\ell}}{(1/T_2)^2_{j,j-1;\ell} + (\chi + \omega_{j-1,\ell} - \omega_{j,\ell})^2}
$$

and

$$g_k = \sum_{\ell=0}^{k} \zeta_{k-\ell}(\ell + 1)(\ell + 2)/2 \quad .$$

In evaluating the $T_{i\leftarrow j}$ of radiative transition rates, a level shift in the ν_3 absorption band caused by excitation in the heat bath modes, equal to 2 cm^{-1} per 1000 cm^{-1}, was assumed, in accord with the spectral measurements of *Nowak* and *Lyman* [6.103] and *Bott* [6.104].

To fit the observed decomposition threshold for SF_6, it was found that a value of H_{anh} large enough to be consistent with arriving in the quasi continuum without serious discrete level losses resulted in too much broadening high in the quasi continuum. In other words, there is no way, based on this model with constant interaction energy, to explain the observed pumping efficiencies for both discrete and dense energy regions. This conclusion was reached for level-dependent widths based on either the random walk model, where widths are proportional to $N_v^{\frac{1}{2}} H_{anh}$, or the *Bixon-Jortner* [6.71] model, where widths are proportional to $N_v \, H_{anh}$. This suggests either that the average interaction energy decreases with increasing vibrational excitation or, more likely, that a decreasing fraction of the total density of states is involved in direct strong coupling to a given state. A resolution of this question poses interesting theoretical problems, again pointing to the need, despite the dearth of reliable potential surface information, for a significant effort in this area to lay down principles for estimating spectroscopic bandwidths and relaxation rates. It was found that good agreement between theory and experiment results if one uses Bixon-Jortner widths and sets an upper limit to the bandwidth around 30 cm^{-1}. It would obviously be desirable, however, to predict such limitations a priori.

In Figs.6.12,13 we compare two model predictions with experiemnts. Figure 6.12 compares the predicted intensity dependence of decomposition (solid curve) with that measured by *Brunner* and *Proch* [6.89] in a collisionless molecular-beam experiment. We have also calculated the average absorbed energy as a function of laser fluence. Figure 6.13 compares the absorption data of *Fuss* and *Cotter* [6.23,105] with the model prediction (solid curve). Another model calculation which also includes discrete level pumping and careful interfacing of discrete levels with the quasi continuum has been carried out by *Ackerhalt* and *Galbraith* [6.52]. Their results show good agreement between theory and experiment.

6.3.5 Laser Frequency Effects

During the late 1950s the isomerization of cyclopropane to propylene played a key role in a dispute over how fast energy is scrambled among a molecule's vibrational degrees of freedom by intramolecular processes. The question was posed in specific terms as whether RRKM theory or Slater's normal mode theory better describes thermally driven unimolecular reactions [6.3,106]. More recently, the same reaction has

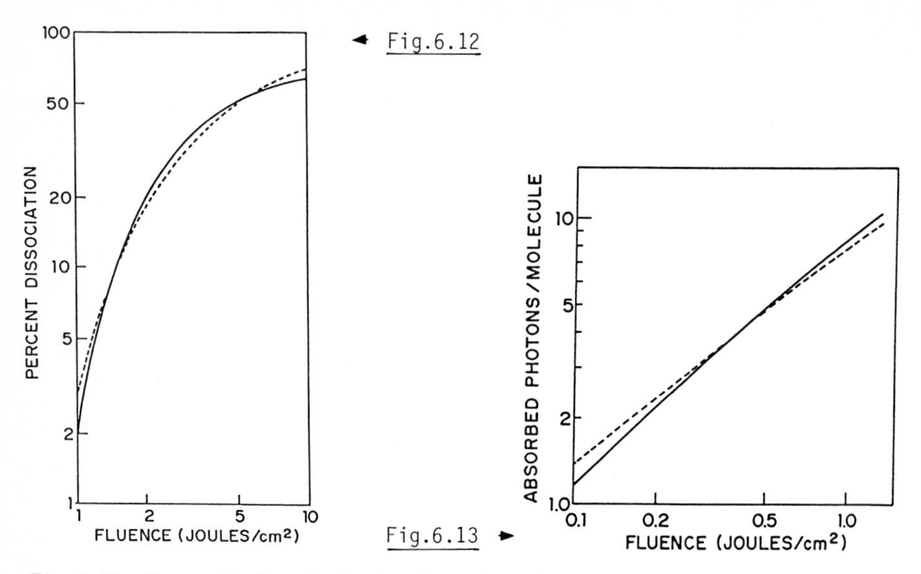

Fig.6.12. Theoretical calculation from [6.68] of decomposition yield (——; $H_{anh} = 0.04\ cm^{-1}$) as a function of laser energy fluence compared with the experimental data of *Brunner* and *Proch* (---) ([6.89] and private communication). Fluence was changed in the calculation by varying the pulse duration at a constant intensity of 20 MW cm^{-2}

Fig.6.13. Theoretical calculation from [6.68] of average number of absorbed photons (*solid curve*, $H_{anh} = 0.04\ cm^{-1}$) as a function of laser energy fluence compared with experimental measurements of *Fuss* (*dashed curve*, from [6.105]). Intensity was fixed at 20 MW cm^{-2} and the pulse duration varied in the calculation to achieve the indicated fluence values

been studied by laser excitation [6.107,108], and in particular *Hall* and *Kaldor* [6.83] have reported some interesting and provocative results for laser excitation at two different frequencies: $\chi = 1050\ cm^{-1}$ matching the vibrational frequency of a CH_2 wag and $\chi = 3100\ cm^{-1}$ matching the vibrational frequency of a C-H asymmetric stretch. Besides the isomerization reaction

$$\text{(cyclo-C}_3) \rightarrow C = C - C$$

for which the activation energy is accurately known, $E_a = 65.0$ kcal mol^{-1} [6.109], products of a fragmentation channel have also been observed in the laser-induced reaction. *Hall* and *Kaldor* [6.83] estimated the activation energy of the fragmentation channel at $E_a \cong 100$ kcal mol^{-1}.

It was found [6.83] that the yield of the low-energy isomerization channel relative to the yield of the high-energy fragmentation channels is quite dependent on the frequency of the exciting laser. At low pressures of neat cyclopropane, almost no fragmentation occurs for $\chi = 3100\ cm^{-1}$, whereas roughly equal yields of isomerization and fragmentation products result for $\chi = 1050\ cm^{-1}$. Also at $\chi = 3100\ cm^{-1}$, with the addition of a buffer gas, fragmentation products are observable,

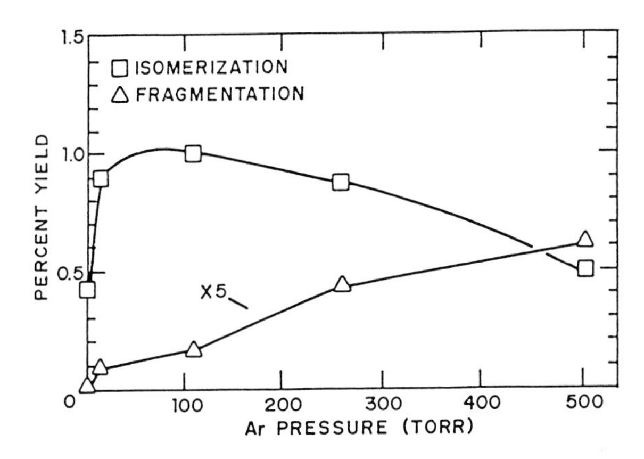

Fig.6.14. Hall and Kaldor's experimental determination of the product yield as a function of buffer gas pressure for two reaction channels in the laser-induced isomerization and dissociation of cyclopropane [6.83]. Percentage yield of major products results from 21,600 pulses, 5 mJ/pulse, of the 3.22 μm laser

becoming comparable in yield to the isomerization product at an Ar pressure of 500 Torr (Fig.6.14). Although as Hall and Kaldor themselves pointed out, the experiments to date do not provide unequivocal proof that the effect is not due to a secondary laser-induced reaction of propylene, the weight of their evidence suggests that this is a genuine selective effect in which the frequency of the exciting laser plays a key role. Whether or not the effect indicates a mode-selective excitation mechanism in which energy is not randomly distributed among vibrational modes seems to us at present to be an open question.

6.4 Effects Specific to Laser Excitation

Many of the questions one can ask about laser excitation are variants of one basic question: how does an ensemble of molecules prepared by laser excitation differ from an ensemble of molecules that are excited by other means and in particular by thermal excitation? Indeed, before the demonstration that laser energy can be fed directly into a molecule's internal degrees of freedom without appreciably heating its translational degrees of freedom, it seemed likely that laser excitation might be nothing more than a generalized heating effect.

Writing for the overall rate of a laser-induced reaction,

$$\text{Rate} = \sum_i N(E_i)k_i \quad , \tag{6.31}$$

where E_i is the energy and k_i the microscopic decay constant of the ith energy shell, one sees immediately the explicit dependence of rate on the energy distribution $N(E_i)$. Not so obvious is an implicit dependence of the k_i's on the distribution of isoenergetic microstates. In particular, if the laser excitation does not distribute microstates randomly on a given energy shell, then RRK or RRKM theory may or may not apply depending on the relative rates of intramolecular relaxation and chemical decay. Also if the shape of the energy distribution for a given average energy or, even better, of the distribution of isoenergetic microstates should prove to be dependent on the details of the excitation process — laser intensity, pulse shape, degree of coherence, or laser frequency — then the possibility of selective chemistry and the systematic probing of dynamical properties such as intramolecular relaxation constants lies ahead.

6.4.1 Energy Distribution: Laser Versus Thermal

We have already discussed in qualitative terms in Sect.6.3.2 how two effects, slow intramolecular relaxation and discrete level bottlenecking, may cause the shape of the energy distribution to depend on the details of the excitation process as well as on the average absorbed energy. In an interesting series of experiments *Bloembergen, Yablonovitch* and co-workers [6.51,55,69] used an optoacoustic technique to determine the product yield for the SF_6 decomposition as a function of average absorbed energy. Figure 6.15 shows the results [6.51] for two laser pulses of widely differing intensities. There is a pronounced laser selective effect at the lower absorbed energies. *Black* et al. [6.51] interpreted these results in terms of a discrete level bottleneck effect that is more pronounced for the lower intensity pulse. Those molecules that make it to the high-energy tail of the distribution and then react are not held back, while some fraction of molecules is retarded by the bottleneck thereby reducing the average energy deposited relative to the population in the tail. A yield curve calculated on the basis of RRKM theory and a thermal distribution is also shown in Fig.6.15. These authors [6.55,110] had previously suggested the possible applicability of a thermal distribution.

A related question of equal interest which has been discussed quantitatively [6.51,111] concerns the nature of the energy distribution under conditions where the details of the excitation process *do* wash out. In other words, when the simple rate equations

$$\frac{dN_j}{dt} = T_{j\leftarrow j-1}N_{j-1} + T_{j\leftarrow j+1}N_{j+1} - T_{j-1\leftarrow j}N_j - T_{j+1\leftarrow j}N_j \tag{6.1}$$

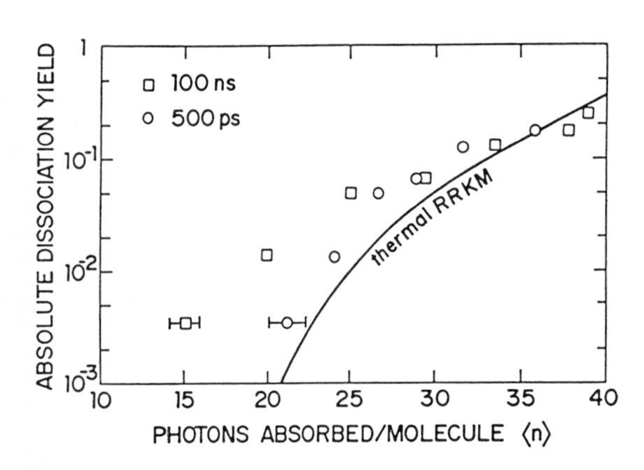

Fig.6.15. Black et al. experimental determination of the dissociation yield per pulse of SF_6 as a function of mean energy of excitation per pulse, $\langle n\rangle\hbar\omega$ [6.51]. This graph gives information relating to the shape of the energy distribution function produced by multiple-photon absorption. The experimental points for two pulse durations are compared with a theoretical curve (thermal RRKM) based on the assumption of a thermal equilibrium distribution of the same mean energy

apply, what is the energy distribution and when might it be a thermal distribution? The possibility of a thermal distribution is supported by the recent interesting result that the solution to these rate equations for an s-fold degenerate harmonic oscillator (used to represent the quasi continuum) gives a time-dependent thermal energy distribution when the absorption cross section is *constant* [6.51], i.e., independent of the degree of vibrational excitation of the molecule. *Anderson* et al. [6.112] determined for a general class of master equations, including (6.1) as a special case, necessary and sufficient conditions for the existence of a thermal solution. We here summarize a simpler derivation [6.111] which applies specifically to rate equations for radiative pumping, and show that the necessary conditions for a thermal solution are quite restrictive.

The formal expressions of the heat bath feedback model for the $T_{j\leftarrow i}$ satisfy the microscopic reversibility condition (6.11), but one can take a somewhat more general approach and define effective degeneracy factors

$$g_0^e = 1 \quad ,$$
$$g_j^e = \frac{T_{j\leftarrow j-1}}{T_{j-1\leftarrow j}} g_{j-1}^e \quad . \tag{6.32}$$

It follows [6.111] that any thermal solution to (6.1) must have the form

$$N_j = g_j^e \, e^{-j\theta(t)} \Big/ \sum_{j=0}^{\infty} g_j^e \, e^{-j\theta(t)} \quad , \qquad j = 0 \ , \ \infty \quad , \tag{6.33}$$

where $\theta(t) = \hbar\omega/kT(t)$, depending on a time-dependent temperature, is determined from the average absorbed energy. Substituting (6.33) into (6.1), one finds after some manipulation — the details are given in [6.111] — that for a thermal solution to be possible, the radiative transition rates must all be related to $T_{1\leftarrow 0}$ by

$$T_{j-1\leftarrow j} = cj \quad ,$$
$$T_{j+1\leftarrow j} = c'j + T_{1\leftarrow 0} \quad , \tag{6.34}$$

where c and c' are arbitrary constants. To show that a thermal solution does exist when the $T_{j\leftarrow i}$'s are restricted in this way, we substitute (6.34) into (6.1) and rewrite the resulting equation in the form

$$\frac{s}{T_{1\leftarrow 0}} e^{-\theta_B} \frac{dN_j}{dt} = (j - 1 + s)\, e^{-\theta_B} N_{j-1} + (j + 1) N_{j+1}$$
$$- [j + e^{-\theta_B}(j + s)] N_j \quad , \tag{6.35}$$

where θ_B and s are defined by

$$\frac{c'}{c} = e^{-\theta_B} \quad , \quad \frac{T_{1\leftarrow 0}}{c} = s\, e^{-\theta_B} \quad . \tag{6.36}$$

Equation (6.35) can be interpreted as the rate equation describing the relaxation (by a Landau-Teller collision mechanism) of an s-fold degenerate oscillator in contact with a heat bath of temperature T_B, where $\theta_B = \hbar\omega/kT_B$. In this context the thermal solution for (6.35) was first given by *Wilson* [6.113].

Using (6.32,34,36) one finds

$$\frac{g_{j+1}^e}{g_j^e} = e^{-\theta_B} \frac{(j + s)}{(j + 1)} \tag{6.37}$$

which implies that

$$g_j^e = e^{-j\theta_B} g_j(s) \quad , \tag{6.38}$$

where the

$$g_j(s) = \frac{(s + j - 1)!}{j!(s - 1)!} \tag{6.39}$$

are, for integer s, the degeneracy factors for an s-fold degenerate oscillator.

From (6.34,36) one easily derives

$$\sigma_j = T_{j+1\leftarrow j} - T_{j-1\leftarrow j} = \frac{T_{1\leftarrow 0}}{s} [s + j(1 - e^{\theta_B})] \quad . \tag{6.40}$$

Equations (6.37,40) express necessary and sufficient conditions for a thermal solution in terms of the physically meaningful effective degeneracy factors g_j^e and microscopic cross sections σ_j. The microscopic cross sections derive their importance from the following relationship for the average absorbed energy per molecule:

$$\frac{1}{\hbar\omega} \frac{d\bar{E}}{dt} = \sum_{j=0}^{\infty} \sigma_j N_j \quad . \tag{6.41}$$

When the σ_j are linear in j, as in (6.40), one computes

$$\frac{1}{\hbar\omega} \frac{d\bar{E}}{dt} = T_{1\leftarrow 0} + \frac{T_{1\leftarrow 0}(1 - e^{\theta_B})}{s\hbar\nu} \bar{E} \quad . \tag{6.42}$$

Therefore the absorption cross section

$$\sigma = (\text{Intensity})^{-1} \frac{dE}{dt}$$

is a linear function of absorbed energy.

The case $\theta_B = 0$ is of special interest. For $\theta_B = 0$ the cross section is constant and the effective degeneracy factors are (for integer s) exactly equal to those of the s-fold oscillator (6.39). This is the case for which *Black* et al. [6.51] recovered the thermal distribution [6.113]. Since it seems more plausible that the

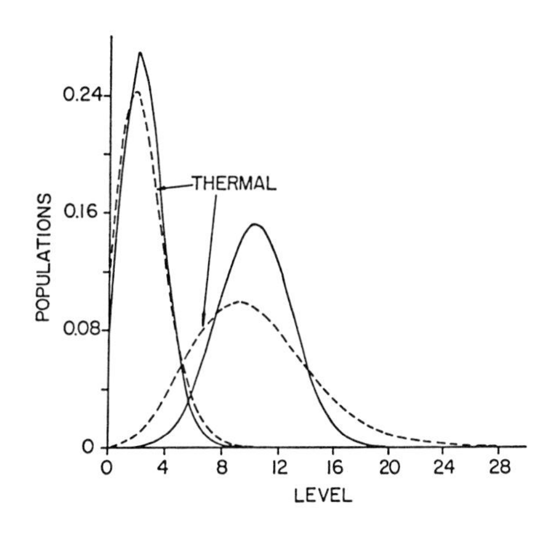

Fig.6.16. Theoretical calculated deviations from thermal behavior [6.111]. Solutions to (6.1) (——) computed with effective degeneracy factors given by those of a 15-fold degenerate oscillator and $\sigma_j = \sigma_0 \exp(-0.154\, j)$ are plotted for two times ($t = 3/\sigma_0$ and $t = 24/\sigma_0$). The thermal distributions (---) have the same average energy

effective degeneracy factors be approximately given by those of an s-fold oscillator than that the cross section be constant (for SF_6 a decrease in cross section of more than one order of magnitude has been observed) we have numerically integrated (6.1) using a cross section which decreases exponentially with energy level. A not atypical case, in which significant departures from the thermal distribution occur, is illustrated in Fig.6.16. The only other possibilities are $\theta_B > 0$, $\theta_B < 0$, and a particular finite level case first discovered by *Anderson* et al. [6.112].

We have also carried out some numerical calculations using a *Whitten-Rabinovitch*-type density-of-states function [6.114] to estimate effective degeneracy factors. In these calculations the cross section was assumed constant so that any departures from the thermal distribution are traceable to the choice of effective degeneracy factors. We examined first the effect of allowing a realistic spread in molecular frequencies and found, somewhat to our surprise, that this refinement had almost no effect on the distribution. A typical result is illustrated by the more sharply peaked curves of Fig.6.17. Another degeneracy factor effect does, however, turn out to be quite important. This second effect arises from the fact that in writing down

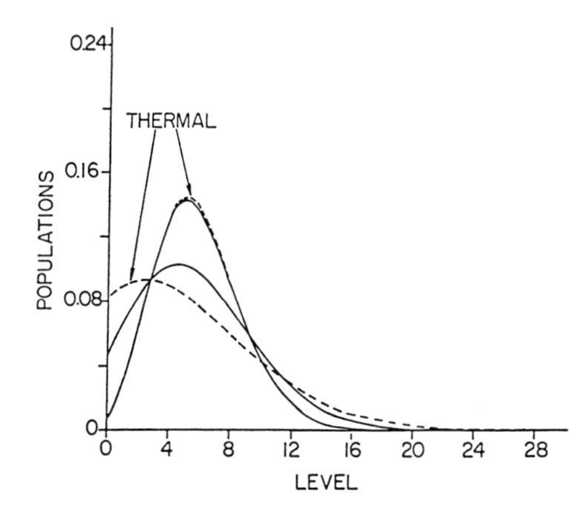

Fig.6.17. Theoretical calculation of effects caused by degeneracy factors that differ from those of the s-fold degeneracy oscillator from [6.111]. Solutions to (6.1), *solid curves*, were computed using a constant cross section and degeneracy factors derived from a density-of-states expression [6.9] in which a spread in vibrational frequencies such that $\langle\omega^2\rangle/\langle\omega\rangle^2 = 1.5$ was assumed. Only in the broad curve, showing a large deviation from thermal behavior, is the laser frequency different from the mean molecular frequency; there, $\chi = \langle\omega\rangle/4$. In both curves $t = 6/\sigma_0$

the rate equation, given by (6.1), one is explicitly choosing a spacing between equivalent levels equal to one quanta of energy at the laser frequency. When the laser frequency departs significantly from the mean molecular vibrational frequency, the effective degeneracy factors may also depart significantly from those giving rise to a thermal distribution. This effect is also illustrated in Fig.6.17.

In summary, since a thermal distribution can apply *only* if both of the conditions given in (6.37,40) hold, and since there is no a priori physical reason for either condition to be true, we regard a thermal distribution of energy to be the exception rather than the rule.

6.4.2 Microstate Formalism

a) Intramolecular Relaxation

It is possible to clarify several questions concerning intramolecular relaxation and the randomness or nonrandomness of the distribution of isoenergetic microstates in a simple formalism using only rate equations. The relevant equation is [6.115]

$$\left(\frac{dP(\mathbf{n},t)}{dt}\right)_{\text{intramolecular}} = \frac{1}{\tau_m}\sum_{\mathbf{n}'} L(\mathbf{n},\mathbf{n}')P(\mathbf{n}',t) - \frac{1}{\tau_m}P(\mathbf{n},t) \quad . \tag{6.43}$$

Here $\mathbf{n} = \{n_1, n_2 \ldots n_s\}$ is a set of approximate quantum numbers that specify the vibrational state of a polyatomic with s degrees of freedom, and the quantity $P(\mathbf{n},t)$ is the probability that at time t, a given polyatomic is in the microstate $\mathbf{n}$. Equation (6.43) describes the time evolution of $P(\mathbf{n},t)$ when intramolecular processes are scrambling vibrational energy among various degrees of freedom. The matrix $L(\mathbf{n},\mathbf{n}')$ gives the probability of transition from microstate $\mathbf{n}'$ to microstate $\mathbf{n}$ upon occurrence of some energy-scrambling event that is assumed, by analogy with collisional relaxation, to occur with frequency $1/\tau_m$ where τ_m is a mean time between intramolecular "collision" events. The probability interpretation of $L(\mathbf{n},\mathbf{n}')$ implies that

$$\sum_{\mathbf{n}} L(\mathbf{n},\mathbf{n}') = 1 \quad . \tag{6.44a}$$

With any approximate description of the energy state, the total energy is not precisely defined. We assume, however, that the energy definition is sufficiently precise to allow one to assign any microstate $\mathbf{n}$ to some energy shell E_n of small but finite width. Conservation of energy requires that only transitions between microstates belonging to the same energy shell be allowed. Equation (6.43) is therefore a closed equation describing the time evolution of just those microstates that belong to a given energy shell. The microcanonical or random distribution

$$P_{\text{micro}}(\mathbf{n}) = \frac{1}{g(E_n)} \tag{6.45}$$

plays a special role. Here $g(E_n)$ is the number of microstates belonging to the energy shell E_n. The "microscopic reversibility" condition

$$L(\mathbf{n},\mathbf{n}') = L(\mathbf{n}',\mathbf{n}) \tag{6.46a}$$

guarantees that the microcanonical distribution is a solution to (6.43).

Interestingly enough there exists for this rate equation formalism a very close analogue to the assumption of ergodicity on the energy shell as it is discussed in statistical mechanics. Specifically, if one makes the *ergodic assumption* that every pair of microstates $\mathbf{n}$ and $\mathbf{n}'$ are connected by at least one allowed sequence of intramolecular transitions, $\mathbf{n} \to \mathbf{n}_1 \to \mathbf{n}_2 \ldots \to \mathbf{n}'$, then it follows that the general solution to (6.43) for microstates belonging to energy shell E_n can be written in the form

$$P(\mathbf{n},t) = \frac{1}{g(E_n)} + Q(\mathbf{n},t) \quad , \qquad \text{where} \tag{6.47}$$

$$\sum_{\mathbf{n}} Q(\mathbf{n},t) = 0 \qquad \text{and} \tag{6.48}$$

$$Q(\mathbf{n},t) \underset{t\to\infty}{=} 0 \quad . \tag{6.49}$$

The function $Q(\mathbf{n},t)$ describing the relaxation of any differences from the microcanonical distribution that exist at time $t=0$ is itself a solution to (6.43). The naming of the ergodic assumption is justified by the fact that the above solution approaches the microcanonical distribution for arbitrary initial conditions. It is known [6.3,106] that besides the assumption of ergodicity, a second fundamental assumption, called the random gap or random lifetime assumption, is built into the statistical (RRK and RRKM) theories of unimolecular decay. As we shall demonstrate in the next section, an analogue of the random lifetime assumption plays a crucial role in determining whether or not the preparation of a nonrandom distribution of microstates by laser excitation will lead to laser selectivity.

b) Relation to RRK and RRKM Theory

It is easy to include the possibility of unimolecular decay from excited vibrational states in the rate equation formalism. One considers the microstates divided into two classes, reactant states and product states, and then allows the intramolecular relaxation matrix to induce transitions from reactant states to product states but not from product states back to reactant states. In other words, we assume

$$L(\mathbf{n},\mathbf{n}') = 0, \ \mathbf{n}\varepsilon R \qquad \text{and} \qquad \mathbf{n}'\varepsilon P \quad , \tag{6.50}$$

where the notation $\mathbf{n}\,\varepsilon\,R$ is read "for microstate $\mathbf{n}$ belonging to class R," R being the class of reactant states and P the class of product states. Retaining the interpretation of the $L(\mathbf{n},\mathbf{n}')$ as conditional probabilities, one has

$$\sum_{\mathbf{n}'} L(\mathbf{n},\mathbf{n}') = \sum_{\mathbf{n}'\varepsilon R} L(\mathbf{n}',\mathbf{n}') + \sum_{\mathbf{n}'\varepsilon P} L(\mathbf{n}',\mathbf{n}) = 1 \quad , \quad \mathbf{n}\varepsilon R \quad . \tag{6.44b}$$

We also assume microscopic reversibility for transitions between reactant states, i.e.,

$$L(\mathbf{n},\mathbf{n}') = L(\mathbf{n}',\mathbf{n}) \quad , \quad \mathbf{n}\varepsilon R \quad , \quad \mathbf{n}'\varepsilon R \quad . \tag{6.46b}$$

The rate equation for intramolecular relaxation and decay on a given energy shell is

$$\left.\frac{dP(\mathbf{n},t)}{dt}\right|_{\substack{\text{intramolecular}\\ \text{decay}}} = \frac{1}{\tau_m} \sum_{\mathbf{n}'\varepsilon R} L(\mathbf{n},\mathbf{n}')P(\mathbf{n}',t) - \frac{1}{\tau_m} \sum_{\mathbf{n}'\varepsilon R} L(\mathbf{n}',\mathbf{n})\, P(\mathbf{n}',t)$$
$$- \frac{1}{\tau_m} \sum_{\mathbf{n}'\varepsilon P} L(\mathbf{n}',\mathbf{n})\, p(\mathbf{n},t) \quad , \quad \mathbf{n}\varepsilon R \quad . \tag{6.51}$$

The first term on the right-hand side of (6.51) accounts for all transitions into state **n** from other reactant states, the second term accounts for all transitions out of state **n** into other reactant states, and the third term accounts for all transitions out of state **n** into product states. Of course, one can use (6.44b) to simplify (6.51) so that it looks exactly like (6.43); however, in (6.51) as it stands, one readily identifies the first two terms as intramolecular relaxation among reactant states and the third term as a unimolecular decay out of the reactant states. To verify this formally, one computes from (6.51)

$$\frac{d}{dt} \sum_{\mathbf{n}\varepsilon R} P(\mathbf{n},t) = - \sum_{\mathbf{n}\varepsilon R} k(\mathbf{n})P(\mathbf{n},t) \quad , \tag{6.52}$$

where the microstate decay constants are given by

$$k(\mathbf{n}) = \frac{1}{\tau_m} \sum_{\mathbf{n}'\varepsilon P} L(\mathbf{n}',\mathbf{n}) \quad , \quad \mathbf{n}\varepsilon R \quad . \tag{6.53}$$

The sum in (6.53) is exactly the probability that a molecule in state **n** will, upon occurrence of an intramolecular "collision", make a transition into any one of the product states. Equation (6.53) is therefore nothing but the basic hypothesis of *Kassel* [6.116], which is an integral part of the present formalism because of the way intramolecular relaxation and unimolecular decay have been linked together as basically one process.

We are now in a position to distinguish between two types of decay. We call the decay mechanism *nondiscriminating* whenever the decay constants defined by (6.53) are the same for all states belonging to the same energy shell, i.e.,

$$k(\mathbf{n}) = k(E_n) \quad , \quad \mathbf{n}\varepsilon E_n \quad . \tag{6.54}$$

We call the decay mechanism *discriminating* whenever the decay constants defined by (6.53) depend on something more than the total vibrational energy. Clearly the question of whether or not the decay mechanism is discriminatory is a central one, since one can have *no laser-specific effects* resulting from a laser excitation of a nonrandom distribution of microstates when the *decay mechanism is nondiscriminating*.

An example of a nondiscriminating mechanism is furnished by the *strong intramolecular relaxation* assumption

$$L(\mathbf{n},\mathbf{n}') = \frac{1}{g(E_n)} \quad , \qquad \mathbf{n}\varepsilon R \quad , \tag{6.55}$$

so named because of the obvious analogy with the better known strong collision assumption [6.117]. With the strong intramolecular assumption one easily computes, from (6.53), the decay constant from a given energy shell,

$$k(E_n) = \frac{1}{\tau_m}\frac{g^p(E_n)}{g(E_n)} \quad , \tag{6.56}$$

where $g^p(E_n)$ is the total number of product states belonging to the energy shell E_n. The relation of (6.56) to RRK and RRKM theory is discussed in [6.115].

We are not suggesting that the strong intramolecular relaxation assumption as given explicitly by (6.55) is a necessary condition for RRK or RRKM theory. It is known, however, that the assumption of a random lifetime distribution is necessary to these theories. In our rate equation formalism the analogue of the random lifetime assumption is the assumption that if one initially has a random or microcanonical distribution of microstates, then the total population of reactant states decays with a single decay constant. In other words, if

$$P(\mathbf{n},t=0) = P_{micro} \quad , \tag{6.57}$$

then

$$\sum_{\mathbf{n}\varepsilon R} P(\mathbf{n},t) = \sum_{\mathbf{n}\varepsilon R} P(\mathbf{n},t=0)\, e^{-\lambda t} \quad . \tag{6.58}$$

Since for a *nondiscriminating mechanism* (6.52) can be immediately integrated to give (6.58) *whatever the initial distribution*, it is obvious that the assumption of a nondiscriminating mechanism implies random lifetimes. The converse, that the assumption of random lifetimes implies a nondiscriminating decay mechanisms at least within the rate equation formalism can also be proven [6.115]. It follows that a basic aspect of RRK and RRKM theory is involved in the question of whether or not a nonrandom distribution of microstates on the energy shell can affect the rate. In the next section we present a simple model, consistent with well-established results from thermal experiments, that does allow laser-specific effects via a discriminating reaction mechanism.

c) A Discriminating Reaction Mechanism

We have in mind a modification of the Kassel model in which the s degrees of freedom are divided into two groups, "A" and "B", of s_1 and s_2 degrees of freedom each,

$s = s_1 + s_2$, in such a way that energy scrambling is rapid within either group, but much slower when one or more quanta of energy must be exchanged between the two groups. For an s-fold-degenerate oscillator with a total energy of n quanta, there are $n + 1$ ways of partitioning the energy with n_1 quanta belonging to the group "A" oscillators and $n - n_1$ quanta belonging to the group "B" oscillators. The intramolecular transition probabilities on a given energy shell allow rapid transitions between each of the

$$g_{n_1}(E_n) = \frac{(n_1 + s_1 - 1)!}{n_1!(s_1 - 1)!} \frac{(n - n_1 + s_2 - 1)!}{(n - n_1)!(s_2 - 1)!} \quad , \qquad n_1 = 0,n \tag{6.59}$$

microstates belonging to a given partition of the total energy, and slower transitions between any pair of microstates belonging to different partitions of the total energy. Retaining the Kassel criterion that reaction occurs whenever m or more quanta accumulate in one particular "breakable" oscillator leads quite naturally to a discriminating reaction mechanism. Assuming that the "breakable" oscillator belongs to the group of s_1 oscillators, there are

$$g^p_{n_1}(E_n) = \frac{(n_1 - m + s_1 - 1)!}{(n_1 - m)!(s_1 - 1)!} \frac{(n - n_1 + s_2 - 1)!}{(n - n_1)!(s_2 - 1)!} \quad , \qquad m \leqslant n_1 \leqslant n \tag{6.60}$$

product states for each partition of the total energy in which m or more quanta belong to the group of s_1 oscillators.

We can now define a set of intramolecular transition probabilities that corresponds in a general way to exactly that type of restriction in the flow of energy between degrees of freedom we have been describing for the Kassel model. Suppose that states belonging to a given energy are divided into subsets of states such that intramolecular relaxation is rapid among states belonging to the same subset, but slower between states belonging to different subsets. Let the transition probabilities on a given energy shell be

$$L(\mathbf{n},\mathbf{n}') = \frac{1 - \alpha(E_n)}{g_I(E_n)} + \frac{\alpha(E_n)}{g(E_n)} \quad , \qquad \mathbf{n} \varepsilon I \quad , \qquad \mathbf{n}' \varepsilon I \quad ,$$

$$L(\mathbf{n},\mathbf{n}') = \frac{\alpha(E_n)}{g(E_n)} \quad , \qquad \mathbf{n} \varepsilon I \quad , \qquad \mathbf{n}' \varepsilon I' \quad , \tag{6.61}$$

$$0 \leqslant \alpha(E_n) \leqslant 1 \quad ,$$

where the index I labels the different subsets, each containing $g_I(E_n)$ states. The entire set of probabilities defines what can be called the *restricted intramolecular relaxation* process. Notice that for $\alpha(E_n) = 1$, these probabilities define the strong intramolecular relaxation process of (6.55), while for $\alpha(E_n) = 0$ they correspond to strong intramolecular relaxation within each of the subsets and no intramolecular relaxation between subsets. For all values of $\alpha(E)$,

$$\Sigma L(\mathbf{n},\mathbf{n}') = 1 \quad ,$$

allowing the $L(\mathbf{n},\mathbf{n}')$ to be interpreted as conditional probabilities. The intramolecular-relaxation-only problem defined by (6.43) can be solved exactly for the restricted intramolecular relaxation process. The solution defines two time scales τ_m and $\tau_m/\alpha(E_n)$. Deviations from a uniform distribution of states within any subset I disappear at a rate $1/\tau_m$, while deviations of the total subset populations

$$\sum_{\mathbf{n}\varepsilon I} P(\mathbf{n},t)$$

from their microcanonical ensemble values, $g_I(E_n)/g(E_n)$, disappear at the potentially much slower rate $\alpha(E_n)/\tau_m$.

When unimolecular decay is included in the restricted intramolecular relaxation mechanism, it is convenient to let I^R denote the set of $g_I^R(E_n)$ reactant states belonging to I, and to let I^P denote the set of $g_I^P(E_n)$ product states belonging to I. For restricted intramolecular relaxation, (6.51) takes the form

$$\frac{dP(\mathbf{n},t)}{dt} = \frac{1-\alpha(E_n)}{\tau_m}\frac{1}{g_I(E_n)}\sum_{\mathbf{n}'\varepsilon I^R} P(\mathbf{n}',t)$$

$$+\frac{\alpha(E_n)}{\tau_m}\frac{1}{g(E_n)}\sum_{\mathbf{n}'\varepsilon R} P(\mathbf{n}',t) - \frac{1}{\tau_m}P(\mathbf{n},t) \quad , \qquad \mathbf{n}\varepsilon I^R \quad . \tag{6.62}$$

Equation (6.62) can be solved by a Laplace transformation technique [6.115], but here we want only to summarize two points made in [6.115]:

1) If one limits the discussion to the modified Kassel model discussed at the beginning of this section, then the relevant statistical factors $g_I^R(E_n)$ and $g_I^P(E_n)$ are completely specified in terms of s_1, s_2, and m.

2) If one includes the solutions to (6.62) in a strong collision theory of thermally driven unimolecular decay, then the slower intramolecular relaxation process occurring at a rate α/τ_m may or may not have an observable effect on the thermal rate constant. When α/τ_m is fast compared to the collision frequency at the pressure where the rate constant begins to fall off from its high-pressure value there is no observable effect. This is important since several studies [6.118,119] made specifically for the purpose of looking for signs of restricted intramolecular relaxation in the rate constant fall off curve have turned up only negative results. Typically, $1/\tau_m$ is of the order of the high-pressure preexponential factors, i.e., $10^{13}s^{-1}$. But α/τ_m may be much smaller, 10^8-$10^{10}s^{-1}$ for moderately complex molecules, and still meet the criterion of not noticeably altering the shape of the fall off curve for the thermal rate constant.

Equation (6.62) can be used to model the case in which radiative pumping is rapid enough to create some distribution of microstates on a highly excited energy shell, which then decays after the laser pulse is over. To describe the more general case where radiative pumping and unimolecular decay compete on the same time scale one must include radiative transition terms in the rate equations for the time evolution of the microstates. For the restricted intramolecular relaxation process, one can take advantage of the fact that any nonrandomness in the distribution of microstates belonging to a subset I of the given energy shell is rapidly washed out on a timescale τ_m, and simplify the rate equations in much the same way that (6.1) for the heat bath feedback model is derived from the more detailed (6.3). The resultant equations have the form [6.115]

$$\frac{dP_I(E_n,t)}{dt} = \left.\frac{dP_I(E_n,t)}{dt}\right|_{\text{radiative}} + \left.\frac{dP_I(E_n,t)}{dt}\right|_{\substack{\text{intramolecular}\\ \text{relaxation}}} + \left.\frac{dP_I(E_n,t)}{dt}\right|_{\text{decay}} \tag{6.63}$$

where

$$\left.\frac{dP_I(E_n,t)}{dt}\right|_{\text{radiative}} = T_{I\leftarrow I-1}P_{I-1}(E_{n-1},t) + T_{I\leftarrow I+1}P_{I+1}(E_{n+1},t) - (T_{I-1\leftarrow I} + T_{I+1\leftarrow I})P_I(E_n,t) \quad , \tag{6.64}$$

$$\left.\frac{dP_I(E_n,t)}{dt}\right|_{\substack{\text{intramolecular}\\ \text{relaxation}}} = \frac{\alpha(E_n)}{\tau_m}\frac{g_I^R(E_n)}{g(E_n)}\sum_I P_I(E_n,t) - \frac{\alpha(E_n)}{\tau_m}\frac{g^R(E_n)}{g(E_n)}P_I(E_n,t) \quad , \tag{6.65}$$

$$\left.\frac{dP_I(E_n,t)}{dt}\right|_{\text{decay}} = -\frac{1-\alpha(E_n)}{\tau_m}\frac{g_I^P(E_n)}{g_I(E_n)}P_I(E_n,t) - \frac{\alpha(E_n)}{\tau_m}\frac{g^P(E_n)}{g(E_n)}P_I(E_n,t) \quad , \tag{6.66}$$

where $P_I(E_n,t)$ is the total population of reactant states belonging to subset I on energy shell E_n, i.e.,

$$P_I(E_n,t) = \sum_{\substack{\mathbf{n}\varepsilon I^R \\ \mathbf{n}\varepsilon E_n}} P(\mathbf{n},t) \quad . \tag{6.67}$$

In writing the radiative transition term in the form given in (6.64) one is assuming that I equals the number of quanta in that group (group A) of modes which are connected to the pump mode by rapid intramolecular relaxation processes. If $\hbar\chi$ is one quanta of laser energy, then

$$E_n = n\hbar\chi = I\hbar\chi + (n - I)\hbar\chi$$

where there are I quanta in the group A modes and n-I quanta in the group B modes. Furthermore, if there are two chemical reaction channels, then one can model the selectivity process by allowing group A modes to decay into one product channel and group B modes to decay into another product channel. The overall model for product selectivity is summarized in Fig.6.18.

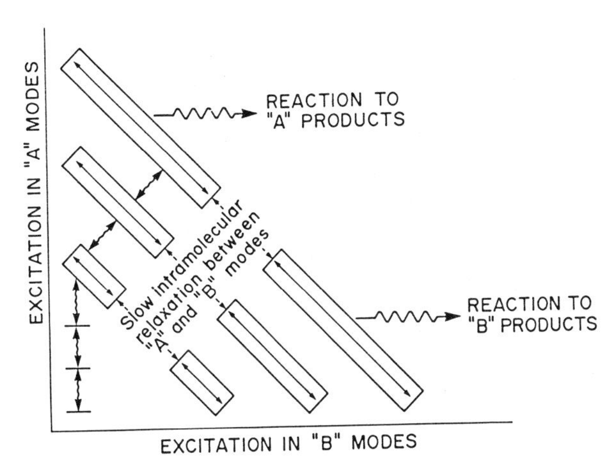

Fig.6.18. Schematic representation of the discriminating reaction mechanism discussed in the text that can lead to product selectivity for laser-induced two-channel reactions. Intramolecular relaxation, in which one or more quanta of vibrational energy are exchanged between group "A" modes and group "B" modes, is assumed slow

If one assumes the Kassel model to compute the statistical factor, then only $\alpha(E_n)$ needs specifying. Taking $\alpha(E_n)$ to be energy independent gives what is basically a one-adjustable-parameter theory of product selectivity. The first term on the right-hand side of (6.66) accounts for a selective decay into just those product states belonging to subset I, while the second term accounts for a nonselective decay into any of the product states. Similarly, when one computes the rate of production of product states belonging to a given subset,

$$\text{Rate into states } I^P = \frac{[1 - \alpha(E_n)]}{\tau_m} \frac{g_I^P(E_n)}{g_I(E_n)} P_I(E_n,t)$$

$$+ \frac{\alpha(E_n)}{\tau_m} \frac{g_I^P(E_n)}{g(E_n)} \sum_I P_I(E_n,t) \quad ,$$

one finds a selective term proportional to $1-\alpha(E_n)$, and a nonselective term proportional to $\alpha(E_n)$.

Clearly, for $\alpha(E_n)=1$ no selectivity is possible, while for $\alpha(E_n)=0$ perfect selectivity is possible. More generally, product selectivity requires that the selective decay rate

$$\frac{[1-\alpha(E_n)]}{\tau_m}\frac{g_I^P(E_n)}{g_I(E_n)}$$

compete effectively with the slow intramolecular relaxation rate $\alpha(E_n)/\tau_m$, since if a microcanonical distribution

$$P_I(E_n,t) \sim g_I(E_n)$$

is established before appreciable reaction can occur then no selectivity will occur. The results mentioned in the previous section show that a relatively small α is, for moderately complex molecules, not inconsistent with the pressure falloff data for thermally driven reactions. Those results do not, however, prove that slow intramolecular relaxation does actually occur. In fact other experiments, such as the careful studies by *Rabinovitch* and coworkers [6.118,120-124] of the reaction of "chemically activated" species, have failed to turn up any evidence of slow intramolecular relaxation. The agreement of the results of *Lee* and coworkers [6.85,125] in laser excitation beam experiments with RRKM theory predictions can also be cited as evidence for no departure from a random distribution of microstates on the energy shell. However, in our opinion, the kind of blanket statement to the effect that "intramolecular relaxation is always fast on a time scale of the order of 10^{-12}s" is certainly not justified on the basis of current evidence.

6.5 Phase Coherence and the Transition to Incoherent Pumping

The physical models and their application to experiments in the preceding sections centered on the use of rate equations to describe the effects of intra- and intermolecular relaxation on the absorption of laser radiation by chemically reactive polyatomic molecules. Only in the specific model calculations for the decomposition of CF_2HCl and SF_6 was the more general Bloch equation used to describe the coherent pumping through the low-lying discrete levels of these molecules. Since the salient properties of the Bloch equation are submerged when it is integrated into the complete formalism used in Sect.6.3, and since the solutions to the Bloch or closely related Boltzmann equation are in themselves of considerable importance, it is appropriate to take a look at certain aspects of the now well-developed mathematical formalism for solving these equations.

It is relevant that solutions to the Boltzmann and Bloch equations exhibit some unexpected properties. These properties, describing the response of an "ideal" resonant absorber to an intense monochromatic radiation field, come under the general heading of phase coherence effects. In Sect.6.5.1 we will focus on the following, to us quite surprising, result of phase coherence. In a domain where inter- or intramolecular energy losses to a heat bath are severe, rate equations, traditionally believed to provide an accurate description of radiative contributions to level population evolutions when the ratio of Rabi to elastic collision frequency is small, are not valid. More generally, rate equations are restricted in two important ways [6.6,61]: (1) the time scale for the approach to the steady state, as predicted by rate equations, is wrong; errors of several orders of magnitude are typical when intensities are large; (2) the rate equations do not predict the correct steady-state populations in either saturating or intermediate ranges of laser intensities. The treatment in Sect.6.5.1 of the so-called generalized master equation (GME) technique shows just which aspects of phase coherence must be taken into account to correct these limitations in rate equations. The relevant Boltzmann equation [6.5,6] can be written in the form

$$\frac{d\boldsymbol{\rho}}{dt} = \left.\frac{d\boldsymbol{\rho}}{dt}\right|_{\text{dynamical}} + \left.\frac{d\boldsymbol{\rho}}{dt}\right|_{\text{intermolecular}} , \tag{6.68}$$

$$\left.\frac{d\boldsymbol{\rho}}{dt}\right|_{\text{dynamical}} = \frac{i}{\hbar}[\rho,H] , \tag{6.69}$$

and in the energy representation

$$\left.\frac{d\rho_{jj}}{dt}\right|_{\text{intermolecular}} = \frac{1}{\tau}\sum_{k\neq j} P_{j\leftarrow k}\rho_{kk} - \frac{1}{\tau}\sum_{k\neq j} P_{k\leftarrow j}\rho_{jj} , \tag{6.70}$$

$$\left.\frac{d\rho_{jk}}{dt}\right|_{\text{intermolecular}} = -\frac{1}{\tau}\rho_{jk} . \tag{6.71}$$

Equation (6.68) is a phenomenological equation describing the time evolution of the density operator $\boldsymbol{\rho}$ for an ensemble of polyatomics interacting with the light field through $d\rho/dt|_{\text{dynamical}}$ and undergoing intermolecular vibrational relaxation via collision with an excess of inert buffer gas molecules. The $P_{j\leftarrow i}$ relaxation terms in (6.70) account for inelastic transitions in which one or more quanta of vibrational energy are gained or lost by the polyatomic through collisions. The $1/\tau$ term in (6.71) accounts for collisional dephasing.

The Bloch equation

$$\frac{d\boldsymbol{\rho}}{dt} = \left.\frac{d\boldsymbol{\rho}}{dt}\right|_{\text{dynamical}} + \left.\frac{d\boldsymbol{\rho}}{dt}\right|_{\text{intramolecular}} \tag{6.72}$$

is almost exactly equivalent in a mathematical sense to the Boltzmann equation. In the Bloch equation the relaxation process is an intramolecular energy exchange and dephasing, between the pump mode and the remaining vibrational degrees of freedom. As pointed out in [6.47], the Bloch equation, as it is usually written, is not adequate to describe pumping through the quasi continuum since it ignores "heat bath feedback" effects.

In Sect.6.5.2 we will discuss in a general way one approach to justifying the use of the phenomenological Boltzmann or Bloch equations based on a generalization of the Fermi Golden Rule argument.

6.5.1 The Generalized Master Equation

Our formalism to treat phase coherence [6.6,61] begins with the Boltzmann or Bloch equations. It is possible to reduce these equations to an exactly equivalent form resembling rate equations, but now the reduced equations take into account the effect of phase coherence in all time domains-transient, posttransient, approach to steady state, and steady state [6.61]. We now describe this reduction for the Boltzmann equation.

The Boltzmann equation (6.68) is a set of coupled equations for off-diagonal and diagonal density matrix elements; the diagonal elements describe level populations and the off-diagonal elements contain phase information. One means to study coherence effects is to determine the contributions that the off-diagonal elements make in the time evolution of the diagonal density matrix elements.

We have developed a systematic technique for the elimination of the off-diagonal elements in the Boltzmann equation while still retaining all essential phase information in the resultant equations for just the level populations [6.6,61]. When dipole matrix elements in the Hamiltonian couple only adjacent levels of the absorbing system, coupling in the Boltzmann equation takes place only between the neighboring groups of density matrix elements as circled in Fig.6.19. Beginning at the corner of the matrix furthest from the diagonal, one expresses in a stepwise manner the furthest elements in terms of their neighboring elements one step nearer the diagonal. This reduction method collapses all N^2-N off-diagonal matrix elements into the diagonal elements; the most complicated single step in the reduction is the inversion of an $(N-1)\times(N-1)$ matrix. A detailed treatment is contained in [6.6].

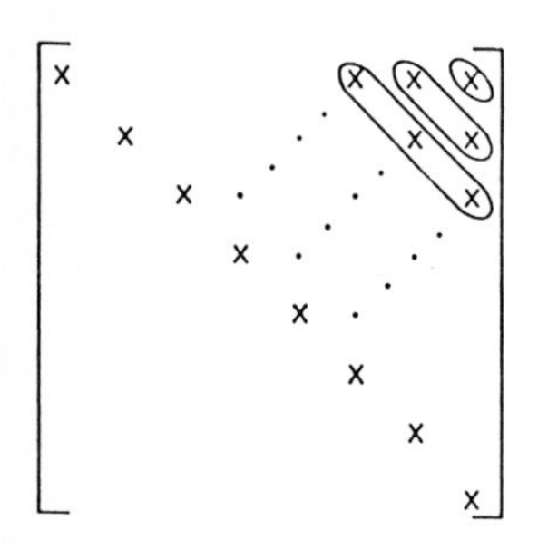

Fig.6.19. Groups of off-diagonal and diagonal density matrix elements involved in the reduction scheme leading to the GME. Coupling of density matrix elements in Boltzmann equation occurs only between adjoining groups

This reduction scheme culminates in the following equation for the time evolution of the energy level populations, ρ_{ii}:

$$\frac{d\rho_{ii}}{dt} = \sum_{j\neq i} W_{i\leftarrow j}(\lambda)\rho_{jj} - \rho_{ii} \sum_{j\neq i} W_{j\leftarrow i}(\lambda)$$

$$+ \sum_{j\neq i} \frac{P_{i\leftarrow j}}{\tau} \rho_{jj} - \rho_{ii} \sum_{j\neq i} \frac{P_{j\leftarrow i}}{\tau} - \frac{2\Gamma_i}{\hbar} \rho_{ii} \quad . \tag{6.73}$$

Based on its obvious resemblance to a master or rate equation, (6.73) has been called a "generalized master equation" [6.6], abbreviated GME. The first two terms on the right-hand side give the rate at which molecules are taken into and out of state i by radiative transitions. We have shown [6.6] that in the limit of sufficiently low laser intensities the radiative transition coefficients $W_{i\leftarrow j}$ $(\lambda = 0)$ are identical to the usual line-broadened perturbation theory results. The third and fourth terms are identified with collisional [6.6] or intramolecular [6.61] energy transfer between levels i and j. The last term gives the loss of molecules from i due to unimolecular chemical decay.

Other formalisms exist in which the same type of coherence information contained in the off-diagonal density matrix elements is retained in diagonal element equations. *Mukamel* [6.70] has applied the Mori projection operator formalism [6.126] to the problem of obtaining equations for a laser-driven polyatomic molecule. Coherence information is contained in this development in a sequence of dipole operator correlation functions.

The effects of phase coherence on level populations can be enumerated by making a direct term-by-term comparison of the GME with an ordinary master equation (OME). We shall do so shortly. First, however, it is important to state that although the GME, containing radiative- and collision-induced transition coefficients, has the same general form as an OME, it is not strictly speaking a rate equation, i.e., it is not a set of differential equations to be solved for ρ_{ii}. Equation (6.73) applies when one assumes $\rho_{ii} = C_i \exp(-\lambda t)$. To solve (6.73) requires the substitution of $-\lambda\rho_{ii}$ for $d\rho_{ii}/dt$. A complete time-dependent solution is obtained by an expansion in terms of eigenvectors derived from the coefficient matrix in (6.73) for each value of λ [6.61]. It should also be noted that the familiar rotating wave approximation [6.60,127,128] is necessary in deriving the GME.

Consider the λ dependence of the radiative transition rates $W_{i\leftarrow j}(\lambda)$ in the GME. If one assumes that the $W_{i\leftarrow j}(\lambda)$ can be replaced by their steady values, $W_{i\leftarrow j}(\lambda = 0)$, then the resulting "rate" equations will necessarily have the correct steady-state $(\lambda = 0)$ behavior in the absence of chemical reaction. *Wilcox* and *Lamb* [6.129] made a similar approximation several years ago when they set the rate of change of all off-diagonal elements to zero in the equations for two- and three-level systems. If, in addition, there is a time domain in the approach to the steady state characterized by a relatively slow variation of the level populations, the eigenvalues λ describing this variation may be small enough (i.e., describe a slow enough variation) so that $W(\lambda) \cong W(\lambda = 0)$. The Wilcox-Lamb approximation is expected to describe accurately the approach to the steady state through such a time domain.

The Wilcox-Lamb rate equations cannot provide an accurate solution in the transient time domain. Here the full N^2-eigenvalue spectrum, with some complex eigenvalues, of the GME produces oscillatory and rapidly decaying effects not described by the N-eigenvalue spectrum of the Wilcox-Lamb approximation. It is also of interest that the complex eigenvalues have a special significance in determining the location and widths of sidebands in the scattered light field [6.7,130,131]. When a chemical decay term $(2\Gamma_i/\hbar)\rho_{ii}$ is present, there is no $\lambda = 0$ eigenvalue and strictly speaking no steady state. The smallest eigenvalue of the GME, which under certain conditions has a physical significance in that it can be identified with a macroscopic rate constant [6.7], must in general be calculated from the GME.

If, in addition to setting $W(\lambda) = W(\lambda = 0)$ in the GME, one also neglects $W_{i\leftarrow j}$ terms that connect nonadjacent levels, a set of rate equations results that has been called the ordinary master equation (OME). At very low laser intensities this assumption is justified, since the nonadjacent $W_{i\leftarrow j}$ are negligible. The adjacent $W_{i\pm1\leftarrow i}$ are, for low intensities,

$$W_{i\pm1\leftarrow i} = \frac{\alpha_{i\pm1}^2 A^2}{2\hbar^2} \frac{(\frac{1}{\tau} - \lambda)}{(\frac{1}{\tau} - \lambda)^2 + (\omega - \chi)^2} \quad . \tag{6.74}$$

Here $1/\tau$ is the dephasing relaxation rate of either the Boltzmann or Bloch equation. Setting $\lambda = 0$ in (6.74) gives the Lorentzian line-shape factors used in the heat bath feedback model (6.4).

It is the assumption that one can neglect the $W_{j\leftarrow i}(\lambda = 0)$ terms connecting nonadjacent levels (this does not even come up in the often-studied two-level system) that is surprisingly inaccurate at even moderate laser intensities. In Fig.6.20 we compare the Wilcox-Lamb approximation $[W(\lambda = 0)]$ to OME and GME solutions for a resonantly driven three-level system. At the saturating field intensity used in this calculation, the GME solutions exhibit oscillatory behavior that cannot, of course, be represented by either rate equation solution. Apart from being unable to account for transient oscillatory behavior, two additional significant differences in GME and OME solutions are illustrated in Fig.6.20. First, the steady-state GME populations $n_0 = 1/2$, $n_1 = 1/6$, $n_2 = 1/3$ [6.61] differ from the populations $n_0 = n_1 = n_2 = 1/3$ predicted by the OME. This result contradicts the commonly held belief that saturation populations for a pumped N-level system should be $n_i = 1/N$ for all levels. The prediction of the OME that all levels are equally populated is true for only a very special class of collision mechanisms [6.6]. Second, as shown in Fig.6.20 the OME ground-state level population approaches the steady state approximately three orders of magnitude faster than the GME population.

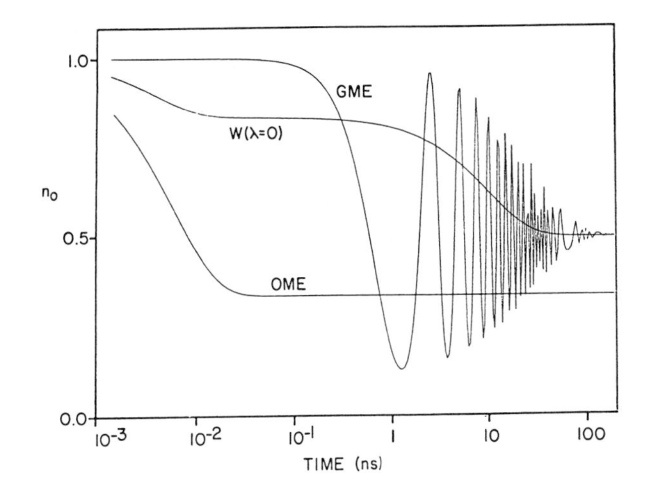

Fig.6.20. Ground-state population evolution for driven three-level system as calculated by the GME, Wilcox-Lamb [$W(\lambda=0)$] and OME for a collision frequency ($3\times10^7 s^{-1}$) small compared with the Rabi frequency ($3\times10^9 s^{-1}$) [6.62]. Saturation steady-state solution of GME is intimately related to inelastic collision model assumed here to be strong collision (instantaneous thermalization) model

More generally, the OME predicts a rate of approach to the steady state on the order of $\Omega^2\tau$, which is, at high enough intensities, much faster than the largest physical response rate in the system, the Rabi frequency Ω. The possibility of evolving with such a rapid rate is implied by radiative transition rate coefficients between adjacent levels. On resonance, $W_{i\leftarrow i+1}=W_{i+1\leftarrow i}=2\Omega^2\tau$. Indeed, it is possible that whenever all the levels are connected through a set of *positive* rate coefficients, each having a characteristically large magnitude R, all the exponential terms $\exp(-\lambda t)$ describing the approach to steady state have characteristic decay rates of approximately R. The same reasoning [6.62] does not apply when the radiative transition coefficients can be either positive or negative. While negative coefficients in a rate equation might appear strange, it can be shown [6.62] that at saturating field intensities some of the coefficients in the GME and the Wilcox-Lamb equations are indeed negative. This fact was also recognized by *Wilcox* and *Lamb* [6.129]. If $\Omega\ll1/\tau$ all the radiative transition coefficients $W_{ij}(\lambda=0)$ in the Wilcox-Lamb equation are of the order $\Omega^2\tau$ [6.62], the same as the OME transition rates. Yet, because near-cancellations between positive and negative contributions to the rate occur in the GME and Wilcox-Lamb equations, their predictions for the approach to the steady state differ substantially from the OME results.

It was pointed out by *Ackerhalt* and *Shore* [6.132] that the Wilcox-Lamb approximation may provide a reasonable coarse-grained approximation to the exact Boltzmann or Bloch equation solutions even when the solution is oscillatory rather than slowly varying. Figure 6.20 illustrates this property of the Wilcox-Lamb approximation. It turns out, however, that there is a systematic discrepancy [6.62] between the Wilcox-Lamb approximation and a true coarse-grained time average over the Boltzmann equation (GME) solutions. This discrepancy is corrected in a new approximation, the renormalized small eigenvalue (RSE) approximation [6.62] which applies a correction factor, determined from the GME, to exponential terms in the Wilcox-Lamb solution.

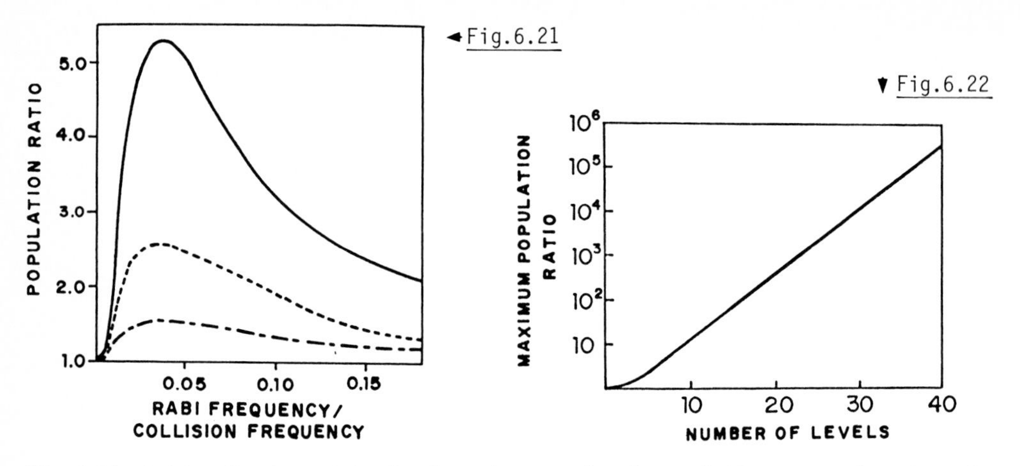

Fig.6.21. Ratio of coherent to incoherent upper-level populations as a function of field strength, from [6.7]. Coherent level populations are derived from the GME (6.73), and incoherent populations are computed using an OME. The three curves are for a six-level system (- ——), an eight-level system (— —) and a ten-level system (——); inelastic collisions governed by Landau-Teller collision model with $P_{0\leftarrow 1} = 0.1$

Fig.6.22. Maximum ratio of upper-level populations including and not including multiple-photon contributions as a function of N, calculated with $P_{i-1\leftarrow i} = 1.0$ [6.133]

The importance of coherence in determining steady-state populations in a driven N-level system is not restricted to saturation laser intensities. In Fig.6.21 we compared GME and OME populations for the highest level of 6-, 8-, and 10-level systems [6.7]. The coherent enhancement of upper-level population is greatest for intermediate intensities when the Rabi to collision frequency ratio $\Omega\tau \cong 0.05$. For a given laser intensity the coherent enhancement is greatest when the inelastic energy loss is most efficient [6.6,133].

It is clear from Fig.6.21 that the coherent enhancement in the upper-level populations increases with N, the number of resonantly coupled levels. In fact, additional calculations show maximum enhancement increasing exponentially with N (Fig. 6.22). It should be emphasized that although the dipole matrix elements in the Hamiltonian defining this model couple only adjacent levels, radiative transitions between nonadjacent levels are present in the GME. These nonadjacent radiative transitions which account for the coherent enhancement are by definition excluded from the OME. The nonvanishing radiative transition coefficients which couple nonadjacent energy levels result from the fact that it is possible for each independent absorber to maintain a phased response with the incident radiation field. We have shown that the conditions most favorable for enhancement in the upper level populations include the predominance of efficient energy transfer to the heat bath when a laser field is present which is intense but below the saturation limit.

In this section we have emphasized the usefulness of the GME in assessing phase coherent effects as they are expressed in the nonzero radiative transition rate $W_{i\leftarrow j}(\lambda)$ between nonadjacent levels. From a strictly computational point of view the GME is important because it effectively reduces the calculation of the full N^2-eigenvalue-eigenvector problem to an iterative N^{th}-order matrix procedure. Furthermore, in examining the "approach-to-steady-state effects" for which the Wilcox-Lamb and RSE approximations are adequate, one needs only $W_{i\leftarrow j}(\lambda = 0)$ and need not compute the eigenvalue spectrum at all. In this and other applications of the GME, a recently developed diagrammatic technique [6.61] permits the functions $W_{i\leftarrow j}(\lambda)$ to be evaluated directly, bypassing certain matrix inversion steps used in the earlier treatment [6.6]. The $W_{i\leftarrow j}(\lambda)$ can now be obtained directly from an appropriate set of diagrams as described in [6.61].

6.5.2 Fermi Golden Rule Considerations

In the previous section we discussed properties of solutions to the Bloch or Boltzmann equations with particular emphasis on the transition from coherent to incoherent pumping. Another question of interest relates to the microscopic basis for these phenomenological equations. A number of derivations [6.134-137], such as REDFIELD's NMR treatment [6.137], arrive at the Bloch equations to describe the time evolution of a properly defined coarse-grained reduced density matrix. Some of these formal treatments have the advantage of relating the $1/T_1$ and $1/T_2$ constants to more fundamental molecular parameters. There is another approach, based on Fermi Golden Rule [6.138] arguments, which also allows one to understand on a microscopic basis the transition from coherent to incoherent pumping and also to estimate in a very general, albeit qualitative, way the magnitude of the T_2 relaxation times even in the absence of detailed knowledge of the molecular Hamiltonian. This approach will also bring out certain limits to the applicability of rate equations as prescribed by the Fermi Golden Rule.

Consider, within the framework of a time-dependent Schrödinger equation, the dynamics of a transition from a discrete state into a set of dense energy levels, as shown schematically in Fig.6.23. A single state at level A is coupled with oscillator strength α_{Ab} to each of the states pictured in the group of states B, the latter spread out in an energy band of width w. All those states which still are effectively being driven in phase as resonant states, will, at any time t, be encompassed in an uncertainty width $\Delta E = \hbar/t$. We assume for convenience that the laser is tuned to resonance at the center of the band, as indicated by the placement of the uncertainty width in Fig.6.23. We distinguish three energy regions. The first region (I) includes all energies further away from exact resonance than the band edge and, thus, does not contain any states to which transitions can be made. The second region (II) has been selected to include the entire band of pumpable states

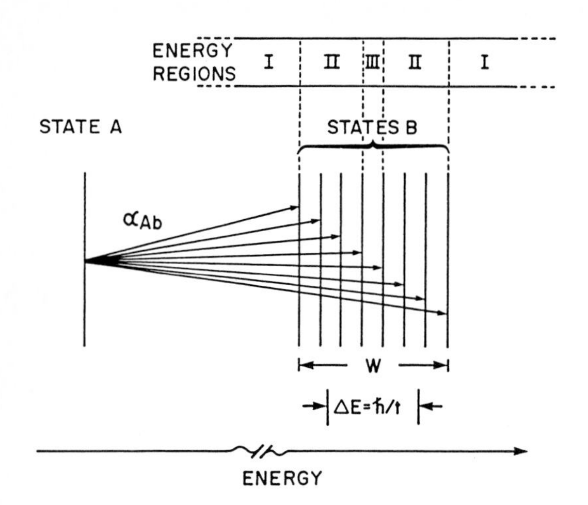

Fig.6.23. Schematic diagram of a group of excited states B near resonance for a laser-induced transition from state A. Energy width of level B is w. Dipole operator matrix element between the initial state and each excited state is α_{Ab}. Population dynamics in the excited level are determined by which states are enclosed in the time-dependent uncertainty width ΔE and depend on whether the uncertainty width limit is in energy region I, II, or III. The energy scale is broken because the spread in states in group B is depicted on a larger scale than the transition energy

except for a small hole, region (III), that is cut out of its center. The central region (III) embraces a total energy width surrounding the exact resonant energy equal to the average spacing between microscopic states so that it will contain on average a single quantum state.

Now, in order to discuss the dynamics, we make the simplifying assumption that each state evolves exactly in phase with the laser field as long as it remains within the uncertainty width, but ceases thereafter to contribute to the total excited-state population. The total population at time t can be found by multiplying the population of a single in-phase state by the number of states n_u remaining within the uncertainty width. If the boundary of the uncertainty region lies in region II, the number of states included within $E = \hbar/t$ decreases with time according to $n_u = (n_{tot}/w)(\hbar/t)$ where n_{tot} is the total number of states in the band; n_u in the other regions is fixed, being given by $n_u = n_{tot}$ throughout region I and by $n_u = 1$ in region III. As a consequence, the time domains which correspond to having uncertainty widths in each region give rise to distinctly different dynamic behavior.

The times delineating the different dynamic regions are significant because of the possibility that population evolution between two levels may switch over from one type of dynamic behavior to another during the course of the transition. For convenience, we have summarized the times bounding each dynamic region in Table 6.1 together with information about how the total excited-state population n_{tot} evolves in each region. To calculate these populations, we have assumed, in agreement with the usual perturbation theory expressions, that each in-phase state has a population equal to

$$n_i = \left(\frac{\alpha_{Ab}A}{2\hbar}\right)^2 t^2 \quad .$$

A constant increase in population where the total population is proportional to t rather than t^2 is predicted by the Fermi Golden Rule, and we see from the Table 6.1

Table 6.1. Time and intensity restrictions governing coherence or incoherence of population dynamics in the dense group of energy levels of Fig.6.23

Region	Width of outer boundary	Time limits	Total excited-state population when $n_{tot} \ll 1$	Intensity range where region dominates
I	∞	$0 < t < T_2$ ($T_2 \equiv \hbar/w$)	$(\alpha_{AB}A/2\hbar)^2 t^2$	$\alpha_{AB}A/2\hbar > 1/T_2$
II	w = bandwidth of dipole-carrying states	$T_2 < t < T_s$ ($T_s \equiv \hbar/s$)	$(\alpha_{AB}A/2\hbar)^2 T_2 t$	$1/T_2 > (\alpha_{AB}A/2\hbar)^2 T_2 > 1/T_s$
III	s = spacing of microscopic states	$T_s < t$	$(\alpha_{Ab}A/2\hbar)^2 t^2$	$1/T_s > \frac{\alpha_{Ab}A}{2\hbar}$

that the Golden Rule applies in Region II. A time-dependent Schrödinger equation applies in both regions I and III, but with an effective macroscopic coupling element

$$\alpha_{AB} = \sum_b \alpha_{Ab}$$

in region I. The importance of each time domain will be determined in large measure by how much population is driven within that domain. The range of intensities required to excite a significant fraction of the population to higher levels within each domain is given in the right-hand column of Table 6.1.

Time-domain restrictions on the Golden Rule were discussed recently by *Quack* [6.75] for both discrete-dense and dense-dense transitions. The result of these restrictions is that if most of the population becomes excited within a single time domain, either a rate equation (for regions II) or a time-dependent Schrödinger equation (for regions I and III) can be applied. If, however, the time required to undergo transitions is roughly comparable to either of the cutoff times T_2 or T_s (Table 6.1) neither equation provides an adequate description. In a multilevel system, both T_2 and T_s are strongly dependent on the vibrational level through level-dependent bandwidths and state densities [see, e.g., (6.10,11)]. Therefore, even if a clean separation of time scales were to exist for each level, different equations, Schrödinger or rate, would still have to be applied to different vibrational levels. The T_2 terms in the Bloch equation produce the same changeover from coherent to incoherent evolution as we have just found for a finite bandwidth $\hbar/T_2$ of driven levels in a time-dependent Schrödinger equation framework. Using a Bloch equation with level-dependent T_2 relaxation rates allows the time cutoff $t > T_2$ to be handled automatically and with appropriate allowance for the level dependence of T_2.

A second relaxation time designated as T_1, and usually included in the Bloch equation, can also be understood in terms of the microscopic energy level picture discussed here. T_1 relaxation corresponds to a transfer of energy from one part of the molecule to another. If the initial state for the transition is a pure state of the infrared-active pump mode, it can be shown [6.32] that the linear combination of dense states which is excited initially before any relaxation can occur, corresponds exactly to an excited state of the pump mode. Population of other states nearby occurs only in a time T_1, of order $\hbar/w$, required for the microscopic states in the band to get out of phase. Thus T_1 energy relaxation is predicted to occur on a time scale commensurate with that of T_2 phase relaxation.

The transition from the Golden rule region into the longest time domain, region III, cannot be mediated by any device analogous to the T_1 and T_2 relaxation times that allowed us to go from region I into the Golden Rule region. Instead, the question is: What is the minimum intensity requirement for excluding region III pumping? It is apparent that this requirement is equivalent to the condition that there be one or more states which satisfy power-broadening requirements. In view of the experimental results of *Bomse* et al. [6.139] which show intensity requirements of less than 100 W cm^{-2} for one of the large molecular ions they investigated, it is of some interest to be able to know from first principles just what determines the ultimate intensity requirements.

6.5.3 Coherent Quasi-Continuum Pumping - an Open Question

The conditions for going from coherent to incoherent pumping were developed in the preceding section. The microscopic picture clearly shows that energy levels begin to evolve out of phase with one another whenever coupling is occurring between bands of states where the power broadening (Rabi frequency) is small compared with the anharmonic broadening (intramolecular T_2 dephasing relaxation times). Maintaining coherence, it would appear, requires that the radiation field couple sparse discrete states only. Is it therefore highly improbable that coherent driving in the quasi continuum can occur? Despite a multitude of current theoretical treatments of the quasi continuum by incoherent rate equations, e.g., (6.1), we regard this question of coherence as still open.

Suppose, for example, that a simplified model of the quasi continuum were to consist of a ladder of states in which each rung of the ladder were characterized by a degeneracy factor increasing with energy. We have shown [6.40] that, not surprisingly, the effective Rabi frequency increases quite markedly with degeneracy, resulting in rapid coherent pumping up the energy ladder. Incoherence is introduced when one accounts for the finite linewidth (T_2 phase relaxation) associated with each individual rung on the energy ladder. The relevant issue, as shown in Sect.6.5.2, is how the phase broadening linewidth compares with the inverse Rabi frequency (dynamic power broadening). Simply stated, the Fermi Golden Rule does *not* address this question. The Golden Rule dictates only that the linewidth be proportional to the density of final states which are intramolecularly coupled to an initial doorway state or states.

Given our present paucity of knowledge regarding a proper description of the eigenstates for the quasi continuum, one should not rigidly assume that the entire density of states in the vicinity of each unperturbed energy state strongly mixes together giving rise to broad anharmonic lineshapes. As experimental techniques evolve to the point of probing lineshapes for transitions in highly excited polyatomics, we may in fact find that the linewidths are much narrower than supposed. Although not entirely analogous, unexpectedly narrow bands were observed in the high-overtone spectrum for benzene [6.140,141]. If, for highly degenerate resonant states, Rabi frequencies turn out to be commensurate with T_2 level widths, coherent pumping in the quasi continuum may be realized at achievable laser intensities. Apart from opening a new area for multilevel spectroscopy, a practical result of phase coherence, discussed in Sect.6.5.1, is that energy level populations and chemical decay rates could be expected to be many orders of magnitude higher for phase coherent, compared with incoherent, pumping [6.133].

6.6 Addendum

Since this paper was originally submitted a number of new results have been obtained. Here we briefly mention some of these. The basic equations of the heat bath feedback model (Sect.6.2.2) have been derived from first principles by *Kay* [6.142] (see also Sect.6.5.2). The random walk model for linewidths (Sect.6.2.3) has been replaced by a theory based on restricted quantum exchange [6.143]. Detailed expressions have been obtained for intramolecular T_1 and T_2 relaxation rates [6.144]. Restricted quantum exchange posits that matrix elements of the coupling potential are strongly dependent and decreasing functions of the number of quanta exchanged between coupled states and predicts linewidth saturation with increasing energy [6.144] (Sect.6.3.4b).

There have been additional studies on the decomposition of CF_2CFCl. The high-pressure falloff (Sect.6.3.3) has been measured at different laser intensities and successfully compared to a theory using restricted quantum exchange expressions for the radiative pumping rates [6.145]. The distribution of energy among reactant products, CF_2 and CFCl, has been directly measured and used as a test for the use of statistical theory (RRK) rate constants in MPD [6.146]. In general, pressure falloff studies have provided useful tests of radiative pumping rate expressions. For example, *Rossi* et al. [6.147] have recently determined a strong laser-frequency dependence in the radiative pumping rates for the decomposition of C_3F_7I.

The second overtone ($3_{\nu 3}$) of SF_6 (Sect.6.3.4a) has been measured and interpreted in high resolution [6.148], thus providing exact data on the relevant importance of anharmonic and rotation-vibrational couplings. *Hodgkinson* and *Robiette* [6.149] have given a clear discussion of the implications for discrete level pumping in SF_6. It should also be mentioned that pressure effects observed in the $\chi = 3100$ cm^{-1} decomposition of cyclopropane that suggested some selectivity (Sect.6.3.5) have not been confirmed in more recent experiments [6.150]. Also relevant to the question of selectivity (Sect.6.4.2d) and energy randomization are the recent criticisms [6.151] showing that the translational energy distribution in products cannot be used to infer energy randomization. Finally a new application for the developed mathematical formalism for solving Bloch equations (Sect.6.5) has been found by *Stone* [6.152]. He gives an analytical method for calculating the time-averaged density matrix.

While the theory reviewed in this article can be classified as "statistical" in most of its aspects, the possibility of nonstatistical excitation cannot be ruled out. The statistical approaches require a dense and fairly smooth distribution in the dipole coupling between exact energy eigenstates (Fig.6.3). Recent theoretical work suggests that this may not always be the case [6.153,154]. In particular, dipole coupling can be much stronger between those quantum states which are related through semiclassical quantization to quasi-periodic classical trajectories [6.153]. Also, the Los Alamos group [6.154] has suggested that in SF_6 symmetry restrictions and restricted quantum limitations reduce the number of states coupled to the 0th-order pump mode state below the density required for a statistical treatment. Either of these situations suggest the possibility of coherent-like pumping in which molecules are pumped to high energies up a ladder of almost equally spaced exact energy eigenstates. This coherent pumping might bring molecules to energies well above the three or four quanta which are usually assumed for the threshold of the quasi continuum. It is even possible that this might provide for excitation above the reaction threshold and thus provide a mechanism for selectivity. Experimentally one should look for this under beam or extremely low-pressure conditions and at moderately low laser intensities, the idea being not to try and beat intramolecular relaxation rates, but rather to avoid power broadening over too many eigenstates.

Acknowledgement. It has been a pleasure to work with Professor David Dows in the Chemistry Department at the University of Southern California, to collaborate with Dr. John Horsley, a member of the Laser Chemistry group at Exxon Corporate Research Laboratories under the direction of Dr. Andrew Kaldor, and to collaborate with Drs. John Stephenson and David King from the Molecular Spectroscopy Division at the National Bureau of Standards. We also want to thank Ms. Sarah Wright for patient and skillful typing of the manuscript and Captain Jack Eppinger and Karin Fouts for figure preparation. This work was supported by a grant from the National Science Foundation CHE 76-84180.

References

6.1 F.V. Bunkin, R.V. Karapetyan, A.M. Prokhorov: Sov. Phys.-JETP **20**, 145 (1965)
6.2 G.A. Askar'yan: Sov. Phys.-JETP **21**, 439 (1965)
6.3 N.B. Slater: *Theory of Unimolecular Reactions* (Cornell Univ. Press, Ithaca 1959)
6.4 M.F. Goodman, H.H. Seliger, J.M. Minkowski: Photochem. Photobiol. **12**, 355 (1970)
6.5 M.F. Goodman, E. Thiele: Phys. Rev. A**5**, 1355 (1972)
6.6 J. Stone, E. Thiele, M.F. Goodman: J. Chem. Phys. **59**, 2909 (1973)
6.7 M.F. Goodman, J. Stone, E. Thiele: J. Chem. Phys. **59**, 2919 (1973)
6.8 K.B. Eisenthal, W.L. Peticolas, K.E. Rieckhoff: J. Chem. Phys. **44**, 4492 (1966)
6.9 N.V. Karlov, Y.N. Petrov, A.N. Prokhorov, O.M. Stel'makh: JETP Lett. **11**, 135 (1970)
6.10 M.C. Richardson, N.R. Isenor: Opt. Commun. **5**, 394 (1972)
6.11 N.G. Basov, E.P. Markin, A.N. Oraevskii, A.V. Pankratov, A.N. Skachkov: JETP Lett. **14**, 165 (1971)
6.12 F.R. Lory, S.H. Bauer, T. Manuccia: J. Phys. Chem. **79**, 545 (1976)
6.13 N.R. Isenor, V. Merchant, R.S. Hallsworth, M.C. Richardson: Can. J. Phys. **51**, 1281 (1973)
6.14 R.V. Ambartzumian, Yu.A. Gorokov, V.S. Letokhov, G.N. Makarov: JETP Lett. **21**, 171 (1975)
6.15 J.L. Lyman, R.J. Jensen, J. Rink, C.P. Robinson, S.D. Rockwood: Appl. Phys. Lett. **27**, 87 (1975)
6.16 R.V. Ambartzumian, Yu.A. Gorokov, G.N. Makarov, A.A. Puretzki, N.P. Furzikov: Chem. Phys. Lett. **45**, 231 (1977)
6.17 S.H. Bauer, J.A. Haberman: IEEE J. QE-**14**, 233 (1978)
6.18 M.J. Berry: J. Chem. Phys. **68**, 4419 (1978)
6.19 A.S. Sudbø, P.A. Schulz, Y.R. Shen, Y.T. Lee: J. Chem. Phys. **69**, 2312 (1978)
6.20 P.L. Houston, C.B. Moore: J. Chem. Phys. **65**, 757 (1976)
6.21 A. Kaldor, P. Rabinowitz, D.M. Cox, J.A. Horsley, R. Brickman: J. Opt. Soc. Am. **68**, 684 (1978)
6.22 S.E. Bialkowski, W.A. Guillory: J. Chem. Phys. **68**, 3339 (1978)
6.23 W. Fuss, T.P. Cotter: Appl. Phys. **12**, 265 (1977)
6.24 D. Gutman: J. Chem. Phys. **67**, 4291 (1977)
6.25 M. Rothschild, W.S. Tsay, D.O. Ham: Opt. Commun. **24**, 327 (1978)
6.26 J.W. Hudgens: J. Chem. Phys. **68**, 777 (1978)
6.27 H.R. Bachmann, R. Rinck, H. Noth, K.L. Kompa: Chem. Phys. Lett. **45**, 169 (1977)
6.28 C.R. Quick Jr., C. Wittig: Chem. Phys. **32**, 75 (1978)
6.29 J.L. Lyman: J. Chem. Phys. **69**, 1196 (1978)
6.30 D.S. King, J.C. Stephenson: Chem. Phys. Lett. **51**, 48 (1977)
6.31 J.C. Stephenson, D.S. King: J. Chem. Phys. **69**, 1485 (1978)
6.32 J. Stephenson, D. King, M.F. Goodman, J. Stone: J. Chem. Phys. **70**, 4496 (1979)
6.33 R.V. Ambartzumian, Yu.A. Gorokhov, G.N. Makarov, A.A. Puretsky, N.P. Furzikov: Kvantovaya Elektron. (Moscow) **4**, 1590 (1977)
6.34 R.V. Ambartzumian, N.P. Furzikov, V.S. Letokhov, A.P. Dyadkin, A.Z. Grasyuk, B.I. Vasilyev: Appl. Phys. **15**, 27 (1978)
6.35 S. Bittenson: J. Chem. Phys. **67**, 4819 (1977)
6.36 T.J. Manuccia, M.D. Clark, E.R. Lory: J. Chem. Phys. **68**, 2271 (1978)
6.37 D.K. Evans, R.D. McAlpine, F.K. McClusky: Chem. Phys. **32**, 81 (1978)
6.38 J.J. Tiee, C. Wittig: Appl. Phys. Lett. **32**, 236 (1978)
6.39 J.J. Tiee, C. Wittig: J. Chem. Phys. **69**, 4756 (1978)
6.40 J. Stone, M.F. Goodman, D.A. Dows: Chem. Phys. Lett. **44**, 411 (1976); J. Chem. Phys. **65**, 5052 (1976); J. Chem. Phys. **65**, 5062 (1976)
6.41 S. Mukamel, J. Jortner: Chem. Phys. Lett. **40**, 150 (1976); J. Chem. Phys. **65**, 5204 (1976)
6.42 N. Bloembergen: Opt. Commun. **15**, 416 (1975)
6.43 D.M. Larsen: Opt. Commun. **19**, 404 (1976)
6.44 D.M. Larsen, N. Bloembergen: Opt. Commun. **17**, 254 (1976)

6.45 N. Bloembergen, C.D. Cantrell, D.M. Larsen: "Collisionless Dissociation of Polyatomic Molecules by Multiphoton Infrared Absorption", in *Tunable Lasers and Applications*, ed. by A. Mooradian, T. Jaeger, P. Stokseth, Springer Ser. Opt. Sci., Vol.3 (Springer, Berlin, Heidelberg 1976)
6.46 D.P. Hodgkinson, J.S. Briggs: Chem. Phys. Lett. **43**, 451 (1976)
6.47 J. Stone, M.F. Goodman: J. Chem. Phys. **71**, 408 (1979)
6.48 R.V. Ambartzumian, V.S. Letokhov: Acc. Chem. Res. **10**, 61 (1977)
6.49 J.L. Lyman, K.M. Leary: J. Chem. Phys. **69**, 1858 (1978)
6.50 A. Kaldor, R.B. Hall, D.M. Cox, J.A. Horsley, P. Rabinowitz, G.M. Kramer: J. Am. Chem. Soc. **101**, 4465 (1979)
6.51 J.G. Black, P. Kolodner, M.J. Shultz, E. Yablonovitch, N. Bloembergen: Phys. Rev. A**19**, 704 (1979)
6.52 J.R. Ackerhalt, H.W. Galbraith: In *Laser-Induced Chemical Processes*, ed. by J.I. Steinfeld (Plenum, New York 1981)
6.53 E.R. Grant, P.A. Schulz, A.S. Sudbø, Y.R. Shen, Y.T. Lee: Phys. Rev. Lett. **40**, 115 (1978)
6.54 J.L. Lyman: J. Chem. Phys. **67**, 1868 (1977)
6.55 N. Bloembergen, E. Yablonovitch: Phys. Today **31**, 23 (May 1978)
6.56 D.M. Brenner: Chem. Phys. Lett. **57**, 357 (1978)
6.57 M.N.R. Ashfold, G. Hancock, G. Ketley: Faraday Discuss. Chem. Soc. **67** (1979)
6.58 D.S. King, J.C. Stephenson: Chem. Phys. Lett. **66**, 33 (1979)
6.59 F. Bloch: Phys. Rev. **102**, 104 (1956)
6.60 T.H. Einwohner, J. Wong, J.C. Garrison: Phys. Rev. A**14**, 1452 (1976)
6.61 J. Stone, M.F. Goodman: Phys. Rev. A**18**, 2618 (1978)
6.62 J. Stone, M.F. Goodman: Phys. Rev. A**18**, 2642 (1978)
6.63 R.V. Ambartzumian: "Dissociation of Polyatomic Molecules by an Intense Infrared Laser Field", in *Tunable Lasers and Applications*, ed. by A. Mooradian, T. Jaeger, P. Stokseth, Springer Ser. Opt. Sci., Vol.3 (Springer, Berlin, Heidelberg 1976)
6.64 R.V. Ambartzumian, Yu.A. Gorokhov, V.S. Letokhov, G.N. Makarov, A.A. Puretzkii/ JETP Lett. **23**, 22 (1976); Sov. Phys.-JETP **44**, 231 (1976)
6.65 C.D. Cantrell, H.W. Galbraith: Opt. Commun. **21**, 374 (1977)
6.66 J.R. Ackerhalt, H.W. Galbraith: J. Chem. Phys. **69**, 1200 (1978)
6.67 H.W. Galbraith, J.R. Ackerhalt: Opt. Lett. **3**, 152 (1978)
6.68 J.A. Horsley, J. Stone, M.F. Goodman, D.A. Dows: Chem. Phys. Lett. **66**, 461 (1979)
6.69 J.G. Black, E. Yablonovitch, N. Bloembergen, S. Mukamel: Phys. Rev. Lett. **38**, 1131 (1977)
6.70 S. Mukamel: Phys. Rev. Lett. **42**, 168 (1979)
6.71 M. Bixon, J. Jortner: J. Chem. Phys. **48**, 715 (1968)
6.72 C. Tric: Chem. Phys. **14**, 189 (1976)
6.73 E.J. Heller, S.A. Rice: J. Chem. Phys. **61**, 936 (1974)
6.74 J.D. Campbell, G. Hancock, J.B. Halpern, K.H. Welge: Chem. Phys. Lett. **44**, 404 (1976)
6.75 M. Quack: J. Chem. Phys. **69**, 1282 (1978)
6.76 M. Quack: J. Chem. Phys. **70**, 1069 (1979)
6.77 P. Kolodner, C. Winterfeld, E. Yablonovitch: Opt. Commun. **20**, 119 (1977)
6.78 J.L. Lyman, J.W. Hudson, S.M. Freund: Opt. Commun. **21**, 112 (1977)
6.79 G.P. Quigley: Opt. Lett. **3**, 106 (1978)
6.80 D.S. King, J.C. Stephenson: private communication
6.81 C.R. Quick, Jr.: "IR Photodissociation of Fluorinated Ethanes and Ethylenes and SF_6: A Time Resolved Probe of the Dissociation Process"; Ph.D. Dissertation, University of Southern California (1979) p.39
6.82 C.R. Quick, Jr., C. Wittig: J. Chem. Phys. **69**, 4201 (1978)
6.83 R.B. Hall, A. Kaldor: J. Chem. Phys. **70**, 4027 (1979)
6.84 Yu.A. Gorokov, V.S. Letokhov, G.N. Makarov: JETP Lett. **21**, 171 (1975)
6.85 M.J. Coggiola, P.A. Schulz, Y.T. Lee, R. Shen: Phys. Rev. Lett. **38**, 17 (1977)
6.86 F. Brunner, T.P. Cotter, K.L. Kompa, D. Proch: J. Chem. Phys. **67**, 1547 (1977)
6.87 M.C. Gower, T.K. Gustafson: Opt. Commun. **20**, 119 (1977)
6.88 H. Kildal: J. Chem. Phys. **67**, 1287 (1977)
6.89 F. Brunner, D. Proch: J. Chem. Phys. **68**, 4936 (1978)

6.90 E.R. Grant, M.J. Coggiola, Y.T. Lee, P.A. Schulz, A.S. Sudbø, Y.R. Shen: Chem. Phys. Lett. **52**, 595 (1977)
6.91 H.S. Kwok, E. Yablonovitch: Phys. Rev. Lett. **41**, 745 (1978)
6.92 D.M. Cox: Opt. Commun. **24**, 336 (1978)
6.93 V.N. Bagratashvili, I.N. Knyazev, V.S. Letokhov, V.V. Lobko: Opt. Commun. **18**, 525 (1976)
6.94 S.S. Alimpiev, V.N. Bagratashvili, N.V. Karlov, V.S. Letokhov, V.V. Lobko, A.A. Makarov, B.G. Sartakov, E.M. Khokhlov: JETP Lett. **25**, 547 (1977)
6.95 A.S. Akhmanov, V.Yu. Baranov, V.D. Pismenny, V.N. Bagratashvili, Yu.R. Kolomiisky, V.S. Letokhov, E.A. Ryabov: Opt. Commun. **23**, 357 (1977)
6.96 R.S. McDowell, H.W. Galbraith, C.D. Cantrell, N.G. Nereson, E.D. Hinkley: J. Mol. Spectrosc. **68**, 288 (1977)
6.97 W.B. Person, K.C. Kim: J. Chem. Phys. **69**, 2117 (1978)
6.98 C.D. Cantrell, H.W. Galbraith: Opt. Commun. **18**, 513 (1976)
6.99 C.D. Cantrell, K. Fox: Opt. Lett. **3**, 151 (1978)
6.100 J.R. Ackerhalt, H. Flicker, H.W. Galbraith, J. King, W.B. Person: J. Chem. Phys. **69**, 1461 (1978)
6.101 K.T. Hecht: J. Mol. Spectrosc. **5**, 355 (1960)
6.102 R.V. Ambartzumian, Y.A. Gorokov, V.S. Letokhov, G.N. Makarov, A.A. Puretskii, N.P. Furzikov: JETP Lett. **23**, 194 (1976)
6.103 A.V. Nowak, J.L. Lyman: J. Quant. Spectrosc. Radiat. Transfer **15**, 945 (1975)
6.104 J.F. Bott: Appl. Phys. Lett. **32**, 624 (1978)
6.105 W. Fuss: Chem. Phys. **36**, 135 (1979)
6.106 P.J. Robinson, K.A. Holbrook: *Unimolecular Reactions* (Wiley, New York 1972)
6.107 M. Lesiecki, W. Guillory: J. Chem. Phys. **66**, 4317 (1977)
6.108 Z. Karny, R. Zare: Chem. Phys. **23**, 34 (1977)
6.109 J. Simons, B. Rabinovitch: J. Phys. Chem. **68**, 1322 (1964)
6.110 J.G. Black, P. Kolodner, M.J. Shultz, E. Yablonovitch, N. Bloembergen: Phys. Rev. A**19**, 704 (1979)
6.111 E. Thiele, J. Stone, M.F. Goodman: Chem. Phys. Lett. **66**, 457 (1979)
6.112 H.C. Andersen, I. Oppenheim, K.E. Shuler, G.H. Weiss: J. Math. Phys. (N.Y.) **5**, 522 (1964)
6.113 D.J. Wilson: J. Chem. Phys. **36**, 1293 (1962)
6.114 G.Z. Whitten, B.S. Rabinovitch: J. Chem. Phys. **38**, 2466 (1963)
6.115 E. Thiele, M.F. Goodman, J. Stone: Opt. Eng. **19**, 10 (1980)
6.116 L.S. Kassel: *Kinetics of Homogeneous Gas Reactions* (Chemical Catalog Co., New York 1932)
6.117 R. Karplus, J. Schwinger: Phys. Rev. **73**, 1020 (1948)
6.118 I. Oref, B.S. Rabinovitch: Acc. Chem. Res. **12**, 166 (1979)
6.119 J. Aspden, N.A. Khawaja, J. Reardon, D.J. Wilson: J. Am. Chem. Soc. **91**, 7580 (1969)
6.120 J.D. Rynbrandt, B.S. Rabinovitch: J. Phys. Chem. **75**, 2164 (1971)
6.121 B.S. Rabinovitch, T.F. Meagher, K.J. Chao, J.R. Barker: J. Chem. Phys. **60**, 2932 (1974)
6.122 F.M. Wang, B.S. Rabinovitch: Can. J. Chem. **54**, 943 (1976)
6.123 A.N. Ko, B.S. Rabinovitch, K.J. Chao: J. Chem. Phys. **66**, 1374 (1977)
6.124 A.N. Ko, B.S. Rabinovitch: Chem. Phys. **30**, 361 (1978)
6.125 E.R. Grant, P.A. Schulz, Aa.S. Sudbø, M.J. Coggiola, Y.T. Lee, Y.R. Shen: "Multiphoton Dissociation of Polyatomic Molecules Studied with a Molecular Beam", in *Laser Spectroscopy III*, ed. by J.L. Hall, J.L. Carlsten, Springer Ser. Opt. Sci., Vol.7 (Springer, Berlin, Heidelberg 1977)
6.126 H. Mori: Prog. Theor. Phys. **33**, 423 (1965)
6.127 I.I. Rabi: Phys. Rev. **51**, 652 (1937)
6.128 P.R. Fontana, P. Thomann: Phys. Rev. A**13**, 1512 (1976)
6.129 L.R. Wilcox, W.E. Lamb: Phys. Rev. **110**, 1915 (1960)
6.130 B.R. Mollow: Phys. Rev. **188**, 187 (1969); A**2**, 76 (1970)
6.131 F. Schuda, C.R. Stroud, M. Hercher: J. Phys. B**7**, L198-L202 (1974)
6.132 J.R. Ackerhalt, B.W. Shore: Phys. Rev. A**16**, 277 (1977)
6.133 J. Stone, M.F. Goodman: Phys. Rev. A**14**, 380 (1976)
6.134 M. Lax: J. Phys. Chem. Solids **25**, 487 (1964); Phys. Rev. **109**, 1921 (1958); Phys. Rev. **129**, 2342 (1963)

6.135 R. Zwanzig: J. Chem. Phys. **33**, 1338 (1960)
6.136 C.D. Cantrell, S.M. Freund, J.L. Lyman: In *Laser Handbook*, Vol.3, ed. by M. Stitch (North Holland, New York 1978)
6.137 A.G. Redfield: In *Advances in Magnetic Resonance*, Vol.1, ed. by J.S. Waugh (Academic, New York 1965)
6.138 L.I. Schiff: *Quantum Mechanics*, 3rd ed. (McGraw-Hill, New York 1968)
6.139 D.S. Bomse, R.L. Woodin, J.L. Beauchamp: J. Am. Chem. Soc. **101**, 5503 (1979)
6.140 D.F. Heller, S. Mukamel: J. Chem. Phys. **70**, 463 (1979)
6.141 R. Bray, M.J. Berry: J. Chem. Phys. **71**, 4909 (1979)
6.142 K. Kay: J. Chem. Phys. **75**, 1691 (1981)
6.143 J. Stone, E. Thiele, M.F. Goodman: Chem. Phys. Lett. **71**, 171 (1980)
6.144 J. Stone, E. Thiele, M.F. Goodman: J. Chem. Phys. **75**, 1712 (1981)
6.145 J. Stone, E. Thiele, M.F. Goodman, J.C. Stephenson, D.S. King: J. Chem. Phys. **73**, 2259 (1980)
6.146 J.C. Stephenson, S.E. Bialkowski, D.S. King, E. Thiele, J. Stone, M.F. Goodman: J. Chem. Phys. **74**, 3905 (1981)
6.147 M.J. Rossi, J.R. Barker, D.M. Golden: J. Chem. Phys. **76**, 406 (1982)
6.148 A.S. Pine, A.G. Robiette: J. Mol. Spectrosc. **80**, 388 (1980); C.W. Patterson, B.J. Krohn, A.S. Pine: J. Mol. Spectrosc. **88**, 133 (1981)
6.149 D.P. Hodgkinson, A.G. Robiette: Chem. Phys. Lett. **82**, 193 (1981)
6.150 R.L. Woodin, R.B. Hall, C.F. Meyer, A. Kaldor: Abstr. Pap.-Am. Chem. Soc. **182**, 100 (1981)
6.151 E. Thiele, J. Stone, M.F. Goodman: Chem. Phys. Lett. **69**, 18 (1980); W.L. Hase: Chem. Phys. Lett. **67**, 263 (1979)
6.152 J. Stone: Phys. Rev. A**26**, 1157 (1982)
6.153 G. Hose, H.S. Taylor: J. Chem. Phys. **76**, 5356 (1982)
6.154 H.W. Galbraith, J.R. Ackerhalt: Chem. Phys. Lett. **84**, 458 (1981)

7. A Method of Laser Isotope Separation Using Adiabatic Inversion

G. L. Peterson and C. D. Cantrell

With 6 Figures

This chapter proposes a laser isotope separation scheme based on the phenomenon of adiabatic inversion. A model of a multilevel system is presented in which population may be transferred from the ground state to higher excited states with nearly 100% efficiency, using laser pulses with field strengths varying slowly compared to the response time of the system. Adiabatic inversion in this model is sharply dependent upon the laser frequency, allowing separation of isotopes by taking advantage of small isotopic shifts in molecular energy level spacings. Furthermore, a criterion that determines if a laser pulse is adiabatic or not also depends sharply on frequency, thereby enhancing the separation. A time-dependent solution of the Schrödinger equation for a laser pulse with a slow rise time and rapid fall time explicitly demonstrates the effectiveness of the method in selectively exciting only one of two different isotopes.

7.1 Background

Adiabatic inversion in a two-level system was predicted by *Treacy* using the Bloch equations [7.1]. By slowly varying the laser frequency from below to above resonance during the laser pulse (or vice versa) he found that population could be transferred from the lower to the upper state with nearly 100% efficiency. In a subsequent paper he and *DeMaria* demonstrated the effect experimentally [7.2].

Grischkowsky and his coworkers later predicted adiabatic inversion on a two-photon transition using variations in field strength alone [7.3,4]. One- and two-photon adiabatic inversion was demonstrated experimentally by *Loy* using an external dc electric field to shift the energy levels of the system during the laser pulse [7.5,6]. Of particular importance in recent years is the theoretical work of *Kuz'min* and *Sazonov* [7.7,8], whose analytical calculations predict inversion by adiabatic changes in field strengths for a number of multilevel systems. They also noted the sharp frequency dependence of adiabatic inversion and recognized its isotope separation possibilities. Adiabatic inversion has been predicted for several multilevel systems by *Peterson* et al. [7.9] using numerical techniques; these authors also applied an adiabatically condition due to *Messiah* [7.10] to arrive at a general expression for the limits of the pulse rise times that must be respected by multilevel systems.

The work presented in this chapter differs in several respects from previous work in the area. First, Messiah's adiabatic and sudden conditions are applied to give a detailed specification of the pulse shape required for effective isotope separation. Second, the important role of the sharp frequency dependence of Messiah's adiabatic condition in enhancing the separation is presented; this effect was not considered by previous authors. Finally, a numerical solution of the time-dependent Schrödinger equation for a specific model of two isotopes explicitly demonstrates the effectiveness, at least in principle, of the technique.

7.2 Physical Principles

The multilevel system illustrated in Fig.7.1 exhibits many of the important features of adiabatic inversion. It consists of a single ground state and three manifolds of closely spaced excited states. Electric dipole transitions are allowed between adjacent manifolds but transitions within a manifold are forbidden. The equations governing such a system under the adiabatic approximation are presented in the next section and in [7.9]. The simple model system shown in Fig.7.1 does not, however, have some of the features of real atomic or molecular systems such as degeneracy of the levels or multiplicity of the ground state. The effects of these features on adiabatic excitation are currently under investigation.

Figure 7.1 shows two frequencies at which adiabatic inversion will occur (there are others). One-photon adiabatic inversion is achieved by tuning the laser frequency so that $\hbar\omega_1$ lies just within the first manifold of states. A computer calculation of the net population in the first manifold versus applied field strength at $\nu_1 = 947.93\ \mathrm{cm}^{-1}$ is summarized by the solid curve in Fig.7.2. The inversion is

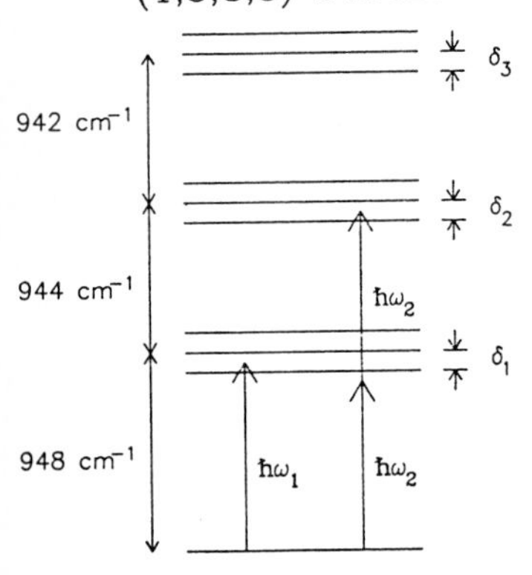

Fig.7.1. Multilevel system used in the calculations in this chapter. Parameter: $\delta_1 = 0.1\ \mathrm{cm}^{-1}$, $\delta_2 = 0.08\ \mathrm{cm}^{-1}$, $\delta_3 = 0.06\ \mathrm{cm}^{-1}$. The band strengths are $\Sigma_{A,B}\ \mu^2_{nA;n+1,B} = (0.4\ \mathrm{esu})^2$

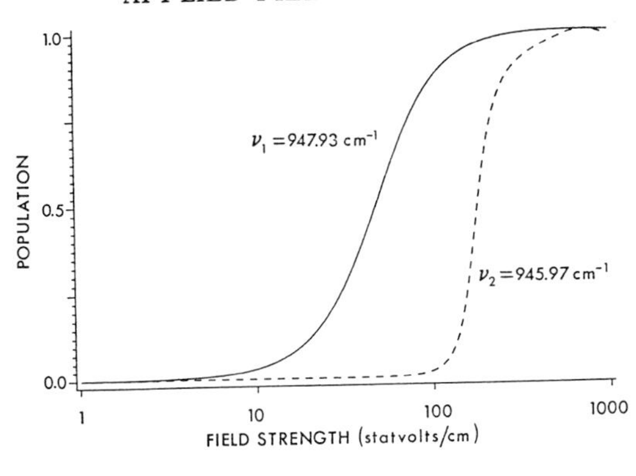

Fig.7.2

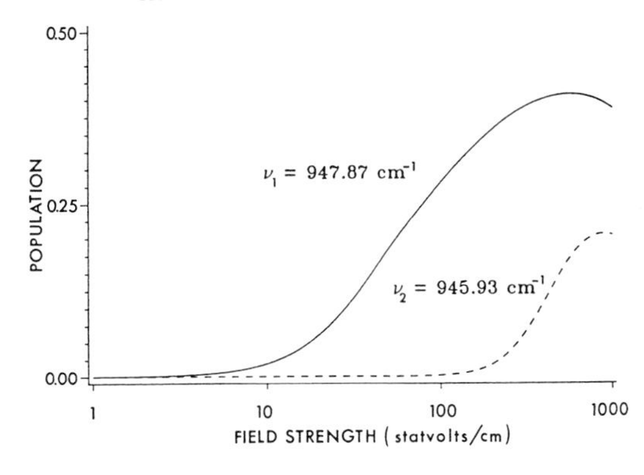

Fig.7.3

Fig.7.2. (——): Net population in the first manifold versus field strength for a one-photon transition with $\nu_1 = 947.93$ cm^{-1}. (---): Net population in the second manifold versus field strength for a two-photon transition with $\nu_2 = 945.97$ cm^{-1}

Fig.7.3. (——): Net population in first manifold versus field strength for a one-photon transition with $\nu_1 = 947.87$ cm^{-1}. (---): Net population in the second manifold versus field strength for a two-photon transition with $\nu_2 = 945.93$ cm^{-1}

nearly complete at a laser electric field amplitude $E_0 = 100$ statvolts cm^{-1} corresponding to an intensity $I = 1.2$ MW cm^{-2}. Two-photon adiabatic inversion is possible when $2\hbar\omega_2$ lies just within the second manifold of states. An example of two-photon inversion is given by the broken curve in Fig.7.2, representing the net population in the second manifold versus field strength when $\nu_2 = 945.97$ cm^{-1}. A higher field strength is required here with complete inversion occurring around $E_0 = 400$ statvolts cm^{-1} corresponding to $I = 9.5$ MW cm^{-2}.

Adiabatic inversion is sharply dependent on the frequency of the laser field. If the laser is tuned below a manifold rather than within it, no inversion will occur, as demonstrated by a calculation summarized in Fig.7.3. At best the population in the first manifold, indicated by the solid curve, is barely above 40% for the one-photon transition with $\nu_1 = 947.87$ cm^{-1}. The two-photon transition, at $\nu_2 = 945.93$ cm^{-1}, is even less successful with less than 25% of the population moving to the second manifold as indicated by the broken curve.

The sharp frequency dependence displayed in this model may be used as a basis for laser isotope separation. If the energy levels of a molecule containing one isotope are shifted up or down with respect to those of the same molecule containing a different isotope, one may tune the laser frequency $\hbar\omega_1$ or $2\hbar\omega_2$ within the manifold of one isotope, and below the manifold of the other. This situation is illustrated for two-photon adiabatic inversion in Fig.7.4. One isotope will then be almost completely inverted while the other will achieve a manifold population of less then 50%, providing a separation between the two.

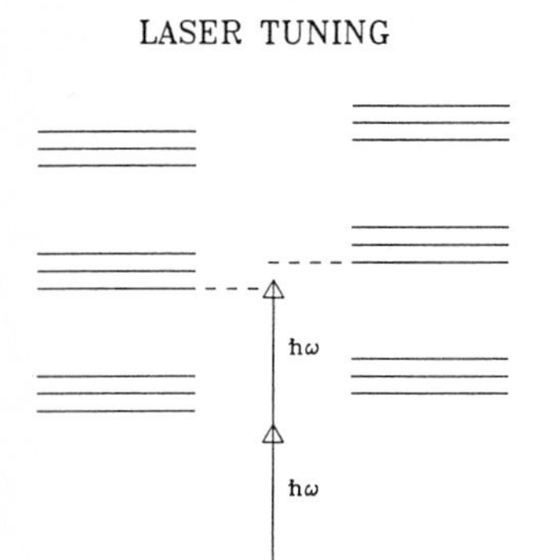

Fig.7.4. Relation of energy levels in isotopes A and B to the laser frequency required for adiabatic laser isotope separation

7.3 Multilevel Systems Under the Adiabatic Approximation

The state of an N-level system such as that in Fig.7.1 may be expanded in terms of the unperturbed energy eigenstates ψ_{mA} to give

$$\Psi(t) = \sum_{mA} c_{mA}(t)\psi_{mA} \quad .$$

The index m labels the manifold (m = 0 for the ground state) while A labels the states within the manifold. Substituting this expansion into the time-dependent Schrödinger equation with $H = H_0 - \mu E_0(t)\cos\omega t$ (μ is the component of the dipole operator parallel to the electric field), making the substitution $\tilde{c}_{mA} = \tilde{c}_{mA}\exp(-im\omega t)$, and neglecting rapidly varying terms (rotating wave approximatin) gives the following equation for the $\tilde{c}$'s:

$$i\dot{\tilde{c}}_{mA} = -\Delta_{mA}\tilde{c}_{mA} - \sum_{B}(\Omega_{mA,(m+1)B}\tilde{c}_{(m+1)B} + \Omega_{mA,(m-1)B}\tilde{c}_{(m-1)B})$$

where

$$\Delta_{mA} = m\omega - \frac{E_{mA}}{\hbar} \quad ,$$

$$\Omega_{mA,nB} = \frac{\mu_{mA,nB}E_0(t)}{2\hbar}\,\delta_{m,n\pm1} \quad ,$$

and $\dot{\tilde{c}}_{mA}$ indicates the time derivative of $\tilde{c}_{mA}$. If the quantities $\tilde{c}_{mA}$ are viewed as the components of a column vector, this equation may be written in the form

$$i\dot{\tilde{c}} = -\tilde{H}\tilde{c} = -(\Delta + \Omega)\tilde{c}$$

with Δ a diagonal matrix of detunings and Ω an off-diagonal matrix involving the field strength. Now, if the laser-pulse envelope $E_0(t)$ varies sufficiently slowly, and the system is initially in its ground state, *Messiah*'s adiabatic condition

[7.10] states that the solution of this equation (apart from an overall time-dependent phase factor) will be one of the eigenvectors of

$$\tilde{H}\mathbf{a}^{(i)} = \lambda_i \mathbf{a}^{(i)} \quad .$$

The states corresponding to these eigenvectors are known as dressed states and may be denoted by $|\lambda_i\rangle$:

$$|\lambda_i\rangle = \sum_{mA} a_{mA}^{(i)} \Psi_{mA} \quad .$$

The particular eigenvector of interest, denoted by $|\lambda_0\rangle$, is the one which evolves continuously from the ground state as $E_0(t)$ rises slowly from zero. This dressed state is said to be correlated with the ground state. The populations in the energy eigenstates of the multilevel system are then given by $|c_{mA}|^2 = |\tilde{c}_{mA}|^2 = |a_{mA}^0|^2$.

7.4 Conditions for Effective Adiabatic Laser Isotope Separation

If the separation scheme discussed in Sect.7.2 is to be effective the shape of the laser pulse must be chosen with some care. One must have a criterion for deciding how slowly the pulse must be turned out to guarantee that the system behaves adiabatically. However, if the inversion is to persist after the pulse has passed, the trailing edge must violate the adiabatic condition, for otherwise the system will evolve adiabatically back into the ground state. Thus a criterion is also needed for deciding how quickly to cut off the pulse in order to trap population in the excited states.

Messiah's conditions for the adiabatic and sudden approximations [7.10], applied to multilevel systems, prescribe the pulse shape one may use to create and maintain adiabatic inversion. Messiah's adiabatic condition places a lower limit on the rise time of a laser pulse if adiabatic inversion is to occur. It requires the relative rate of variation $\dot{E}_0/E_0 = 1/\tau_R$ to satisfy

$$\tau_R = \left(\frac{\dot{E}_0}{E_0}\right)^1 \gg \left(\sum_{i \neq 0} \left|\frac{|\langle\lambda_i|\Delta|\lambda_0\rangle|}{(\lambda_i - \lambda_0)^2}\right|^2\right)^{\frac{1}{2}}$$

as the pulse rises from zero. Here E_0 is the field strength of the laser pulse, $|\lambda_i\rangle$ is the ith dressed state with $|\lambda_0\rangle$ being the dressed state correlated with the ground state, and Δ is the matrix of detunings, which is not diagonal in the dressed-state basis. In the unperturbed energy eigenstate basis $\Delta_{mA,mA} = m\omega - E_{mA}/\hbar$. For the system of Figs.7.1-3 this condition requires $\tau_R > 10$ ns when the laser frequency is tuned within a manifold of states for both one- and two-photon transitions.

Messiah's sudden condition places an upper limit on the fall time of the pulse if adiabatic inversion is to persist after it has passed by. Applying this condition to the adiabatically inverted state $|\lambda_0\rangle$ gives

$$\tau_F \ll \frac{2}{(\langle\lambda_0|\Omega^2|\lambda_0\rangle - \langle\lambda_0|\Omega|\lambda_0\rangle^2)^{\frac{1}{2}}}$$

where $\Omega = \mu E_0/2\hbar$. For the given system this reduces to $\tau_F \ll 10^{-1}$ ns when the laser frequency is tuned within a manifold of states. Thus a pulse which has a rise time of more than 10 ns and a fall time of less than 10^{-1} ns will trap the system in an inverted condition.

Messiah's adiabatic condition has a sharp frequency dependence which may be used to enhance isotope separation based on adiabatic inversion. For the previous system, while Messiah's adiabatic condition for a laser tuned inside the manifold is $\tau_R > 10$ ns, the condition for a laser tuned outside the manifold is $\tau_R > 10^{-2}$ns. Thus if a laser is tuned within the manifold of one isotope but below the manifold of another, a pulse of rise time 10 ns will be adiabatic for both. However, a fall time of less than 10^{-1}ns will be sudden for the isotope with the laser tuned inside the band, trapping the inversion, but may be adiabatic for the isotope with the laser tuned below the band, allowing what little population was transferred to higher manifolds to adiabatically revert back to the ground state, enhancing the separation.

A numerical solution of the time-dependent Schrödinger equation explicitly demonstrates laser isotope separation using pulses with slow rise times and rapid cutoffs. Figure 7.5 displays the pulse applied to the (1,3,3,3) systems of Fig.7.4. This pulse has a rise time of the order of 20 ns, adiabatic for both isotopes, and

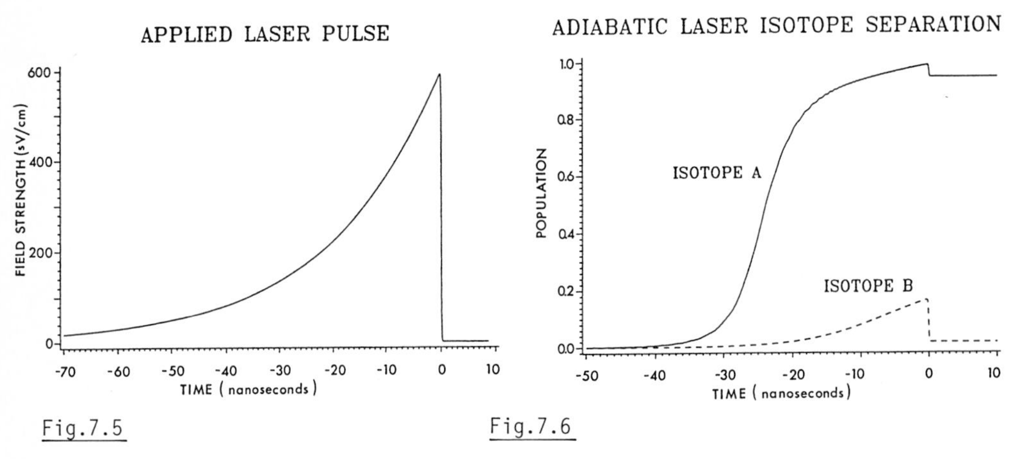

Fig.7.5 Fig.7.6

Fig.7.5. Laser pulse applied to the multilevel systems of Fig.7.4, with $\tau_R = 20$ ns and $\tau_F = 0.05$ ns

Fig.7.6. (——): Net population in the second manifold versus time for isotope A. (---): Net population in the second manifold versus time for isotope B. Isotope B is identical to A except the second manifold of states is raised 0.1 cm^{-1} in B

a fall time of about 0.05 ns = 50 ps, which is sudden for isotope A but adiabatic for isotope B. Its frequency is $\nu = 945.97\ \mathrm{cm}^{-1}$ which is within the second manifold of isotope A, but below the second manifold for isotope B. Figure 7.6 summarizes the populations of the second manifold for isotopes A (solid curve) and B (broken curve). Isotope A is inverted to nearly 100% while B is nearly unexcited. Note the enhancement of separation by designing the pulse to be rapid for isotope A but adiabatic for isotope B which is returned to the ground state as the pulse falls to zero.

References

7.1 E.B. Treacy: Phys. Lett. **27**A, 421 (1968)
7.2 E.B. Treacy, A.J. DeMaria: Phys. Lett. **29**A, 369 (1969)
7.3 D. Grischkowsky, M.M.T. Loy, P.F. Liao: Phys. Rev. A**12**, 2514 (1975)
7.4 D. Grischkowsky, M.M.T. Loy: Phys. Rev. A**12**, 1117 (1975)
7.5 M.M.T. Loy: Phys. Rev. Lett. **32**, 814 (1974)
7.6 M.M.T. Loy: Phys. Rev. Lett. **41**, 473 (1978)
7.7 M.V. Kuz'min, V.N. Sazonov: Zh. Eksp. Teor. Fiz. **79**, 1759 (1980) [English transl.: Sov. Phys.-JETP **52**, 889 (1980)]
7.8 M.V. Kuz'min, V.N. Sazonov: Zh. Eksp. Teor. Fiz. **83**, 50 (1982) [English transl.: Sov. Phys.-JETP **56**, 27 (1982)]
7.9 G.L. Peterson, C.D. Cantrell, R.S. Burkey: Opt. Commun. **43**, 123 (1982)
7.10 A. Messiah: *Quantum Mechanics*, Vol.II (North-Holland, Amsterdam 1963) pp. 739-759

8. Three-Level Superfluorescence

F. P. Mattar, P. R. Berman, A. W. Matos, Y. Claude, C. Goutier, and C. M. Bowden

With 30 Figures

This chapter reviews recent work where dynamic pump depletion and quantum fluctuations are examined in the buildup of three-level superfluorescence (SF) with the use of a unidirectional semiclassical model in conjunction with either an average Langevin force or quantum fluctuations as an initiation process. Using the computational methods previously developed for pulse propagation studies, the dynamic evolution of SF from an optically pumped system has been studied. The code used has been extensively applied to self-induced transparency (SIT) and two-level SF and is well documented in the literature. Results generated by it have been validated by quantitative agreement with experiments by several groups on Na, Ne, I, HF, Cs, and ^{87}Rb.

The full longitudinal and transverse reshaping associated with the concomitant propagation of two light beams in a three-level medium has been evaluated for the first time. Neither the mean field theory nor the adiabatic following nor even the rate equation or the weak overlap approximations have been used to simplify this analysis. Instead, the full Maxwell-Bloch equations with phase and diffraction coupling effects included have been solved rigorously, using self-consistent numerical methods.

Specification of certain initial pump beam conditions at a given frequency results in specific SF characteristics at another frequency, as recently observed in CO_2-pumped CH_3F and Ba. That is, the output characteristics of the collective spontaneous emission of the SF pulse (delay time, pulse width, peak intensity, shape, etc.) can be controlled, deterministically, by appropriately selecting certain initial and boundary conditions for the injected pump pulse. In particular, it has been shown that the injected coherent pump pulse initial characteristics, such as on-axis area, temporal and radial width and shape, can have significant deterministic effects on the SF pulse evolution.

Pump dynamics have been studied in terms of the gain Fresnel number associated with the Beers absorption length F_g and the ratio of cooperation time τ_R to pulse duration τ_p for a given atomic density N. For large optical thickness αl, strong overlap between the SF and the pump occurs leading to two-photon effects. For large Fresnel numbers and/or comparable τ_R and τ_p, the SF exhibits phase wave characteristics (i.e., temporal distortion) only when the pump Langevin force is present.

This means that temporal incoherence in the pump will manifest itself as SF distortion whenever the diffraction is not sufficiently strong. Additional dispersion effects associated with detuning of the pump or off-resonance superfluorescence that affect the evolution of self-action phenomena are presented.

8.1 Background

Spontaneous emission by an excited atom is a random process in which the stored energy is emitted in the natural lifetime τ_{sp} of the excited state. In 1954, *Dicke* [8.1] predicted that under certain conditions all the energy could be released cooperatively in a much shorter time, τ_{sp}/N, where N is the effective number of excited atoms [8.1-3]. In this process, which he called superradiance, the atoms would decay cooperatively, instead of independently. The emission intensity would then be proportional to N^2, instead of N as expected for incoherent radiation.

The concept of N^2 emission in an array of driven electrical or magnetic dipoles is well known. For instance, when the dipoles are confined to a volume with linear dimensions smaller than the wavelength of the driving field, they oscillate in phase and the resulting emission is proportional to the square of the number of dipoles. Such N^2 intensities have been recognized in the field of NMR, for example, in spin echoes [8.4-6]. There the pulses are much shorter than the lifetime divided by N because the lifetime of an individual spin is enormously long and there are always some broadening mechanisms. Experimental verification of Dicke superradiance only became feasible with the advent of sources of intense laser radiation, albeit with volumes large compared to a cubic wavelength, and resultant qualifications, i.e., the pulse intensity, delay, and width are proportional to N^2, 1/N, and 1/N, respectively. *Friedberg* and *Hartmann* [8.5] noted a high gain requirement to insure that coherent decay processes predominate over incoherent decay $\alpha L \geqslant 1$ and that superradiance occurs.

The first experiment to demonstrate Dicke superradiance was reported in 1973 by *Skribanowitz* et al. in rotational transitions of HF gas [8.7] Like all subsequent superradiance experiments [8.8-27], a long, optically thick vapor of two-level atoms (a HF rotational transition in this case) was prepared in a state of total inversion by indirectly (i.e., incoherently) pumping to the upper level with a short light pulse, in this case via a coupled ground-state transition (Fig.8.1). There was no optical cavity and stray feedback was negligible. In the first part of the paper, as in most other theoretical treatments, the dynamical effect of the pump was not considered; only the relaxation process from the state of complete inversion in a two-level manifold was treated. After a relatively long delay, proportional to 1/N, an intense pulse of radiation was emitted at the superradiant transition, which totally deexcited the sample. The peak intensity was found to be

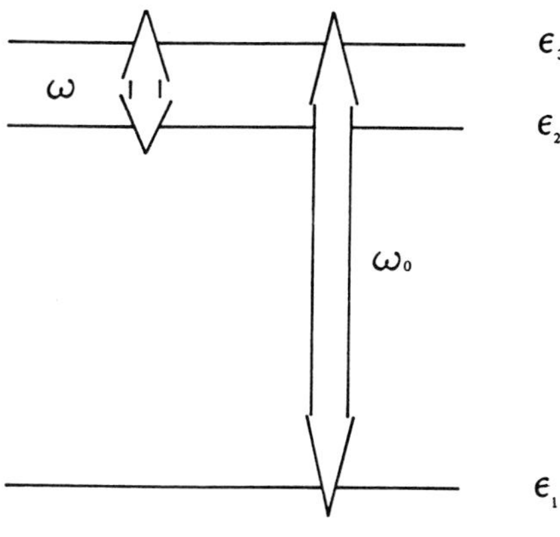

Fig.8.1. Model three-level atomic system and electric field tunings under consideration. For the results reported here, the injected pulse is tuned to the 1 ↔ 3 transition and the SF emission radiates at the 3 ↔ 2 transition

proportional to N^2, and the emission process was completed in a time six orders of magnitude smaller than the radiative lifetime. Under many experimental conditions, the emitted pulses exhibited strong ringing. This interesting feature, not predicted by theories developed at that time [8.28-30], was attributed to propagation effects [Refs.8.2, pp.27-33; 8.3,7,8]. These results were interpreted by means of a propagation model based on the coupled Maxwell-Schrödinger ("semiclassical") formalism [8.7,8,31]. For simplicity, the superradiant pulse was assumed to be in the form of a plane wave ("one-dimensional" model). The theory was found to predict all of the observed features, and to be in reasonable agreement with the data.

A noteworthy feature of these experiments is that the atoms are prepared in an inverted state, and hence initially do not possess a macroscopic polarization. Classically, the sample would not radiate, except in the presence of blackbody radiation. The triggering of the superradiant emission process is in fact a fundamental quantum phenomenon, brought about by random spontaneous emission events and/or blackbody radiation. Hence, the consequent fluctuations in delay time, peak intensity, and pulse shape provide information about quantum statistics [8.22,23]. In the semiclassical propagation model, these fluctuations are simulated by a small initiating pulse of area (tipping angle) θ_0 [8.7,8], or by a polarization source distributed throughout the medium [8.31]. Dicke superradiance emitted by an initially inverted sample is often called superfluorescence (SF) [8.32]. All observations of Dicke superradiance to date have been of this type.

Following the HF studies, SF was studied in several atomic [8.9-14] and molecular [8.15-17] systems, and the general features of the initial observation were confirmed: pulse intensities proportional to N^2, pulse delays proportional to 1/N, and ringing.

In 1976, *Gibbs* et al. carried out a careful set of quantitative observations on the $7P_{3/2}$-$7S_{1/2}$ transition of atomic Cs at 2.33 μm in a 3 kg magnetic field [8.18]. The transition chosen was free of M-degeneracies and the experimental parameters obeyed a set of conditions formulated in the mean field theory (MFT) approximation of *Bonifacio* et al. [8.32]. Mean field theories inherently excluded propagation, and generally predicted "single-pulse" (i.e., no ringing) emission. The Cs experiment established the existence of a regime of single-pulse emission, in qualitative agreement with the MFT, but the observed pulse widths and delays did not agree [8.19-21]. The disagreement was quantitative to at least a factor of 2, i.e., the pulse width was found to be at least twice as large as foreseen by *Bonifacio* and *Lugiato* [8.32b] and closer to the estimate of *MacGillivray* and *Feld* [8.29,33-35]. Independently, *Ressayre* and *Tallet* concluded that SF cannot be described accurately by a one- or two-mode model, i.e., geometrical and propagational effects are essential for a quantitative description [8.36].

Furthermore, the SF output was shown to fluctuate from shot to shot, i.e., when the sample was repeatedly prepared with the atoms in the excited state. These fluctuations in both delay time and shape are of quantum origin (as has been shown by *Haake* and co-workers [8.37-40] and *Vrehen* and co-workers [8.41,42]); they correspond to the initial quantum uncertainties in the state of the field and the atomic system. Thus, SF offers the unique possibility of studying amplified microscopic quantum fluctuations in the time domain, as was noted by *Vrehen* et al. [8.20].

These SF experiments raised the question: Why is the output pulse sometimes smooth, but at other times exhibits multiple structure or ringing? *MacGillivray* and *Feld* proposed that the lack of ringing was due to the non-plane-wave nature of the evolving SF pulse [8.29,33-35], but this idea was not pursued. Another explanation put forward was that the initial tipping angle θ_0 is of the order of 10^{-2}rad, which is much larger than the generally accepted value of $\leqslant 10^{-4}$rad. The propagational model predicted that such a large value would greatly curtail the ringing and reduce the decay. However, a direct measurement of θ_0 by *Vrehen* and *Schuurmans* [8.22] showed that such a large value could be excluded.

In 1977, two quantized field treatments which rigorously incorporated propagation effects and described the quantum initiation process, were developed, one by *Glauber*, *Haake* and co-workers [8.37-40] and the other by *Polder*, *Schuurmans*, and *Vrehen* [8.41,42]. These treatments were developed to describe the observed statistical fluctuations and, hopefully, to account for the lack of ringing. The predicted fluctuations agreed with the Cs observations but, due to the plane-wave nature of the theory, the lack of ringing could not be accounted for. These quantized field treatments confirmed the validity of using the semiclassical approach, and showed that the quantum initiation process could be properly included in the propagational model by means of a set of initiating polarization sources or fields randomly distributed throughout the medium [8.43,44].

By this time, it was generally accepted that the lack of ringing was due to the non-plane-wave nature of the evolving SF pulse. *Bonifacio* et al. [8.45] examined transverse effects in the MFT approximation, while *Feld* and *MacGillivray* [8.46] and *Bullough* et al. [8.47] included a loss term in the Maxwell equation to describe diffraction. The results of these treatments gave qualitative support to the role of diffraction in reducing the ringing, but quantitative agreement with single-pulse observations was poor.

Mattar and co-workers [8.48-52] tackled the problem of relaxing the various limiting approximations to rigorously assess the validity of the propagational model. They found that transverse effects largely removed the strong ringing so prominent in the plane-wave calculations. Additionally, inclusion of broadening and/or initial fluctuations achieved agreement with experimental data, and further reduce the tail. All refinements of the propagational model have made for improved quantitative agreement with published experimental results and predict conditions for further experiments, as well as providing a reliable basis for interpretation and fundamental understanding of phenomena. In contrast, the mean field theory *cannot* come close to obtaining quantitative agreement with the experimental observations.

This thorough study of the propagational model which combines *all* experimental parameters has a profound fundamental physical significance. In particular, the simultaneous treatment of statistics and transverse effects is important because the calculation displays the amplified quantum fluctuations which are unique to this particular system. It is the first study to evaluate their interplay (that of diffraction and fluctuation) and to assess their significance. They are macroscopic manifestations of quantum fluctuations, even though they are proportional to the inverse of N, the number density of atoms. One usually does not expect to be able to see these small fluctuations for N greater that 10. However, in SF one can easily (macroscopically) see this 10% fluctuation for N as large as 10^8.

Mattar's model, which included a transverse Laplacian term in the reduced Maxwell equation, has been developed to describe transverse (i.e., non-plane-wave) field variations [8.53] in self-induced transparency (SIT) [8.54]; its validity has been confirmed by several SIT experiments in sodium, neon and iodine [8.55-58]. The results of the non-plane-wave SF analysis provided the first complete explanation of the absence of ringing and, for the first time, gave quantitative agreement (within measurement uncertainties) with the definitive Cs experiments.

Transverse effects influence the SF pulse evolution process in two major ways, one of which is spatial averaging. In SF experiments the initial inversion density is radially dependent, since the pump light pulse which inverts the sample has typically a Gaussian-like profile. In the absence of diffraction, the resulting cylinder of inverted atoms can be thought of as a set of concentric cylindrical shells, each with its own density, initial conditions, and delay time. The radiation is a sum of plane-wave intensities; when the entire output signal is viewed, the ringing averages out, resulting in an asymmetric pulse with a long tail [8.33-35]. This characteristic, which was reported by *Mattar* et al. [8.49] in their computational analysis, was recently confirmed in Rb observations by *Heinzen* et al. [8.59].

A second transverse effect, diffraction, is unavoidable in any SF situation. Diffraction becomes important in samples with small Fresnel number $F = A/\lambda L$, with L the cylinder length and A the cross section. This effect causes light emitted by one shell to affect the emission from adjacent shells. It couples together atoms in various parts of the cylindrical cross section so that they tend to emit at the same time. This coupling mechanism causes transverse energy flow and, in addition, slightly increases the delay and reduces the tail and asymmetry [8.33-35].

Therefore, SF is an inherently transverse-effect problem, even for large F samples, since the off-axis modes develop and diffraction couples them together. That work was a collaboration of *Mattar* with *Gibbs*, *McCall*, and *Feld* [8.48,49] and was the first one to correctly include this crucial, long-sought-for element [8.60].

With the exception of *Bowden* and *Sung*'s [8.61] theoretical analysis in the mean field theory, all theoretical studies have dealt exclusively with the relaxation process from a prepared state of complete inversion in a two-level manifold of atomic energy levels, and thus do not consider the dynamic effects of the pumping process. Yet, all reported experimental work has utilized optical pumping on a manifold of a minimum of three atomic [8.7-14,18-23] or molecular [8.15-17,24-27] energy levels by laser pulse injection into the nonlinear medium, which subsequently superfluoresces. (Note that the two-level analysis is only valid for $\tau_{RSF} \gg \tau_P$, where τ_{RSF} is the characteristic SF time and τ_P is the pump pulse temporal width, and this has not been realized over the full range of reported data.) This study addresses the issue of a finite pump, which gets depleted and affects the upper-level population of the SF transition.

Our analysis extends *Bowden* and *Sung*'s analytical treatment [8.62]. That is, we do not confine our solution to the linearized short-time regime but have modified the semiclassical model advanced by *Feld* and co-authors [8.7,8,31,33-35], where both transient and propagation effects are rigorously studied, to include quantum fluctuations [8.37-42,62,63]. This is compared with a classical deterministic (not random) tipping angle concept used to initiate the polarization to simulate the fluorescence initiation. The latter method is well established for both two- and three-level [8.48-52,64-70] propagation calculations. The pump depletion results extend the pumpless analysis by *Gibbs* et al. [8.18-21] of the Cs experiment. Since transverse effects are also considered, the present results also generalize the plane-wave studies of CO_2-pumped CH_3F by *Rosenberger* et al. [8.71].

The work described in this chapter includes the additional effect on the SF emission of the pumping that produces the initial population inversion. In all experiments to date, a long cylindrical sample is pumped along its axis, which gives rise to a front of inversion density that moves down the sample at the speed of light. Furthermore, as found by *Bowden* and *Sung* [8.58], even in the mean field approximations and the linearized regime, when τ_R is comparable to or larger than τ_P the process of coherent optical pumping on a three-level system can have dramatic effects on the SF. In particular, as the SF pulse grows and propagates down the medium, its delay (relative to the pump pulse) decreases. Therefore, in a sufficiently long sample, overlap will occur between the pump and evolving pulses, resulting in a strong interaction which depletes the pump pulse and reshapes the evolving pulse in a way similar to the Raman process. Thus the full nonlinear copropagational aspects of the injected pump pulse, together with the SF which evolves, must be treated explicitly in the calculation, as presented by *Mattar* and *Bowden* [8.69,70].

To analyze the pump dynamics, the physical model must encompass three-level atoms instead of two-level ones. Here the λ-configuration is considered. The $1 \leftrightarrow 3$ transition is induced by a coherent injection pulse of frequency nearly tuned to the identical transition. The properties of this pumping pulse are specified initially in terms of the initial and boundary conditions. The transition $3 \leftrightarrow 2$ evolves by spontaneous emission at frequency ω. It is assumed that the energy-level spacing is such that $\varepsilon_3 > \varepsilon_2 \gg \varepsilon_1$ (Fig.8.1) so that the fields at frequencies ω_0 and ω can be treated by separate wave equations. However, the nonlinear material closely intertwines the two fields. The energy levels 1 and 2 are not coupled radiatively due to parity considerations. We neglect spontaneous relaxation in the $3 \leftrightarrow 1$ transition whereas in the $3 \leftrightarrow 2$ transition it is simulated by the choice of a small but nonzero initial transverse polarization characterized by the parameter $\psi_0 \sim 10^{-3}$, as described in [8.64,66-68,72]. This small-signal initiation is distinct from quantum initiation [8.37,42] since the latter is a linear process whereas Eqs. (A-16) - (A-18) in [8.69] are nonlinear. The two processes are different. The initial condition is consistent with having nearly all the population in the ground state; other atomic variables are chosen consistently according to the initial equilibrium properties of the system as discussed by *Feld* and co-workers [8.73]. The initiation is described in terms of a deterministic Langevin force in the SF transition. Once the on-axis 2π pump creates a population inversion in both the $1 \leftrightarrow 3$ and $3 \leftrightarrow 2$ transitions, the SF begins to evolve. This explains why SF arises after the pump peak occurs, i.e., after the evolution of the first half of the pump pulse, whose areas is π.

The full statistical treatment of the quantum initiation process with resulting temporal fluctuations utilizes a bivariate distribution in the driving Langevin force in the material Bloch equation, as was recently derived by *Bowden* and *Sung* [8.62]. In particular, herein we show that the injected coherent pump-pulse initial characteristics, such as on-axis area, temporal and radial width and shape, can have significant deterministic effects on the SF pulse delay time, peak intensity, temporal width and shape. Thus, by specifying certain initial properties of the injected pump pulse, the superfluorescent pulse can be shaped and altered for initialization by both an average initial tipping angle [8.69,70] and an ensemble of (quantum) fluctuations [8.74]. The full nonlinear set of results predicts the conditions under which an injected light pulse of a given frequency can be used to generate, shape, and control a second light pulse of a different frequency via a nonlinear medium, even with random initiation, thus demonstrating a new aspect of the phenomenon of light control by light. Three specific regions are encountered; the SF buildup, the full development of SF with the pump depleted, and the highly nonlinear regime where the SF and the pump overlap significantly, if not totally.

This phenomenon, known as pump dynamics [8.69,70], is interesting because the characteristics of the superradiant pulse emitted at one frequency, such as on-axis area, temporal and radial widths (and associated gain length Fresnel number), and shape, can be altered, and thus controlled, by specifying certain characteristics of the pump pulse injected at a different frequency [8.69,70]. The effects of changing the effective gain [8.75] of either the SF or the pump transition and the density of active atoms are also studied. For sufficiently large effective gain and/or large input pump area, the two light pulses overlap and the two-photon processes [resonant coherent Raman (RCR)] make large contributions to the mutual pulse development.

Dependencies of this type have been recently observed in methyl fluoride [8.15-17] and in barium [8.27]. Also, under other conditions, we obtained a SF pulse of temporal width much less than that of the pump pulse, even though the two pulses temporally overlapped. This calculation agrees qualitatively with the results of recent experiments in mode-locked CO_2-pumped CH_3F [8.71]. Thus, a new aspect of light control by light was demonstrated [8.69,70].

Recently the quantum fluctuations for initiation of three-level systems were derived by *Bowden* and *Sung*. In particular, they presented in [8.62] a more comprehensive treatment of SF in the linearized regime of SF initiation by combining (i) coherent pump dynamics on the three-level system and (ii) simultaneous as well as subsequent quantum mechanical initiation of the SF emission. Their attention, as is ours, was confined to situations satisfying $\tau_P < \tau_{RSF}$, which is the condition where the effects of SF quantum fluctuations are expected to be most important in terms of subsequent temporal fluctuation. Using their initiation through a Langevin force as additional driving forces in the material (generalized three-level Bloch equations), we studied numerically the nonlinear regimes which include propagation, transverse, and diffraction effects and found that the light control by light remained [8.74]. Furthermore, when a second Langevin force was included at the pump transition, temporal distortion in the superfluorescence output power developed. The resulting pulse exhibits temporal distortion, namely, a first peak (maxima) which is not the highest peak. Such unusual trajectories can be interpreted as phase waves. This is in accordance with the two-level SF calculations carried out by *Watson* et al. [8.64,65]. Similar temporal distortions were discussed by *Hopf* [8.76] in the uniform-plane-wave regime. The apparition of phase waves means that the superfluorescence pulse cannot be shielded from fluctuations occurring in the pump transition. However, when the diffraction coupling strengthens (i.e., the

Fresnel number decreases), the communication improves, thus, the larger the mutual influence between the various shells and the less likely the phase wave output appears. Moreover, when the pump duration shortens for a given cooperation time, the phase wave characteristic is most likely to be exhibited. Furthermore, it must be emphasized that two transverse dimensions are required to describe the more complex case when quantum fluctuation initiation is present, since this destroys the azimuthal symmetry in samples with Fresnel numbers larger than unity.

More recently, we have analyzed the effect of dispersion on the SF buildup within a deterministic initiation process. The interplay of dispersion with diffraction is thus reported in coherently pumped three-level deterministic SF. Due to the presence of detuning, the Beer length is modified, thus altering the effective gain that the SF experiences. The resulting SF pulse width, pulse delay, and pulse intensity vary from one detuning to another. The spectral variation of the SF buildup from a deterministic initiation (that is, a deterministic Langevin force or, equivalently, a deterministic tipping angle) has been rigorously calculated for the first time. Moreover, the phase that the nonresonant pump acquires can compensate the phase evolving from the diffraction coupling. Thus, any self-focusing or self-defocusing the pump may experience can now be subdued, compensated for, and potentially annihilated leading to a coherent quasi-trapping [8.77] situation where the beam radius essentially stabilizes into some nonlinear channeling process. Universal curves are presented to display the interplay of diffraction, dispersion, and Raman action.

8.2 Pump Dynamics Effects in Three-Level Superfluorescence

In this section we present a model and calculational results and analysis for the effects of coherent pump dynamics, pump depletion, quantum initiation, copropagation, transverse and diffraction effects on SF emission from a collection of a density N of optically pumped three-level systems. The full, nonlinear, copropagational aspects of the injected pump pulse, together with the SF which evolves, are explicitly treated in the calculation.

The model upon which the calculation is based is the semiclassical version [8.63] of the fully quantum mechanical model used to derive the effects of quantum initiation in the linearized region of SF in a coherently pumped three-level system [8.62] (i.e., during and after the pump-pulse time frame). The semiclassical aspects amount to representing the SF and pump fields by classical fields determined by the Maxwell equations, including the transverse contribution, and representing the Langevin force fluctuation terms responsible for spontaneous relaxation and quantum initiation by complex-valued c-numbers [8.63]. The amplitudes and phases of the complex-valued fluctuation terms are determined by Gaussian and uniform statistics, respectively, which are derived from the fully quantum model [8.62,63].

The first model for the study of dynamical effects of coherent pumping on SF evolution was the three-level model proposed by *Bowden* and *Sung* [8.61]. The model is composed of a collection of identical three-level atoms, each having the energy-level scheme such that the $1 \leftrightarrow 3$ transition is induced by a coherent electromagnetic field pulse of frequency ω_0 and wave vector k_0. The transition $3 \leftrightarrow 2$ evolves by spontaneous emission at a much lower frequency ω. It is assumed that the energy-level spacing is such that $\varepsilon_3 > \varepsilon_2 \gg \varepsilon_1$, and spontaneous relaxation in the pump transition $1 \leftrightarrow 3$ is retained for generality. The energy levels ε_2 and ε_1 are not coupled radiatively due to parity considerations. The injected pump field is treated as a coherent state [8.78]. The Hamiltonian which describes the system and the corresponding equations of motion have been discussed in [8.62,63,69,70], and the reader is referred to those equations.

The normally ordered Heisenberg equations of motion for the SF fluorescence field obtained from the Hamiltonian for the system [8.62,63] are formally integrated, and then separated into the contribution due to the self-field of the atom, the vacuum contribution, and the contribution due to the presence of all the other atoms (i.e., the extended dipole contribution). The first-mentioned separated field leads to natural atomic relaxation γ^{-1} for the $3 \leftrightarrow 2$ transition in the normally ordered Heisenberg equations, and the vacuum contribution leads to Langevin force terms $f(\tau)$ which satisfy the ensemble average over the vacuum fluctuations (i.e., which are delta correlated).

$$\langle f_{a,b}(\tau) f^{\dagger}_{a,b}(\tau)\rangle = \frac{1}{N_i \tau_{RP,RSF}} \delta(\tau - \tau') \quad , \tag{8.1}$$

$$\langle f^{\dagger}_{a,b}(\tau) f_{a,b}(\tau')\rangle = 0 \quad , \tag{8.2}$$

where N is the total number of atoms and τ_{RP}, τ_{RSF} are the characteristic pump and SF times (i.e., the time in which, on average, one cooperative photon is emitted) given by

$$\tau_{RP,RSF} = \frac{L}{c}\left(\frac{2\pi|\mu_{P,SF}|^2}{\hbar} N_i\right)^{-1} \quad . \tag{8.3}$$

Here, N is the atomic density, L is the longitudinal length of the medium, and μ_P, μ_{SF} are the atom-field coupling in the neighborhood of the resonance for the pump and the SF transition, respectively. The Maxwell equations in retarded time coordinates are derived in a manner similar to that leading to (2.14) of [8.69]. The details of the derivations were presented elsewhere.

Both field operator variables $\bar{E}_a$ and $\bar{E}_b$ and the atomic (material) operator variables $\bar{P}_a$,$\bar{P}_b$ (polarization), $\bar{Q}$ (quadrupole moment), $\bar{W}_a$, and $\bar{W}_b$ [the population differences in transition a ($3 \leftrightarrow 1$) and transition b ($3 \leftrightarrow 2$)] are to be understood as slowly varying functions of z and τ, and the fluctuating force terms f_b and f_a, which are responsible for quantum initiation, can be shown to be functions of retarded time τ only [8.62].

The Langevin terms f_a corresponding to the $1 \leftrightarrow 3$ transition obey relations identical to those of f_b [8.76,79], but with τ_{RSF} replaced by τ_{RP}. The Langevin force terms in the Heisenberg equations of motion [8.62] give rise to Gaussian random quantum initiation statistics in both allowed transitions [8.63]. The pumping field [8.63] envelope E_A is taken as a pulse propagating to the right that is injected into the medium with specified initial and boundary conditions and in general is described by classical Maxwell equations.

8.3 Semiclassical Equations of Motion and Computational Method

The calculation of SF pulse evolution in the nonlinear regime is necessarily a calculational problem if propagation is included explicitly. We use an algorithm presented elsewhere [8.80] and the model defined by the equations of motion [8.62,63] to analyze the effects of coherent pump dynamics, propagation, transverse and diffraction effects on SF emission. To facilitate numerical calculation, the equations of motion are taken in their factorized, semiclassical form [8.69,70] with the field operator replaced by its classical representation, which is described by the Maxwell equations. The pump field and fluorescence field operator are determined dynamically and spatially in retarded time by initial and boundary conditions and the equations

$$E_{a,b} = \mathrm{Re}\{\tilde{E}_{a,b} \exp[i(\omega_{a,b}t - k_{a,b}z)]\} \quad , \tag{8.4}$$

$$P_{a,b} = \mathrm{Re}\{i\tilde{P}_{a,b} \exp[i(\omega_{a,b}t - k_{a,b}z)]\} \quad , \tag{8.5}$$

$$Q = \mathrm{Re}\left\{\tilde{Q} \exp\{i[(\omega_a - \omega_b)t - k_c t]\}\right\} \quad , \tag{8.6}$$

$$E_{a,b} = (2\mu\tilde{E}_{a,b}/\hbar)\tau_{Ra,b} \quad , \tag{8.7}$$

$$P_{a,b} = \tilde{P}_{a,b}/\mu_{a,b} \quad , \tag{8.8}$$

$$\tau = 2(t - z/c)/(\tau_{Ra} + \tau_{Rb}) \quad , \tag{8.9}$$

$$\eta_{a,b} = z/l \quad . \tag{8.10}$$

If one selects l in terms of the unit nanosecond pulse duration $l = c\tau_p$, one recovers the SIT normalization [8.77,81,82], whereas if one selects $l = c\tau_R$, the two normalizations given by (8.10 and 15) become equivalent.

Here $E_{a,b}$ and $P_{a,b}$ are the slowly varying complex amplitudes of the electic field and polarization, respectively; τ is the retarded time; $\eta_{a,b}$ is the normalized axial a,b coordinate; $\mu_{a,b}$ is the transition dipole moment matrix element; and τ_{Ra}, τ_{Rb} are the radiation times for transitions a and b, respectively.

For equal pulse lengths $\tau_{pa} = \tau_{pb} = \tau_p$, one has $l_a = c\tau_{pa}$, $l_b = c\tau_{pb}$, $l = l_a = l_b$ thus $\eta = \eta_a = \eta_b$.

$$- \frac{i}{4F_{a,b}} \nabla_T^2 E_{a,b} + \frac{\partial E_{a,b}}{\partial \eta} = (\alpha_{Ra,b} l) d \int P_{a,b}(\Delta\Omega) g(\Delta\Omega) d(\Delta\Omega) \quad , \tag{8.11}$$

with

$$\alpha_{Ra,b} = \frac{\omega_{0a,b}\mu_{a,b}^2}{n\hbar c} N\tau_{Ra,b} \quad , \tag{8.12}$$

where ω_{0a} and ω_{0b} are the frequencies of the active transitions a $(3 \leftrightarrow 1)$ and b $(3 \leftrightarrow 2)$ of the three-level atomic system a,b,c; N is the atomic number density (assumed longitudinally homogeneous); and n is the refractive index, which is assumed to be identical for each transition wavelength. Also

$$g_{a,b} = \frac{\alpha_{Ra,b} T_2}{\tau_{Ra,b}} \quad , \tag{8.13}$$

where T_2 is a relaxation time and the geometric Fresnel number is

$$F_{a,b} = \frac{\pi r_p^2}{\lambda_{a,b} L} \quad . \tag{8.14}$$

Using

$$\zeta_{a,b} = z\alpha_{Ra,b} \tag{8.15}$$

as the normalized axial a,b coordinate instead of $\eta_{a,b}$, one obtains

$$- \frac{i}{4F_{ga,b}} \nabla_T^2 E_{a,b} + \frac{\partial E_{a,b}}{\partial \zeta_{a,b}} = d \int P_{a,b}(\Delta\Omega) g(\Delta\Omega) d(\Delta\Omega) \quad ; \tag{8.16}$$

the quantity $F_{ga,b}$ is discussed below.

Equivalently, one solves numerically

$$- i\, F_{ga}^{-1} \nabla_T^2 E_a + \partial_{\zeta a} E_a = dP_a \quad , \tag{8.17}$$

$$- i \left(\frac{\alpha_{Rb}}{\alpha_{Ra}}\right) F_{gb}^{-1} \nabla_T^2 E_b + \partial_{\zeta a} E_b = d \left(\frac{\alpha_{Rb}}{\alpha_{Ra}}\right) P_b \quad , \tag{8.18}$$

where $\partial_{\zeta a} \equiv \partial/\partial\zeta_a$.

Diffraction is taken into account by the transverse Laplacian

$$\nabla_T^2 = \frac{1}{\rho} \frac{\partial}{\partial \rho} \left(\rho \frac{\partial}{\partial \rho}\right) \quad , \tag{8.19}$$

with $\rho = r/r_p$, for cylindrical geometry. Furthermore, diffraction is also explicitly taken into account by the boundary condition that $\rho = \rho_{max}$ (or $\xi = \xi_{max}$ and $\zeta = \zeta_{max}$) corresponds to completely absorbing walls (i.e., $\partial_\rho E_{a,b} = 0$). To insure that (1) the entire field is accurately simulated, (2) no artificial reflections are introduced at the numerical boundary $\rho_{max} \gg r_p$, and (3) fine diffraction variations near the axis are resolved, the sample cross section is divided into nonuniform stretching cells as shown in [8.73,83].

If the on-axis effective gain is

$$\alpha_R = \frac{\omega_{0a,b}}{n\hbar c} N\tau_R \tag{8.20}$$

(where α_R^{-1} is the effective Beer length as defined by *Gibbs* and *Slusher* [8.84] in SIT for a sharp-line atomic system, but with T_2 replaced by τ_R, i.e., $\alpha_R = 2\pi\mu^2\omega_0 N\tau_R/\hbar c$), then

$$F_{ga,b} = \frac{\pi r_p^2}{\lambda_{a,b}\alpha_{a,b}^{-1}} = Z_{da,b}\alpha_{a,b} \quad , \qquad \text{with} \tag{8.21}$$

$$Z_{da,b} = \frac{\pi r_p^2}{\lambda_{a,b}} = \kappa_{a,b}^{-1} \quad . \tag{8.22}$$

Here Z_d is the diffraction length, also known as the Rayleigh length. The gain length Fresnel number $F_{ga,b}$ is related to the usual geometric Fresnel number $F_{a,b}$ by

$$F_{ga,b}/F_{a,b} = \alpha_{Ra,b}L \tag{8.23}$$

i.e., the total gain of the medium. One can think of $F_{ga,b}$ as the ratio of effective gain $\alpha_{Ra,b}^{-1}$ to diffraction loss $\kappa_{a,b}^{-1}$; whereas $F_{a,b}$ can correspondingly be thought of as the reciprocal of the strength of the diffraction loss $\kappa_{a,b}$ for a length L [8.66-70].

$$F_{ga,b} = \alpha_{Ra,b}/\kappa_{a,b} \qquad \text{and} \tag{8.24}$$

$$F_{a,b} = (\kappa_{a,b}L)^{-1} \quad . \tag{8.25}$$

Mattar's model automatically includes the effects of both spatial averaging and diffraction coupling. The first calculations described a geometry with cylindrical symmetry (two spatial dimensions). Subsequent calculations for the two-level case have been extended to the more complex case where azimuthal symmetry is absent and two transverse dimensions are required [8.52,64,65]. The latter model is needed to describe short-time-scale phase and amplitude fluctuations which result in multiple transverse mode initiation and lead to multidirectional output with hot spots. This effect is only important in samples with large Fresnel numbers, since diffraction singles out a smooth phase front in small F samples.

However, for the three-level case we have for the moment limited our analysis to cylindrical geometry and left the Cartesian calculation for future work.

$$\tau_{Ra,b} = \frac{\hbar\lambda_{a,b}}{4\pi^2\mu_{a,b}^2 N_0^0 L} = \frac{8\pi\tau_0}{3N_0^0\lambda_{a,b}^2 L} \quad , \tag{8.26}$$

where N_0^0 is the on-axis atomic density $N_0(\rho=0)$. From the population relaxation times $T_1^{a,b}$, the polarization dephasing times $T_2^{a,b,c}$, and the Doppler broadening time T_2^*, the normalized relaxation times are obtained as follows:

$$\tau_1^{a,b,c} = \frac{2T_1^{a,b,c}}{\tau_{Ra} + \tau_{Rb}} \quad , \tag{8.27}$$

$$\tau_2^{a,b} = \frac{2T_1^{a,b}}{\tau_{Ra} + \tau_{Rb}} \quad , \tag{8.28}$$

$$\tau_2^* = \frac{2T_2^*}{\tau_{Ra} + \tau_{Rb}} \quad . \tag{8.29}$$

$$\Delta\Omega_{a,b} = (\omega_{0a,b} - \omega_{a,b})(\tau_{Ra} + \tau_{Rb})/2 \quad , \tag{8.30}$$

$$\int_{-\infty}^{\infty} g(\Delta\Omega)d(\Delta\Omega) = \left(\frac{\pi}{\tau_2^*}\right)^{\frac{1}{2}} \int_{-\infty}^{\infty} \exp\{-[(\Delta\Omega)\tau_2^*]^2\}d(\Delta\Omega) = 1 \quad , \tag{8.31}$$

$$d = \exp[-(\rho/\rho_N)^m] = \frac{N(\rho)}{N(0)} \quad , \tag{8.32}$$

where ρ_N is the (1/e) radial width of the atomic density distribution N. For uniform density, $d = 1$, i.e., $\rho_N \to \infty$. For $m > 0$, the radial population density distribution is variable. For, say, an atomic beam,

m = 2 for a Gaussian density profile, m = 4 for a super-Gaussian density profile, m = 6 for a hyper-Gaussian density profile.

The relationships of α_R and τ_R lead directly to

$$\alpha_R L = 1 \quad . \tag{8.33}$$

As noted previously in the plane-wave regime, a universal scaling of the equation exists. The gain length Fresnel number F_g quantifies the competition between diffraction (i.e., transverse effects) and the nonlinear gain associated with a given atomic number density of the SF system, as it did in SIT [8.53]. *Drummond* and *Eberly* [8.85] also recently recognized this scaling.

The following energy consideration holds for a sharp line $g(\Delta\Omega) = \delta(\Delta\omega)$:

$$-iF_{ga}^{-1}\nabla_T(E_a\nabla_T E_a^* - E_a^*\nabla_T E_a) + \partial_{\zeta a}|E_a|^2 = d(E_a^* P_a + E_a P_a^*) \quad , \tag{8.34}$$

$$-i\left(\frac{\alpha_{Rb}}{\alpha_{Ra}}\right)F_{gb}^{-1}\nabla_T(E_b\nabla_T E_b^* - E_b^*\nabla_T E_b) + \partial_{\zeta a}|E_b|^2 = d\left(\frac{\alpha_{Rb}}{\alpha_{Ra}}\right)(E_b^* P_b + E_b P_b^*) \quad . \tag{8.35}$$

Summing up for $\alpha_{Rb} = \alpha_{Ra}$,

$$\nabla \cdot J_{Tot} = -2d[\partial_\tau W_a + (W_a - 1)/\tau_1^a + \partial_\tau W_b + (W_b - 1)/\tau_1^b] \quad , \tag{8.36}$$

with J_{Tot} having both longitudinal $J_z^{a,b}$ and transverse $J_T^{a,b}$ (or equivalently, $J_x^{a,b}$ and $J_y^{a,b}$) components for each color. When using the polar representation of the complex envelope, we have

$$E_{a,b} = A_{a,b} \exp(i\Phi_{a,b}) \tag{8.37}$$

$$J_z^{a,b} = A_{a,b}^2 \quad \text{and} \tag{8.38}$$

$$J_T^{a,b} = 2iF_{ga,b}^{-1} A_{a,b}^2 \nabla_T \Phi_{a,b} \quad . \tag{8.39}$$

For cylindrical geometry

$$J_T^{a,b} = 2iF_{ga,b}^{-1} A_{a,b}^2 \partial\Phi_{a,b}/\partial\rho \quad . \tag{8.40}$$

The components $J_z^{a,b}$ and $J_T^{a,b}$ represent the longitudinal and transverse energy current flow. Thus, the existence of transverse energy flow is clearly associated with the radial variation of the phase $\Phi_{a,b}$ of the complex field amplitude. When $J_T^{a,b}$ is negative (i.e., $\nabla_T\phi_{a,b} > 0$), self-induced focusing dominates diffraction spreading.

One may rewrite the energy continuity equation (8.36) in the laboratory frame to recover its familiar form

$$\nabla \cdot J_{Tot} = - d\left[\frac{\partial}{\partial t}\left(2(W_a + W_a) + \frac{1}{c\tau_{RA}\alpha_{RA}} A_a^2 + \frac{1}{c\tau_{Rb}\alpha_{Rb}} A_b^2\right) - \frac{(W_a - W_a^0)}{\tau_1^a - (W_b - W_b^0)} \tau_1^b\right] \quad . \tag{8.41}$$

It is noteworthy that the two scalar field equations appear to be uncoupled. Nevertheless, the two fields are coupled through the material equations [8.76,86] where the cross-coupling appears explicitly. The inertial response of the resonant three-level atoms tightly interweaves the two waves through the parametric terms.

The Hamiltonian that describes the atomic system has been discussed by *Bowden* and *Sung* [8.61] and the corresponding Heisenberg equations will not be set down again here. Instead, we write out the c-number Bloch equations for three-level SF as follows [8.63]:

$$\frac{\partial P_a}{\partial\tau} + \beta_a P_a = E_a W_a - \frac{1}{2} E_b Q - f_b Q + f_a W_a \quad , \tag{8.42}$$

$$\beta_a = \frac{1}{\tau_2^a} + i\Delta\omega_a \quad , \tag{8.43}$$

$$\frac{\partial P_b}{\partial\tau} + \beta_b P_b = E_b W_b - \frac{1}{2} E_a Q^* - f_a Q^* + f_b W_b \quad , \tag{8.44}$$

$$\beta_b = \frac{1}{\tau_2^b} + i\Delta\omega_b \quad , \tag{8.45}$$

$$\frac{\partial Q}{\partial \tau} + \beta_c Q = \frac{1}{2}\left(E_a P_b^* + E_b P_a^*\right) + f_b P_a + f_a P_b \quad , \tag{8.46}$$

$$\beta_c = \frac{1}{\tau_2^c} + i\Delta\omega_c \quad , \tag{8.47}$$

$$\frac{\partial W_a}{\partial \tau} + \gamma_a(W_a - W_a^0) = -\frac{1}{2}\left(E_a^* P_a + E_a P_a^*\right) - \frac{1}{4}\left(E_b^* P_b + E_b P_b^*\right) - 2\left(f_a^* P_a + f_a P_a^*\right)$$
$$- \left(f_b^* P_b + f_b P_b^*\right) \quad , \tag{8.48}$$

$$\frac{\partial W_b}{\partial \tau} + \gamma_b(W_b - W_b^0) = -\frac{1}{2}\left(E_b^* P_b + E_b P_b^*\right) - \frac{1}{4}\left(E_a^* P_a + E_a P_a^*\right) - \left(f_a^* P_a + f_a P_a^*\right)$$
$$- 2\left(f_b^* P_b - f_b P_b^*\right) \quad , \tag{8.49}$$

where W_a^0 and W_b^0 are the equilibrium population differences.

$$\gamma_a = \frac{1}{\tau_1^a} \quad , \qquad \gamma_b = \frac{1}{\tau_1^b} \quad .$$

In general, γ_a need not be equal to γ_b. For simplicity they may be taken as equal, but one is not restricted to this.

$$\frac{\partial W_b}{\partial \tau} + \gamma_b(W_b - W_b^0) = -\frac{1}{2}\left(E_b^* P_b + E_b P_b^*\right) - \frac{1}{4}\left(E_a^* P_a + E_a P_a^*\right) - \left(f_a^* P_a + f_a P_a^*\right)$$
$$- 2\left(f_b^* P_b + f_b P_b^*\right) \quad , \tag{8.50}$$

where W_a^0 and W_b^0 are the equilibrium population differences. The amplitudes and phase of the complex-valued fluctuation terms are determined by Gaussian and uniform statistics, respectively, which are derived from the fully quantum model [8.62].

In (8.42-50),

$$f_{a,b} = |f_{a,b}| \exp(i\phi_{a,b}) \quad , \tag{8.51}$$

$$0 \leqslant \phi_{a,b} \leqslant 2\pi \quad , \tag{8.52}$$

$$|f_a| = f_P \quad , \tag{8.53}$$

$$P(f_P^2) = \pi^{-1}\sigma_P^{-1} \exp\left[-\left(\frac{f_P}{\sigma_P}\right)^2\right] \quad , \tag{8.54}$$

$$|f_b| = h_{SF} \quad , \tag{8.55}$$

$$P(h_{SF}^2) = \pi^{-1}\sigma_{SF}^{-1} \exp\left[-\left(\frac{h_{SF}}{\sigma_{SF}}\right)^2\right] \quad , \tag{8.56}$$

$$\sigma_P = \frac{\tau_n^2}{(\tau_{RP} L/c) N_i} = \frac{\tau_n^2}{\tau_{RP}\tau_E N_i} = \frac{\tau_n^2}{N_i \tau_c} \quad , \tag{8.57}$$

$$\tau_c^2 = \tau_{RP}\tau_E \quad , \tag{8.58}$$

where τ_E is the escape time and τ_c the cooperation time.

$$\sigma_{SP} = \frac{\tau_n^2}{(\tau_{RSF} L/c) N_i} \quad , \tag{8.59}$$

$$\tau_{RP} = \frac{L}{c}\left(\frac{2\pi|\mu_P|^2}{\hbar} N_i\right)^{-1} \quad , \quad \mu_P \equiv \mu_a \quad , \tag{8.60}$$

$$\tau_{RSF} = \frac{L}{c}\left(\frac{2\pi|\mu_{SF}|^2}{\hbar} Ni\right)^{-1} \quad , \quad \mu_{SF} \equiv \mu_b \quad , \tag{8.61}$$

$$N_i = N(\rho_i) = \exp[-(\rho_i/r_N)^m][\pi(\rho^2_{i+1/2} - \rho^2_{i-1/2})/B] \quad , \tag{8.62}$$

$$B = \pi\rho^2_{i/2} + \sum_{i=2}^{N\rho-1} \exp[-(\rho_i/r_N)^m][\pi(\rho^2_{i+1/2} - \rho^2_{i+1/2})]$$

$$+ \{\exp[-(r_{N\rho}/r_N)^m]\pi(\rho^2_{N\rho} - \rho^2_{N\rho-1/2})\} \quad . \tag{8.63}$$

The Langevin force contributions to the semiclassical equations of motion give rise to initiation of fluorescence in the $3 \leftrightarrow 2$ transition when the $3 \leftrightarrow 1$ transition is coupled by the pumping field E_a. Normally, one can ignore superfluorescence in the $3 \leftrightarrow 1$ pump transition, i.e., ignore contributions from f_a. It is important to realize that the polarization P_a is driven not only by the pump field component E_a but also by E_b, since the product $E_b Q$ corresponds to an oscillation at the pump frequency also. Similarly, the product $E_a Q^*$ contributes to the temporal rate of change of the P_b polarization as in parametric oscillators. The nonlinear coupling of E_a and E_b field components just described involves the quadrupole component Q. Note that, due to the absence of a corresponding dipole moment, Q does not give rise to an electric field component E_c radiating at the difference of the two colors, since it is a forbidden transition due to parity considerations.

Furthermore, the pump and SF pulses are also coupled via the population of level 3. During the early stage linear regime ($N_1 = 1$, $N_{2,3} = 0$), the two pulses evolve independently of one another. For visualization of the physics, let us assume SF builds up from a small field E_b whose ratio to the pump field E_a is δ, where $\delta \ll 1$. To zeroth order in δ, the pump evolves as a self-induced transparency (SIT) pulse. The SF builds up as first order in δ, whereas the feedback of the SF field into the pump is of second order (δ^2), etc. Of course, there is no cross talk or cross-correlation between the pump and the SF pulse in the linear regime, but as soon as the nonlinearities become effective, the two pulses lose their independence. As the most obvious manifestation of the interdependence we must expect both pulses to overlap for large αl, i.e., to have their intensity maxima at rather close times. On the other hand, the maximum of the two intensities cannot develop any strong correlation.

Thus, by utilizing the relations (8.42-63), a complete ensemble simulation can be constructed, and in this way the manifestations of amplified quantum initiation can be calculationally analyzed over the full range of dynamical SF evolution. This amounts to generating the calculational results, using the semiclassical representation of the Heisenberg equations of motion [8.69,70] for specified initial and boundary conditions for each of the values selected for $|f_a|$, $|f_b|$, ϕ_a, and ϕ_b according to the statistical distributions (8.54-61). In the computation, N in (8.12) is taken as N_i given by (8.62) for each volume element or cell in each computational state, as suggested by Feld. It is noted that

$$\tau_R L/c = \tau_c \quad , \tag{8.64}$$

which is invariant with volume. One must then take the ensemble averages of the different trajectories and associated variances. For the results presented here, we have ignored superfluorescence in the pump transition. We first consider a Langevin driving force in the SF transition only; subsequently we consider a second Langevin force acting in the pump transition. The material parameters chosen for these calculations are arbitrary, but correspond roughly to those for optically pumped metal vapors.

The initial and boundary conditions are such that all the atomic population is in the ground state ε_1 at $\tau = 0$. The pumping pulse that pumps the $1 \leftrightarrow 3$ transition is injected at $z = 0$ and propagates toward the right, and its initial characteristics are specified at $z = 0$. The SF pulse subsequently evolves in z, ρ, and τ due to the initiation of fluorescence instigated by $|f|$ and ϕ as indicated in the equations of motion [8.15-17] in their semiclassical form discussed above [8.1-3]. The pump pulse, whose initial characteristics are specified at injection, and the SF pulse copropagate and interact via the nonlinear medium.

8.4 Deterministic Effects of Pump Dynamics in the Nonlinear Regime of Superfluorescence

We have computed the effects on SF pulse evolution of various initial conditions of the injected (pump) pulse. The results presented here demonstrate many facets of the control and shaping of the SF signal by control of the input signal initial characteristics.

Although the simulation inherently yields numerically accurate results for a particular experimental design, the results reported here must be taken as qualitative because the material parameters chosen here are arbitrary. Our main purpose here is to demonstrate and analyze specific correlations between the initial and boundary conditions associated with the injected pump pulse and the characteristics of the SF pulse which envolves. In many of the cases which follow, rules are established through the analysis which can be used to predict quantitative results for any particular experimental conditions. Our choice of particular initial and boundary conditions has been motivated in part by processes which may have been operative in experiments which have been reported [8.7-27] and in part by the feasibility of experimental selection or specification. In connection with the latter, we demonstrate the control of one light signal by another via a nonlinear medium, which permits nonlinear encoding for information transfer and pulse shaping of the SF from specific initial and boundary conditions associated with the pump injection signal.

Since the average values of τ_D and the peak SF intensity are important quantities for interpreting experimental results with theories of SF [8.1-3,29,32-42] manner in which the pump pulse coherence and initial on-axis area affect these quantities is seen to be of extreme importance in any analysis.

The simulation parameters (unless otherwise noted) are as follows: The injected pulses are initially Gaussian in ρ and τ with widths (FWHM) $r_P = 0.24$ cm and $\tau_D = 4$ ns respectively; the level spacings are such that $(\varepsilon_3 - \varepsilon_1)/(\varepsilon_3 - \varepsilon_2) = 126.6$. The effective gain for the pump is $\alpha_{RP} = 0.364$ $(g_P = 17)$ cm^{-1} that for the SF transition is $\alpha_{RSF} = 0.171 (g_{SF} = 291.7)$ cm^{-1}. The ratio $\beta = g_{SF}/g_P = 17.16$. The gain length Fresnel numbers for the two transitions are $F_P = 359.72 (F_{gP} = 16800)$ and $F_{SF} = 1.334 (F_{gSF} = 2278)$. The relaxation and dephasing times are assumed to be identical for all transitions and are taken as $T_1 = 80$ ns and $T_2 = 70$ ns, respectively (i.e., homogeneously broadened).

Note the different transition gains (i.e., different oscillator strengths) considered to ensure that the SF overlaps with the pump pulse. The SF can be thought of as a weak pulse (of very small optical area) that experiences a net gain which results from the population transferred into level 3. The population is initially at the ground level (1) of the three-level atomic system. The strong pump of on-axis area 2π sees an absorbing medium $(1 \leftrightarrow 3)$ and it evolves as an SIT pulse. Its energy is depleted at a rate different from the one at which the SF builds up. The SF experiences an optical thickness much larger than that seen by the pump pulse.

For the sake of completeness, the role of the two processes that lead to transverse reshaping is first examined for a two-level scheme, that is, for nonoverlapping pump and superfluorescence pulses. It is assumed that no pump depletion takes place. The set of equations to be solved reduces to a single Maxwell equation for one laser and the density matrix simplifies to one polarization, one population difference and no pseudo quadrupole. However, the Maxwell-Bloch equations are solved rigorously without the linearization assumption of nondepletion of the two-level population difference. Furthermore, one obtains the nonlinear pendulum expression as an analytical solution in the uniform-plane-wave regime with infinite relaxation times. Following *Mattar* et al.'s analysis [8.49], a radial profile for the initial inversion density is assumed: that is, a radial variation $r = \rho r_P$ in N_0 and/or in θ_0 is allowed for the computations.

Figure 8.2 shows isometric graphs versus ρ and τ of three physical situations for the propagation of a small-area SF pulse through a sample with cylindrical

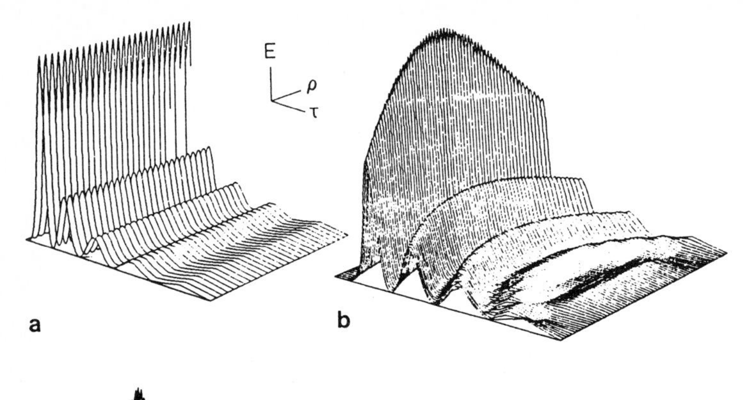

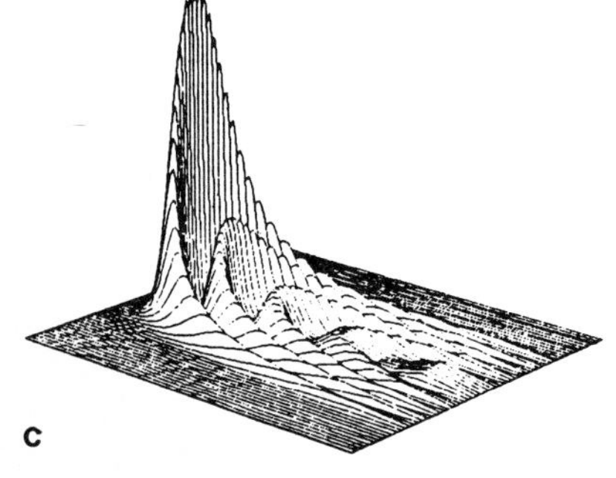

Fig.8.2a-c. Isometric representations of the field energy versus ρ and τ for (**a**) uniform inversion and preexcitation; (**b**) radially distributed tipping angle; and (**c**) Gaussian inversion profile

geometry: (a) Perfect uniform plane-wave calculation with uniform inversion (gain) and uniform preexcitation (uniform effective tipping angle); (b) uniform small field and radially distributed tipping angle, i.e., the shell model; and (c) radial independence of both the initial tipping angle and the inversion density for each shell. The various shells are coupled by diffraction, more specifically, graph (b) represents the case of unit Fresnel number ($F=1$).

The radial variations of intensity peaks, delay, and ringing illustrate how different gain shells contribute independently to the net power. Each shell exhibits a different Burnham-Chiao ringing pattern. Accordingly, their contributions to the net signal interfere and reduce the ringing. However, the central portion of the output pulse should exhibit strong plane-wave ringing. In fact, the ringing observed in the HF-gas experiments may have been just that, since the detector viewed a small area in the near field of the beam.

In Fig.8.3, the propagational character of the SF is clearly displayed by plotting for different optical thicknesses the SF energy and J_T, the transverse energy current, at the right and left of the figure, respectively. The SF buildup with increasing αl confirms the validity of *Friedberg* and *Hartmann*'s [8.5] high-gain requirement in conjunction with *MacGillivray* and *Feld*'s [8.31] semiclassical description. *Bonifacio* et al.'s [8.32] mean field theory, which neglects the propagation effects and predicts a space-independent polarization in the SF sample, is clearly not valid under these conditions.

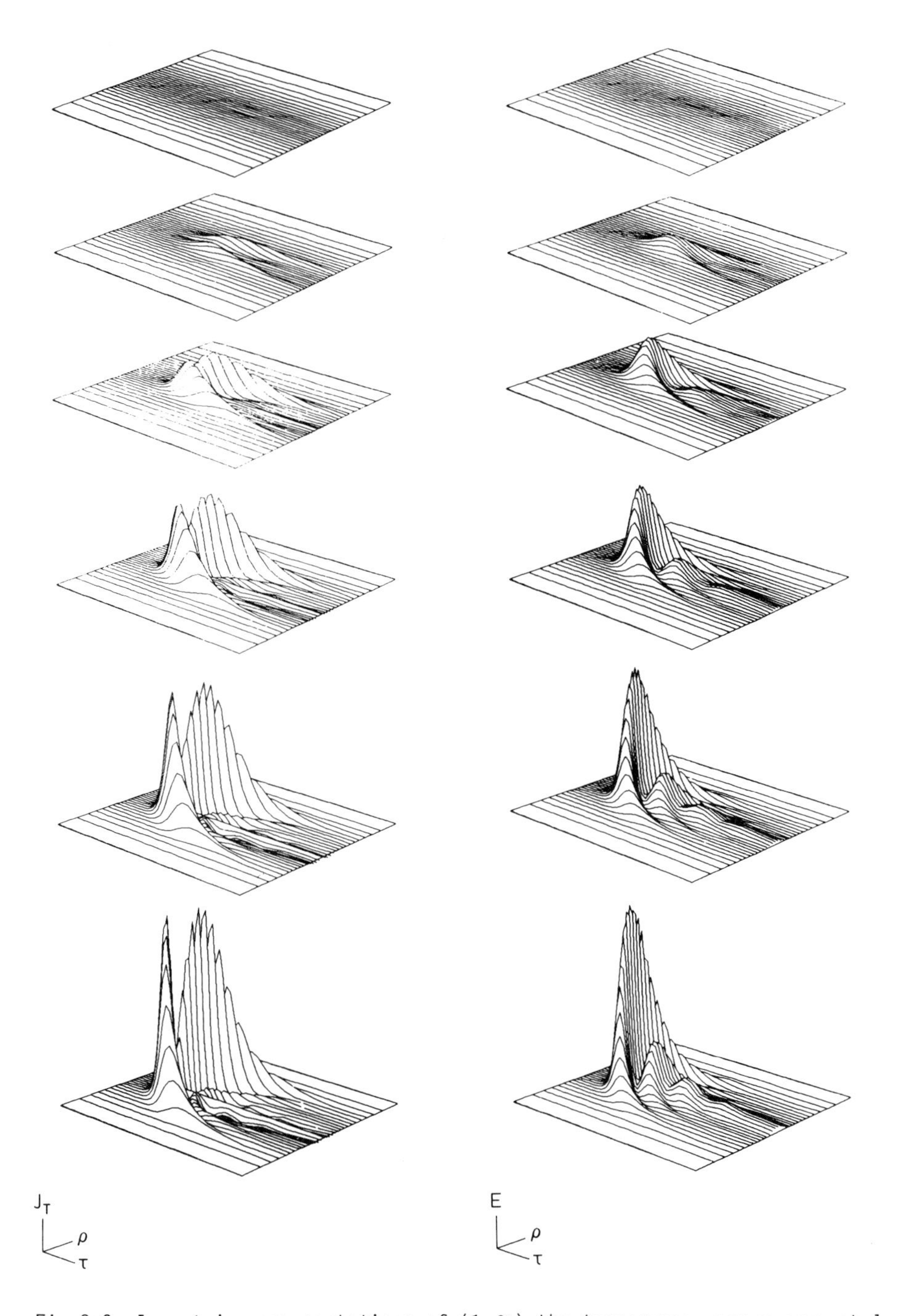

Fig.8.3. Isometric representations of (*left*) the transverse energy current J_T and (*right*) the SF energy for various penetration depths illustrating the propagational character of the SF

The mean field theory is *not* relevant for a quantitative explanation of the HF [8.17], Cs [8.18], and Rb [8.59] observations.

We have studied the role of the pump pulse in SF buildup in the plane-wave regime. To obtain the same SF delay in the three-level configuration as in the two-level system, a larger atomic density (that is, a larger gain in the SF transition) is required. The ratio of the two Beer's lengths (SF to pump) must be larger than unity (i.e., $\beta = \alpha_{RSF}/\alpha_{RP} = \alpha_b/\alpha_a > 1$) to allow the SF pulse evolving from the small deterministic seed or induced by the initial tipping angle to build up at a faster rate than that at which the pump gets depleted. The SF emission experiences amplification and an advance in time (decrease of temporal delay) as the propagation distance increases, whereas the pump pulse evolves as a SIT pulse in resonant absorbers and suffers a time retardation (i.e., an increase of the temporal delay). The difference of oscillator strengths in the SF from the pump transition causes the SF and pump pulses to strongly overlap. A Raman mutual influence arises and an energy exchange of the two beams occurs. The same temporal oscillatory variation, that is, ringing of the nonlinear pendulum, is obtained for the SF pulse. This result corresponds to evolution of the on-axis shell in a cylindrical geometry without coupling to the adjacent shells.

The output power OPOWR detected as a function of time at the end of the nonlinear cell is the radially integrated field energy. Depending on the domain of integration over the transverse cross section, one obtains the on-axis (no radial integration), or the near-axis (only a few radial mesh, i.e., small detector), or the total beam (i.e., large detector) power, reproducing the uniform plane HF observation which displays ringing (i.e., temporal oscillations) or the full-beam single-pulse Cs observation. By modifying the detector widths, one can rigorously reconcile the two experimental observations and provide supporting physical insights. For the sake of completeness, the output power has been calculated in three ways: first, by integrating over the entire beam, $\mathrm{OPOWR}_{a,b}(\tau) = \int_0^{\rho_{max}} |E_{a,b}|^2 \rho \, d\rho$; second, by a partial integration from the beam center with the upper limit varying to enclose a successive number of shells; and third, using the representation of a pinhole of fixed width [e.g., $(\Delta\rho)$, $2(\Delta\rho)$, $3(\Delta\rho)$...,i.e., a fraction of r_p] which is moved across the beam, as in the recent rubidium experiment by Feld and co-workers. The output power is thus evaluated as a shell output power with varying lower and upper limits of integration. The extent of the radial interval which reflects the pinhole width is defined as

$$\mathrm{OPWAIm},\ \mathrm{OPWBIm}(\rho,\tau) = \int_{\rho}^{\rho+m(\Delta\rho)} |E_{a,b}|^2 \rho \, d\rho \quad .$$

The different output powers at a given η output plane have different dimensions; the first one varies only with τ, the second varies with τ and ρ, and the third with τ, ρ, and the detector width $m(\Delta\rho)$. Two transverse effects modify the departure from the uniform-plane-wave behavior. The first one is the "shell model"

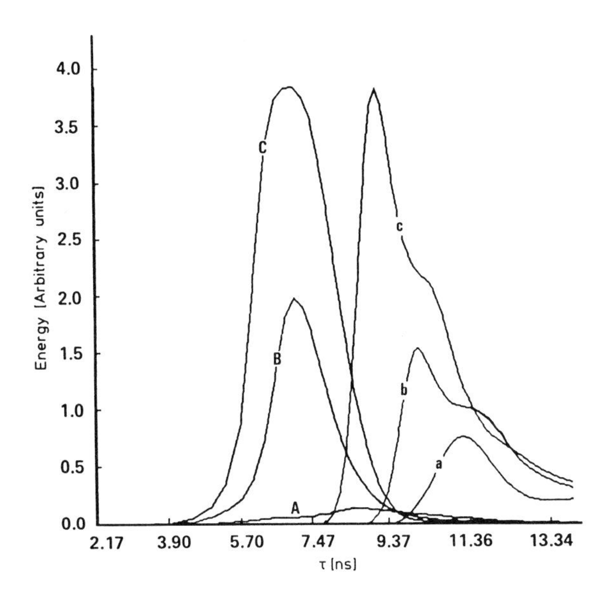

Fig.8.4. Radially integrated normalized energy profiles for the SF (*a-c*) and injected (*A-C*) pulses at z=5.3 cm penetration depth for three different values of the initial on-axis injection pulse area θ_p. The injected pulses are initially Gaussian in r and τ with widths (FWHM) r_0=0.24 cm and τ_p= 4ns, respectively. The level spacings are such that $(\varepsilon_3-\varepsilon_1)/(\varepsilon_3-\varepsilon_2)$=126.6. The effective gain for the pump transition α_p=17 cm^{-1} and that for the SF transition α_{SF}=291.7 cm^{-1}. $\beta=\alpha_{SF}/\alpha_p$=17.16. The gain length Fresnel numbers for the two transitions are F_{gP}=16800 and F_{gSF}=2278. The relaxation and dephasing times are assumed to be identical for all transitions and are taken as T_1=80 ns and T_2=70 ns, respectively. The values of θ_p are (*A*) π, (*B*) 2π, and (*C*) 3π

predominant for large Fresnel numbers, while the second phenomenon is the "diffraction" which is dominant in the small-Fresnel-number regime.

Figure 8.4 shows results of the numerical calculation for the transverse integrated intensity profiles for the copropagating SF and injected pulses at a penetration depth of z = 5.3 cm in the nonlinear medium. These profiles correspond to what would be observed with a wide-aperture, fast, energy detector. The pumping pulses are labeled by capital latters, and the corresponding SF pulses are labeled by the corresponding lowercase letters. Each set of curves represents a different initial on-axis area for the pump pulse, i.e., curve A is the reshaped pump pulse at z = 5.3 cm that had its initial on-axis area specified as $\theta_p = \pi$, and curve a is the resulting SF pulse which has evolved. All other parameters are identical for each set of pulses. The initial conditions for the atomic medium are that nearly all the population is in the ground state ε_1 at $\tau = 0$, and a small but nonzero macroscopic polarization exists between levels ε_3 and ε_2. These two conditions are specified by two parameters, ε and δ, respectively, and we have chosen $\delta = \varepsilon = 10^{-3}$ self-consistently as specified (by Feld and co-workers) in [Ref.8.69, Appendix]. These initial conditions are uniform for the atomic medium and are the same for all results reported here. Notice that we have neglected spontaneous relaxation in the pump transition $1 \leftrightarrow 3$ relative to the SF transition $3 \leftrightarrow 2$. This is justified owing to our choice of relative oscillator strengths (see the caption of Fig.8.4).

These results clearly indicate the coherence effect of the initial pump-pulse area on the SF signal that evolves. Notice that the peak intensity of the SF pulses increases monotonically with initial on-axis area for the pump pulse. This is caused by self-focusing due to transverse coupling and propagation. For instance, a 2π injection pulse would generate a very small SF response compared to an initial π injection pulse for these conditions at relatively small penetration z, or for the corresponding case in one spatial dimension. Even so, the peak SF intensity is approximately proportional to the square of the initial on-axis area of the pump pulse, whereas the delay time τ_D between the pump-pulse peak and the corresponding SF pulse peak is very nearly inversely proportional to the input pulse area. The temporal SF pulse width (FWHM) τ_{SF} is approximately invariant with respect to the injection-pulse area.

Since the average values of τ_D and the peak SF intensity are important quantities for interpreting experimental results in terms of theories of SF [8.1-6], the manner in which the pump-pulse coherence and initial on-axis area affect those quantities is seen to be of extreme importance in any analysis.

As discussed earlier, the pump pulse sees an absorbing transition $(3 \leftrightarrow 1)$ and populates the upper level (level 3), which leads to a population inversion in the transition $(3 \leftrightarrow 1)$. Consequently, the evolving SF experiences a net gain (amplification). Following Mattar and Newstein's study of diffraction effects in SIT, the SIT pulse propagation depends on a number of pertinent parameters, namely τ_p, τ_1, τ_2, r_p, μ, λ, $\Delta\omega$, and θ_p. These parametric dependences are displayed in [Ref.8.53, Figs.11-20]. In particular, the propagation is altered by the choice of the Beer-length Fresnel number F_g through its different constituents for a given pulse and beam profile:

1) The temporal pulse length τ_p or, effectively, the reciprocal ratio of τ_p to the material relaxation times T_1 and T_2.
2) The spatial beam width of the Gaussian profile r_p.
3) The Gibbs and Slusher's sharp-line effective Beer absorption length normalized to the pulse length α/τ_p, which depends on μ, the dipole moment associated with the transition, and N, the atomic density, and thus reflects the nonlinear medium.
4) The carrier wavelength λ.
5) The length l of the absorption cell, which appears in the (geometric) Fresnel number F and the optical thickness αl.
6) The frequency offset between the carrier and the atomic transition $\Delta\omega$.
7) The on-axis time-integrated pulse area θ_p.

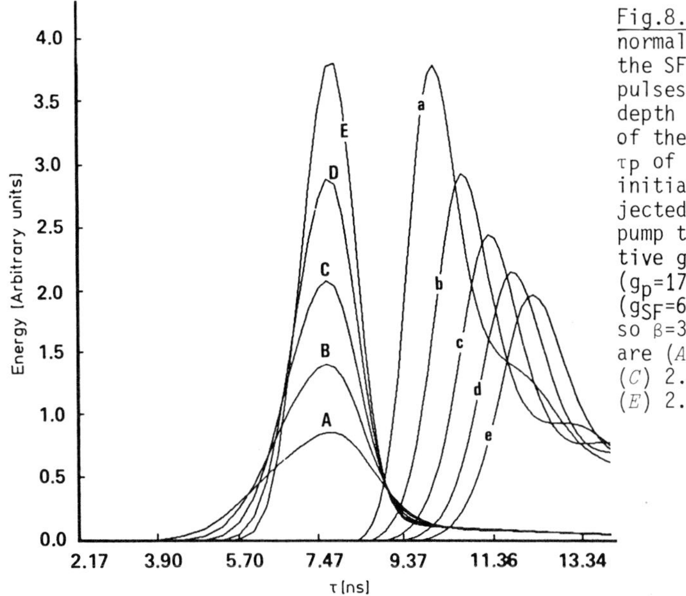

Fig.8.5. Radially integrated normalized energy profiles for the SF (*a-c*) and injected (*A-C*) pulses at z=5.3 cm penetration depth for five different values of the initial temporal width τ_P of the injected pulse. The initial on-axis area of the injected pulse is $\theta_P=\pi$, and the pump transition and SF effective gains are $\alpha_{RP}=0.376\ cm^{-1}$ ($g_P=17.5$) and $\alpha_{RSF}=0.376\ cm^{-1}$ ($g_{SF}=641.7$), respectively, so $\beta=36.67$. The values of τ_P are (*A*) 4 ns, (*B*) 3.3 ns, (*C*) 2.9 ns, (*D*) 2.5 ns, and (*E*) 2.2 ns

These parameters are expected to affect the pump pulse reshaping and its transmission characteristics.

Figure 8.5 shows the effect on the SF evolution of varying the initial temporal width of the injected pulse τ_P. As τ_P decreases, the SIT effective Beer length experienced by the pump becomes smaller. This reduction in the absorption nonlinearity affects the pump pulse reshaping and reduces the pump delay associated with the coherent exchange of energy between the pump pulse and the absorbing transition. The degree of coherence in the interaction is enhanced because the pulse length relative to the relaxation times τ_P/T_1 and τ_P/T_2 is reduced. This lower limit of τ_P emphasizes the necessity of describing the atomic dynamics using the density matrix rather than adopting the simplifying rate equation approximation.

As the pump pulse length and its effective absorption length decrease, the injected energy depletion diminishes and the SF buildup is weaker and SF occurs with a larger delay τ_D. However, the SF temporal width τ_{SF} remains very closely fixed.

It is clear from these results that there exists an approximately linear relationship between the time delay τ_D (between the peak SF intensity and the corresponding pump pulse intensity) and the initial temporal width τ_P of the pump pulse. This linear relationship is shown in Fig.8.6, where the time delay τ_D is plotted versus the corresponding pump-pulse initial temporal width from Fig.8.5. These results generate the following empirical formula for τ_D as a function of τ_P:

$$\tau_D = 0.375\ \tau_R[\ln(4\pi/\psi)]^2 - 4\tau_R\gamma_1(\gamma_R/4\gamma_1 - 1)\tau_P \qquad (8.65)$$

(with $\gamma_1=\gamma_a=\gamma_b$), where [8.6]

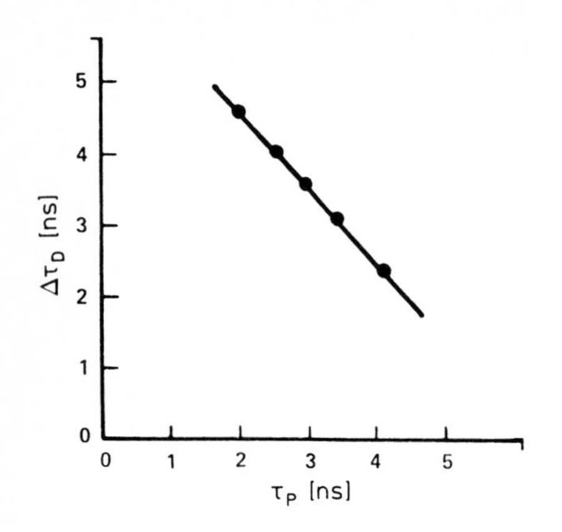

Fig.8.6. Delay time $\Delta\tau_D$ of the SF peak energy from the corresponding pump-pulse peak energy versus the pump-pulse initial full temporal width at half maximum energy, τ_P, according to Fig.8.5

$$\tau_R = \frac{2T_2}{g_{SF}z} \quad , \tag{8.66}$$

or equivalently

$$\frac{\alpha_{RSF,RP}}{g_{SF,P}} = \frac{\alpha_{RSF,RP}}{T_2} \quad , \tag{8.67}$$

is the characteristic superfluorescence time [8.1-3,32] and ψ is a parameter adjusted to give a best fit to the calculational results. For the case treated here, $\tau_R = 41$ ps, $T_2 = 70$ ns, $\gamma = 10^{-8}$, and the geometric Fresnel number $F = 1.47$.

The relation (8.65) is at best in qualitative agreement with the analytical prediction made in [Ref.8.61b, Eq. (5.1)] based upon mean field theory. The first term in (8.65) was chosen to conform with the quantum mechanical SF initiation result [8.73] obtained within the propagational model. The quantity ψ can be interpreted as the "effective tipping angle" for an equivalent π initial impulse excitation, i.e., for $\tau_P \to 0$, which initiates subsequent SF. Note that the value of ψ is dependent upon the choice of δ; however, τ_D varies by less than 25% for order of magnitude changes in δ for $|\delta| < 10^{-2}$. The choice of δ is simply an artificial way of instigating the semiclassical numerical calculation and reasonable variations in its value do not greatly affect the results. The physical parameter is, then, ψ, which, interpreted on the basis of (8.65), is generated through the dynamics caused by the pumping process and represents quantum SF initiation.

The full statistical treatment of three-level SF building up from a Langevin force in the SF transition, with pump dynamics included, was presented at the Fifth Coherence and Quantum Optics Conference [8.74]. Subsequently, we carried out additional calculations which involve the Langevin contribution to the pump transition as well.

These results emphasize the importance of the initiating pulse characteristics in SF pulse evolution, and the effect of SF pulse narrowing with approximate pulse shape invariance with increasing initial temporal width of the injected pulse. It

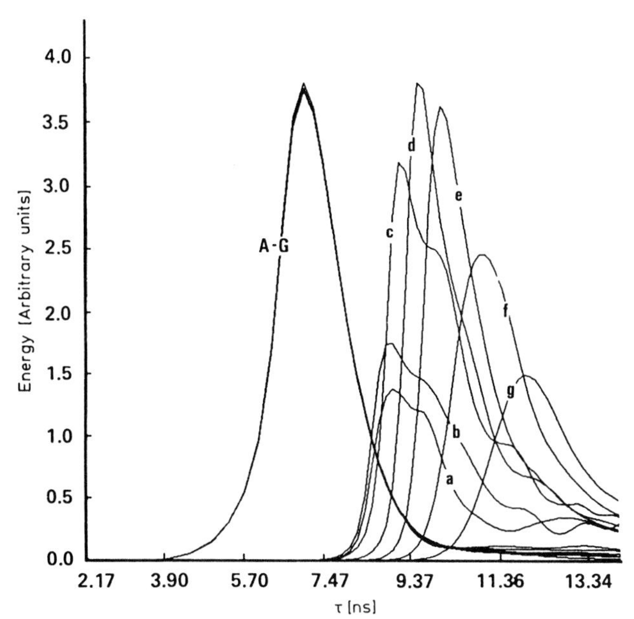

Fig.8.7. Radially integrated normalized energy profiles for the SF (*a-g*) and injected (*A-G*) pulses at z=5.3 cm penetration depth for seven different values of the injected (pump) pulse initial radial width at half maximum, r_P. Initial on-axis area of the injection pulse $\theta_P=2\pi$; SF effective gain $\alpha_{RSF}=0.444$ cm^{-1} ($g_{SF}=758.3$); and the pump transition effective gain $\alpha_{RP}=0.313$ cm^{-1} ($g_P=14.6$)($\beta=51.94$). The values of r_P and the corresponding geometrical Fresnel numbers SF are (*a*) 0.57 cm, 8.46; (*b*) 0.43 cm, 4.79; (*c*) 0.24 cm, 1.47; (*d*) 0.18 cm, 0.85; (*e*) 0.15 cm, 057; (*f*) 0.11 cm, 0.35; and (*g*) 0.09 cm, 0.21

is emphasized that all other parameters, including the initial value of the injected pulse on-axis area, are identical for these sets of curves.

The initial radial width r_P of the injected pulse was varied and the effect upon the SF pulse evolution is shown in Fig.8.7. There is a clearly indicated optimum value for r_P for which the SF peak intensity is a maximum and the SF temporal width τ_S is a minimum. If (8.23) is used in conjunction with the values of the parameters given in Fig.8.7 and its caption, it is seen that optimization occurs for a value of unity of the conventional (geometric) Fresnel number F_{SF} for the SF transition $F_{SF}\approx 1$. Thus from (8.23) and $F_{SF}=1$, we have

$$F_{gSF} = \alpha_{RSF} z_{max} \tag{8.68}$$

for the gain length Fresnel number. Since from (8.14) $F_{SF}\sim 1/z$, the implication is that (8.68) gives the penetration depth z_{max} at which the SF peak intensity reaches a maximum in terms of the ratio F_{gSF}/α_{RSF}. Since this takes both transverse effects and diffraction explicitly into account as well as propagation, this is indeed a profound statement [8.63].

Further insight into the implications of (8.68) can be obtained by considering a one-spatial-dimension analogy. If the linear field loss is taken to be entirely due to diffraction, then the one-dimensional linear loss κ corresponding to the two-dimensional case specified by F_{SF} is given by

$$\kappa_{SF} = \frac{\lambda_s}{2\pi r_0^2} \quad . \tag{8.69}$$

It is also known as the Rayleigh length or even the reciprocal diffraction length. Then, from (8.36),

$$F_{gSF} = \frac{\alpha_{RSF}}{\kappa_{SF}} \quad , \tag{8.70}$$

or equivalently [Ref.8.69, Eq. (3.5)]

$$F_{gSF} = \frac{g_{SF}}{\kappa_{SF}}$$

is the effective gain α_{RSF} to loss κ_{SF} ratio. From (8.68),

$$z_{max} = (\kappa_{SF}^{-1}) \quad , \tag{8.71}$$

i.e., z_{max} is the penetration depth at which the SF peak intensity is a maximum and corresponds to one effective diffraction length, as defined by (8.69). Carrying the one-dimensional analogy one step further, (8.70) used in (8.23) gives

$$F = (\kappa z)^{-1} \quad . \tag{8.72}$$

From (8.69 and 70), we have exhibited the significance of the Fresnel numbers F_g and F in terms of diffraction loss, i.e., F_g can be thought of as a gain to loss ratio, (8.70), whereas F can correspondingly be thought of as the reciprocal of the strength of the diffraction loss in a beam of unit length (8.72).

Figure 8.8 illustrates the Fresnel dependence of the SF buildup by displaying isometrically, versus τ and ρ, the SF energy at the output z plane of the nonlinear cell. The initial inversion may have a radial profile peaking near the beam center. As r_p, the initial beam width of the injected Gaussian pump pulse, increases, both the associated Rayleigh diffraction length z_d and the (geometric) Fresnel number increase, which reduces the linear loss experienced by the pump. Moreover, communication between the adjacent concentric cells representing the beam decreases: the various sections across the beam profile superfluoresce in a quasi-independent manner instead of acting simultaneously as a unit (as occurs when the diffraction coupling is strong). In this large-Fresnel-number regime, the delay associated with the peak of the radially integrated energy decreases. For smaller Fresnel numbers, the delay of the output power increases and the on-axis shell superfluoresces with the peripheral shell, even though each sees a different excitation. The resultant small-Fresnel-number sample SF pulse becomes more symmetrical. Thus, the pump beam

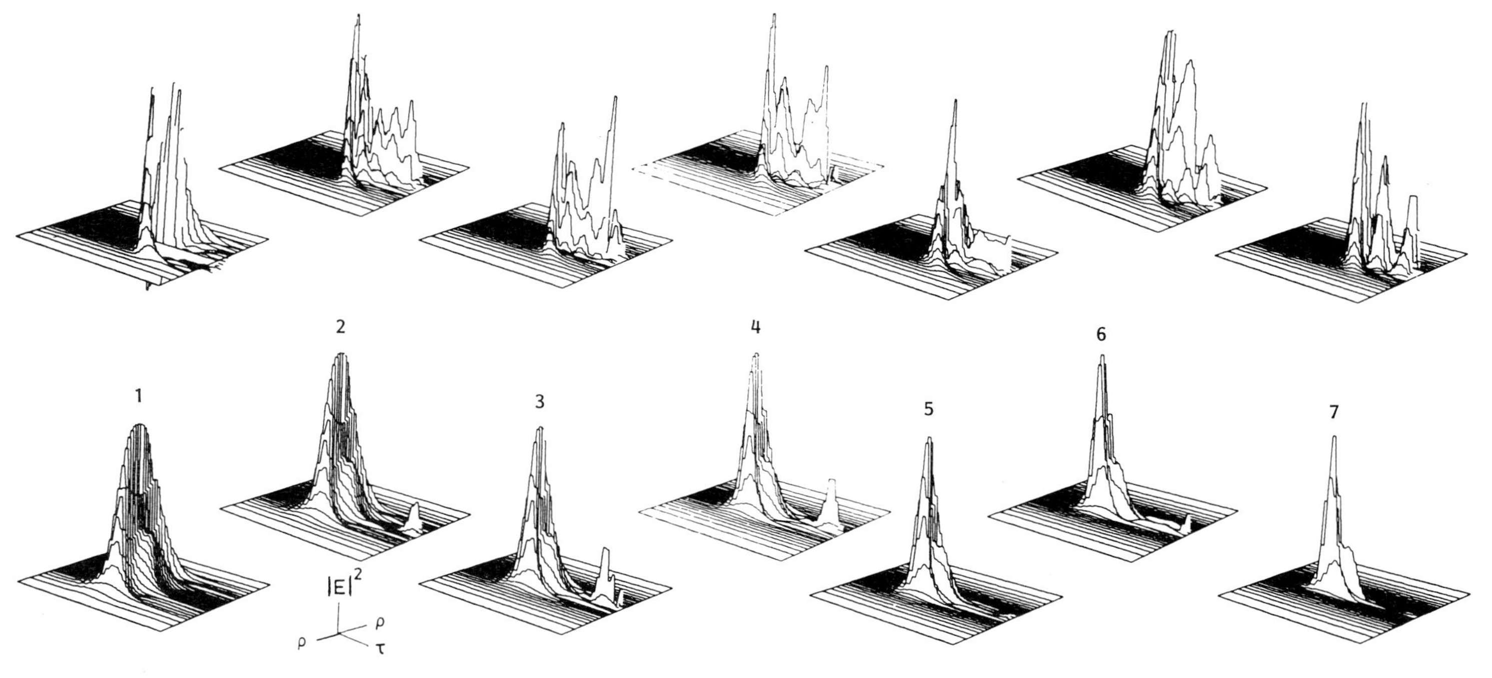

Fig.8.8. Comparison of SF (*top row*) and pump (*bottom row*) energy at a penetration depth z=5.3 cm for different values of the pump geometric Fresnel number (associated with the initial spatial width of the injected signal). The isometric plots of energy are against τ and ρ. Values of F_{gP} from left to right: (1) 4.0, (2) 2.26, (3) 1.0, (4) 0.69, (5) 0.40, (6) 0.27, (7) 0.10

radial variation controls the reshaping of the evolving SF. One should also note that for a given profile and atomic number density N, little diffraction prevents phase encoding, inhibits on-axis energy enhancement, and leads to beam spreading, while too much diffraction may retard and subdue the SF buildup. The SF buildup results from the competition between diffraction loss and gain associated with the preexcitation. The pump dissipates its energy as loss instead of leaking it to the SF emission. An equilibrium between pump beam diffraction and cross SF gain from the coupled preexcited transition for a given N can be reached. The resultant emission would have a stabilized solitary profile.

The effect on SF pulse evolution of variation of the initial radial shape of the initiating pulse is shown in Fig.8.9. The shape parameter v is defined in terms of the initial condition for the pump transition field amplitude $\omega_R(r)$,

$$\omega_R(r) = \omega_R(0)\exp[-(r/r_p)^v] \quad . \tag{8.73}$$

Thus for $v=2$, the initial amplitude of the injected pulse is radially Gaussian, whereas for $v=4$, it is radially super-Gaussian. We see from the results presented in Fig.8.10 that as the initial radial shape of the injected pulse becomes broader, i.e., with larger values of v, the peak intensity of the SF pulse generated becomes larger, and the width τ_{SF} and delay time τ_D diminish. It is emphasized again that all other parameters, including the initial values of the radial and temporal widths, are invariant for these sets of curves.

Thus, if the initial radial shape of the injected pulse is modulated from one injection to the next, the SF OPOWR temporal width and delay time τ_D are correspondingly modulated, as well as the SF peak intensity. Accordingly, the temporal ini-

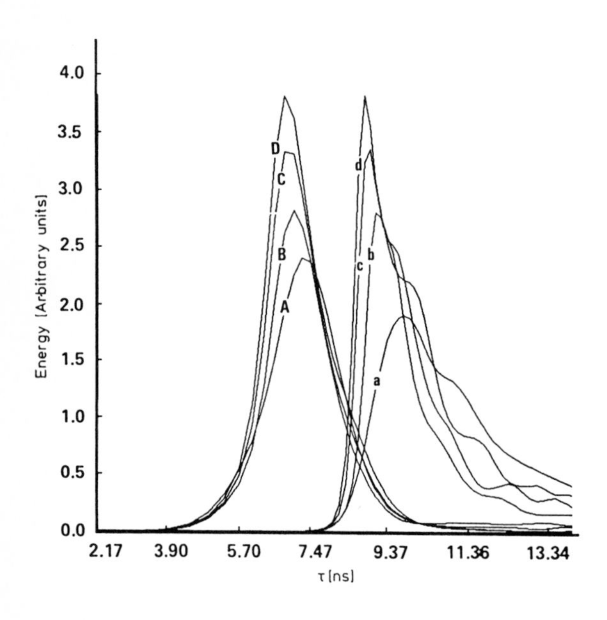

Fig.8.9. Radially integrated normalized energy profiles for the SF (*a-d*) and injected (*A-D*) pulses at z=5.3 cm penetration depth for four different values of the injected pulse initial radial shape parameter v (see text). Initial on-axis area of the injected pulse $\theta_P=2\pi$; SF effective gain $g_{SF}=758.3$; effective gain for the pump transition $g_P=14.6$; $\beta=51.94$; $F_{gSF}=2960$; and $F_P=7017$. All other parameters are the same as for Fig.8.4. The values of v are (*A*) 1, (*B*) 2, (*C*) 3, and (*D*) 4

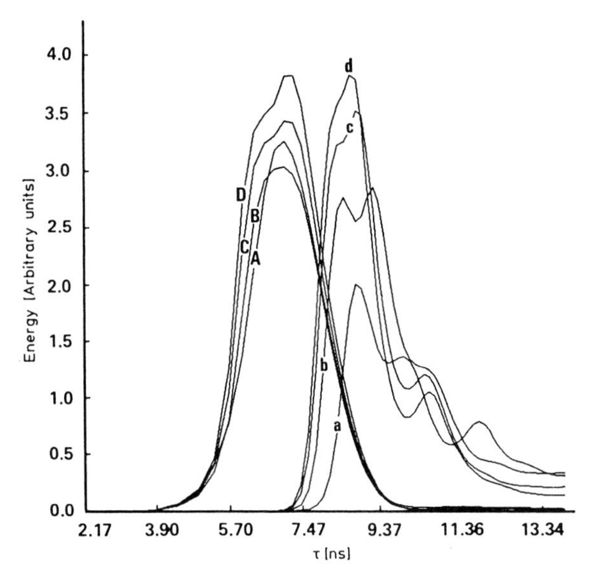

Fig.8.10. Radially integrated normalized energy profiles for the SF (*a-d*) and injected (*A-D*) pulses at z=5.3 cm penetration depth for four different values of the injected pulse initial radial shape parameter v (see text). The initial on-axis area of the injected pulse is $\theta_P=3\pi$. All other parameters are the same as for Fig.8.9

tial radial shape of the pump pulse cannot, with validity, be ignored in interpretating SF experiments in terms of τ_{SF} and τ_D.

Whereas the initial on-axis area for the pumping pulse was $\theta_P = 2\pi$ for the results shown in Fig.8.9, in Fig.8.10 the identical conditions and parameters were imposed, but the initial on-axis pump pulse area was changed to $\theta_P = 3\pi$. It is seen that the major effect of the change is to cause more ringing in the SF pulses and to modify the pump pulse temporal reshaping as is noted by comparing Figs.8.9 and 10.

Figure 8.11 displays isometric graphs contrasting the energy of the pump and the SF for different v (as in Fig.8.9) to illustrate the importance of the spatial profile (e.g., Gaussian and super-Gaussian).

The response of SF pulse evolution to changes in the initial temporal shape of the injection pulse is shown in Fig.8.12, which compares the effect of a Gaussian initial temporal shape for the pump pulse, identified by the temporal shape parameter $\sigma = 2$, with that of a super-Gaussian, identified by $\sigma = 4$. As for the radial distribution discussed previously, the temporal shape parameter σ is defined in terms of the initial condition for the pump transition field amplitude $\omega_R(\tau)$,

$$\omega_R(\tau) = \omega_R(0)\exp[-(\tau/\tau_P)^\sigma] \quad . \tag{8.74}$$

Again, it is seen that the broader initial pump pulse causes an increase in the peak SF intensity and a reduction in the delay time τ_D and SF pulse width τ_{SF}.

The results of Fig.8.13 correspond to the same conditions and values for the parameters as those for Fig.8.12, except that the initial on-axis area for the injection pulse is $\theta_P = 3\pi$ instead of 2π.

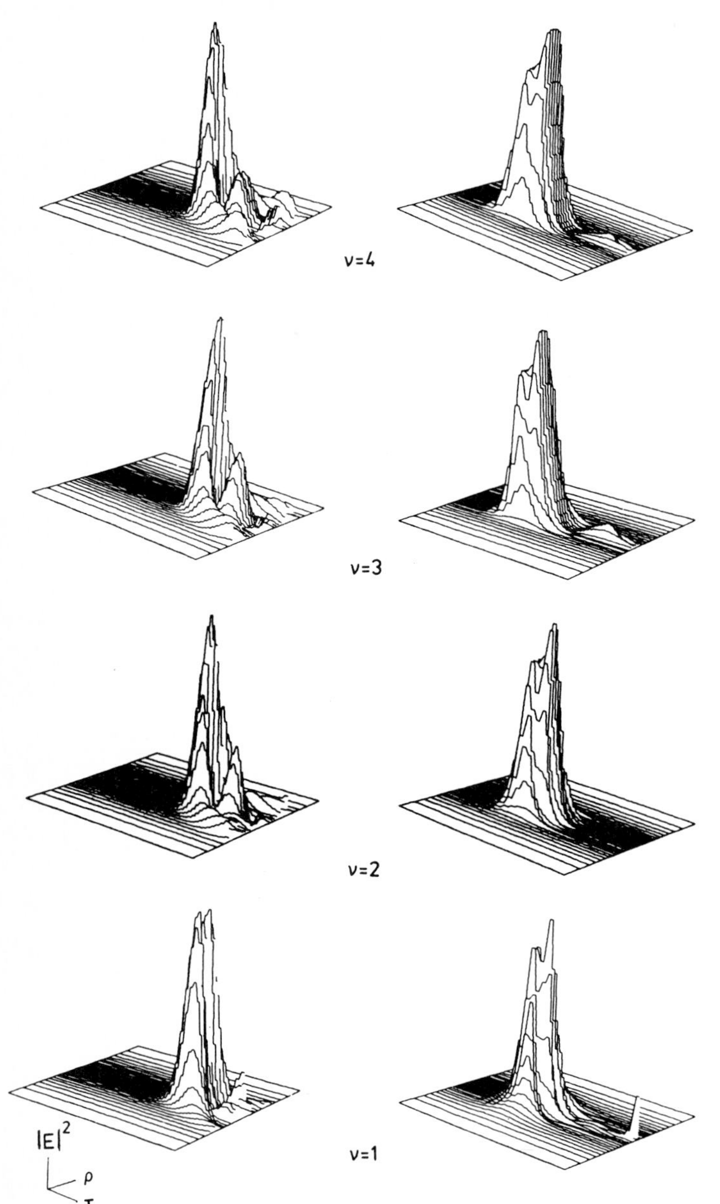

Fig.8.11. Isometric plots of SF energy against τ and ρ at a penetration depth z=5.3 cm for four different values of input radial shape parameter ν. This figure complements Fig.8.9

The effect of changing the effective gain of the SF transitions α_{RSF}, and hence the relative oscillator strength between the SF transition and the pump transition, is demonstrated in the results of Figs.8.14-17. Each of these figures corresponds to a different on-axis initial area θ_P for the injection pulse. Consistent among the entire set of results is that increasing the effective gain α_{RSF} results in a nearly linear increase in the SF peak intensity, as well as a decrease in the delay time τ_D. Also, the smaller-area initiating pulse causes a narrower SF pulse to evolve and with apparently less ringing.

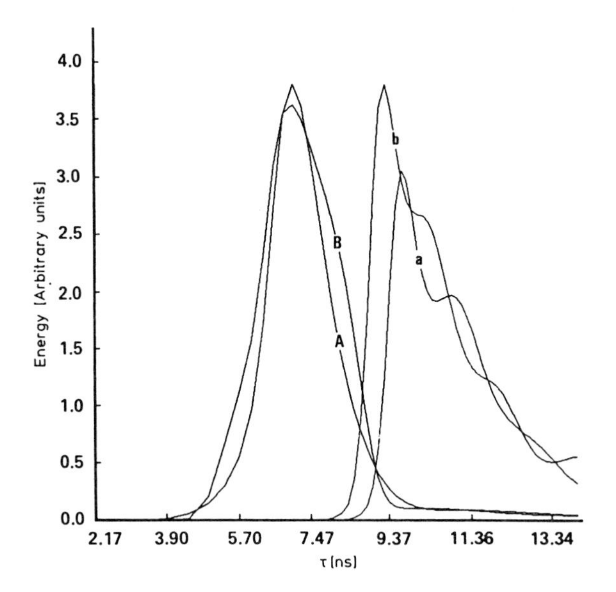

Fig.8.12. Radially integrated normalized energy profiles for the SF (a,b) and injected (A,B) pulses at z=5.3 cm penetration depth for two different values for the injected pulse initial temporal shape parameter σ (see text). The initial on-axis area of the injected pulse is $\theta_P=2\pi$, and the SF effective gain $\alpha_{SF}=641.7$, $\beta=43.95$. All other parameters are the same as for Fig.8.7c. The values of σ are (A) 2 and (B) 4

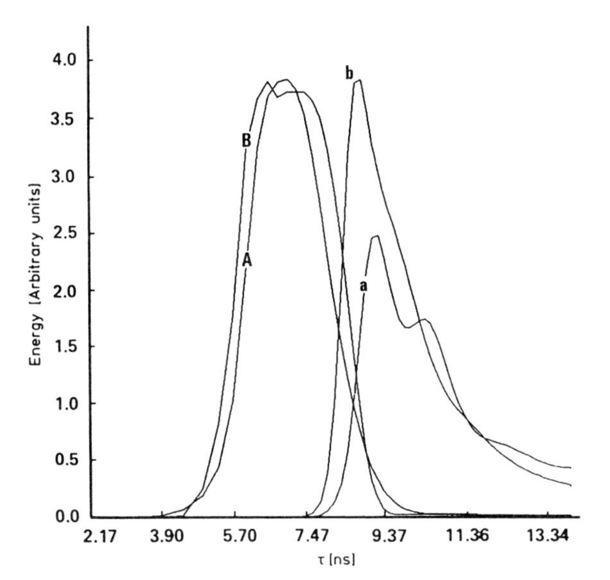

Fig.8.13. Radially integrated normalized energy profiles of SF (a,b) and injected (A,B) pulses at z=5.3 cm penetration depth for two different values of the injected pulse initial temporal shape parameter σ (see text). The initial on-axis area of the injected pulse is $\theta_P=3$, $\beta=43.95$. All other parameters are as for Fig.8.12. The values of σ are (A) 2 and (B) 4

Figure 8.18 shows the effect of variation of the density N of active atoms. Table 8.1 gives the gain values and gain length Fresnel numbers of the sets of curves in the figure. The effective gains α_{RSF} and α_{RP} are changed proportionally, corresponding to a density variation N. The ratio of the SF intensities is $I_c/I_b = 1.76$ and $I_b/I_a = 2.06$; these ratios are larger than the corresponding density ratios squared, $(N_c/N_b)^2 = 1.40$ and $(N_b/N_a)^2 = 1.49$. This difference from the predictions from previous theories of SF [8.1-8] may be due to transverse flux self-focusing, especially since the values of the effective gains used in this case

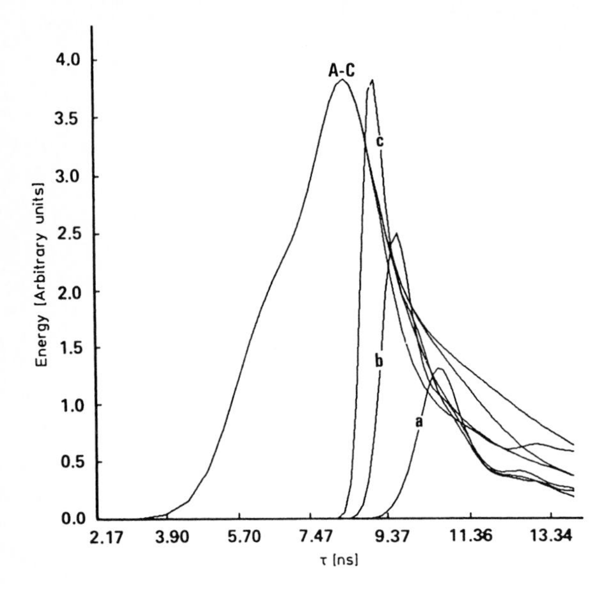

Fig.8.14. Radially integrated normalized energy profiles for the SF (a-c) and injected (A-C) pulses at z=5.3 cm penetration depth for three different values of the SF transition effective gain α_{RSF} (g_{SF}). The on-axis initial area for the injected pulse is $\theta_p=\pi$. All other parameters are as for Fig.8.7c. The values of α_{RSF}, g_{SF}, and β respectively, are

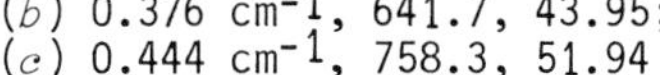

(a) 0.3075 cm^{-1}, 525.0, 35.96; (b) 0.376 cm^{-1}, 641.7, 43.95; (c) 0.444 cm^{-1}, 758.3, 51.94

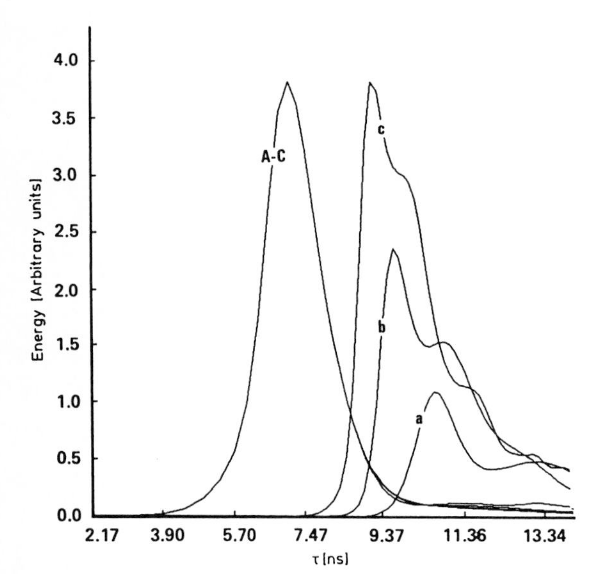

Fig.8.15. Radially integrated normalized energy profiles for the SF (a-c) and injected (A-C) pulses at z=5.3 cm penetration depth for three different values of the SF transition effective gain α_{RSF} (g_{SF}). The on-axis initial area for the injected pulse is $\theta_p=2\pi$. All other parameters are same as for Fig. 8.14

are quite high. However, the ratio of the temporal widths τ_{SF}, FWHM, and thus the optical thicknesses, are within 15% of the corresponding inverse ratios of the densities; the same is true for the delay time τ_D of the SF intensity peak with respect to the pump intensity peak. These results compare qualitatively reasonably well with the mean field predictions for SF in two-level systems initially prepared in a state of complete inversion [8.32].

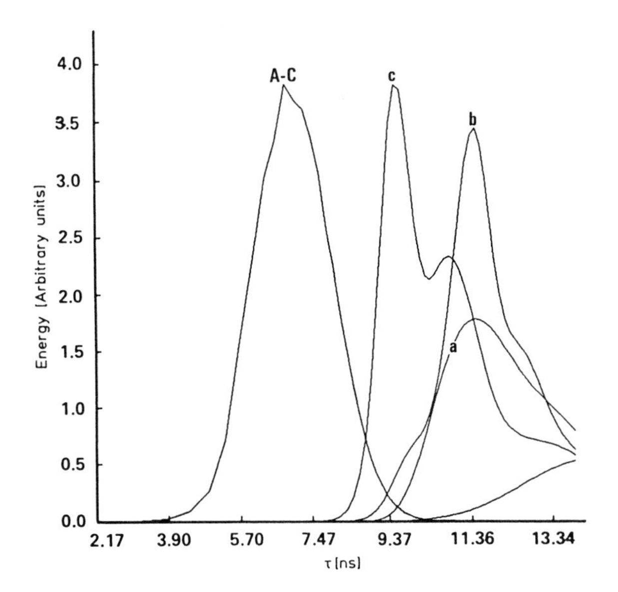

Fig.8.16. Radially integrated normalized energy profiles for the SF (a-c) and injected (A-C) pulses at z=5.3 cm penetration depth for three different values of the SF transition effective gain α_{RSF} (g_{SF}). The on-axis initial area for the injected pulse is $\theta_P=3\pi$. All other parameters are as for Fig.8.14

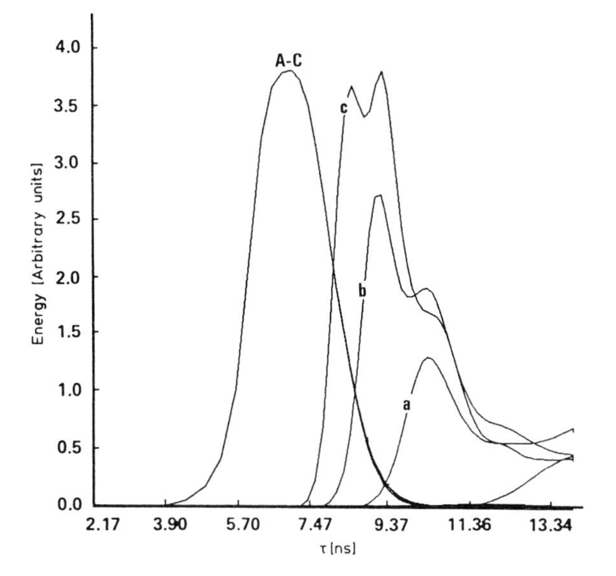

Fig.8.17. Radially integrated normalized energy profiles for the SF (a-c) and injected (A-C) pulses at z=5.3 cm penetration depth for three different values of the SF transition effective gain α_{RSF} (g_{SF}). The on-axis initial area for the injected pulse is $\theta_P=4\pi$. All other parameters are as for Fig.8.14

A comparison of the effects upon the injection pulse of variation in oscillator strengths between the SF and pump transition (variation of α_{RSF}) with effects due to a density variation (variation of both α_{RP} and α_{RSF} proportionally) is given in Figs.8.19 and 20 with gain values and Fresnel numbers given in Tables 8.2,3. It is seen that the respective effects in the pump-pulse reshaping are quite distinct. The variation in oscillator strengths (Fig.8.19) essentially causes "hole burning" in the following edge of the pump pulse, whereas the variation in density (Fig. 8.20) affects the whole pump pulse. This comparison shows the difference between

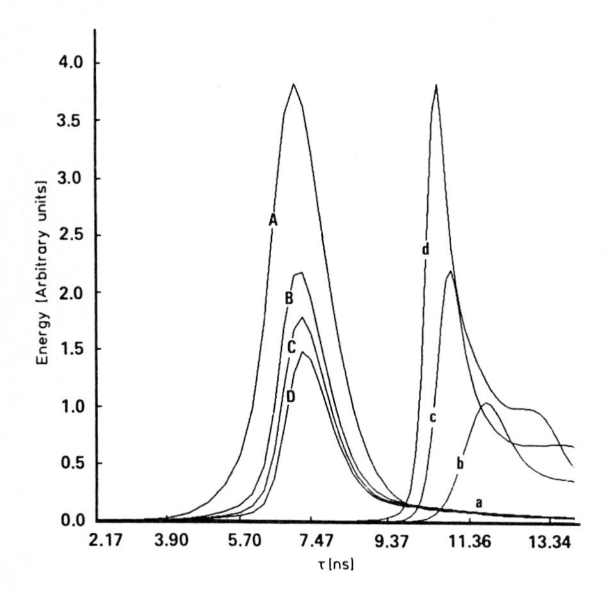

Fig.8.18. Radially integrated normalized energy profiles for the SF (*a-d*) and injected (*A-D*) pulses at z=5.3 cm penetration depth for different values of the density N of atoms. The on-axis initial area for the injected pulse is $\theta_P=2\pi$. Except for the effective Beer (gain) and Fresnel numbers, see Table 8.1, the values of all other parameters are as for Fig.8.7c

Table 8.1. Gain values and Fresnel numbers of the curves in Fig.8.18

Curve	g_P	g_{SF}	β	F_{gp}	F_{gSF}
a	6.25	125	20.0	132.4	974.3
b	26.3	525	19.96	25992	4100
c	32.1	641.7	19.99	31724	5010
d	37.9	758.3	20.01	37456	5922

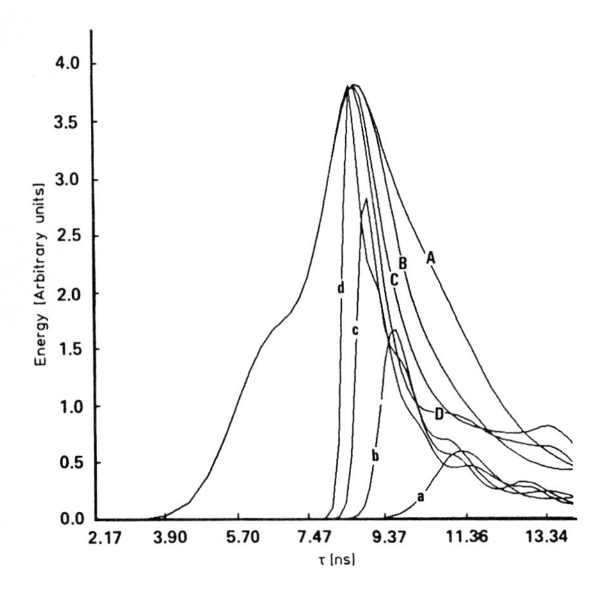

Fig.8.19. Radially integrated normalized energy profiles for the SF (*a-d*) and injected (*A-D*) pulses at z=5.3 cm penetration depth for four different values of the SF transition effective gain $\alpha_{RSF}(g_{SF})$. The initial on-axis area of the injected pulse is $\theta_P=\pi$, and the effective gain of the pump transition $\alpha_{RP}=0.375$ ($g_P=17.5$) cm^{-1}. Except for the effective gain $\alpha_{RSF}(g_{SF})$, see Table 8.2, all other parameters are the same as those for Fig. 8.7c

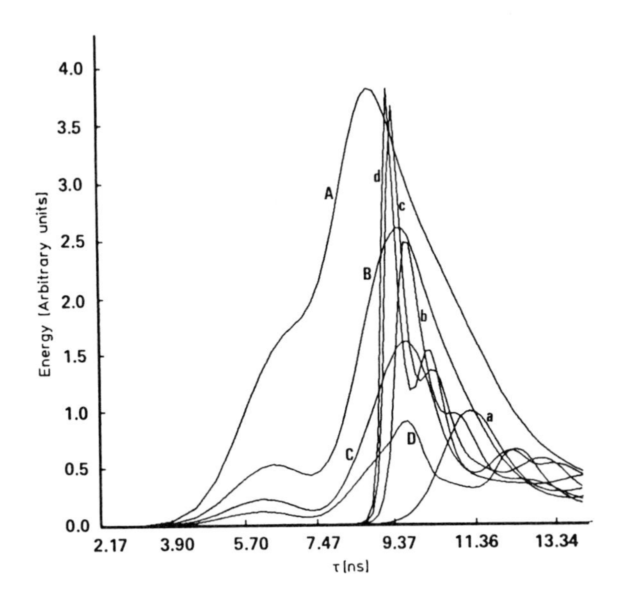

Fig.8.20. Radially integrated normalized energy profiles for the SF (*a-c*) and injected (*A-C*) pulses at z=5.3 cm penetration depth for four different values of the density N of atoms. The on-axis initial area of the injected pulse is $\theta_P=\pi$. Except for the effective Beer (gain) lengths and Beer (gain) length Fresnel numbers, see Table 8.3, the values for all other parameters are as for Fig.8.7c

Table 8.2. Gain values of the curves in Fig.8.19

Curve	g_{SF}	g_P=const	β
a	291.7	17.5	16.67
b	408.3	17.5	23.33
c	525.0	17.5	30.0
d	641.7	17.5	36.67

Table 8.3. Gain values and Fresnel numbers of the curves in Fig.8.20

Curve	g_{SF}	g_P	β	F_{gP}	F_{gSF}
a	291.7	17.5	16.67	17296	2278
b	408.3	24.5	16.67	24212	3188
c	525.0	31.5	16.67	31130	4100
d	641.7	38.5	16.67	38048	5010

inhomogeneous (Fig.8.19) and homogeneous (Fig.8.20) effects on the pump pulse. This effect might be used for the purpose of pulse shaping (i.e., pulse tailoring) under suitable conditions.

Shown in Fig.8.21 is the transverse-integrated SF pulse energy versus retarded time τ (curve 2) together with the transverse-integrated pump-pulse energy versus τ (curve 1) for a gain and propagation depth chosen so that the pulses overlap in time. Under these conditions the two pulses strongly interact with each other via the nonlinear medium, and the two-photon processes resonant coherent Raman(RCR) , which transfer populations directly between levels ε_2 and ε_1, make strong contributions to the mutual pulse development [8.61]. The importance of the

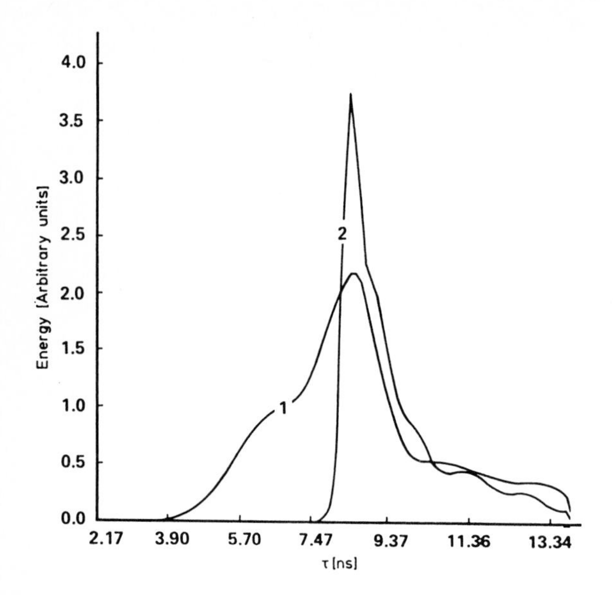

Fig.8.21. Radially integrated energy profiles in units of Rabi frequency, for the SF(*1*) and injected (*2*) pulses at a penetration depth of z=5.3 cm. The effective gain for the pump transition and the SF transition are α_{RP}=0.3643 cm^{-1} (g_P=17) and α_{RSF}=0.376 cm^{-1} (g_{SF}=641.7), respectively. F_gP=16800, F_gSF=2278. The initial on-axis area of the injected pulse is $\theta_P=\pi$. All other parameters are as for Fig.8.4

RCR processes in SF dynamical evolution in an optically pumped three-level system was pointed out for the first time in [8.61]. Indeed, the SF pulse evolution demonstrated here has greater nonlinearity than SF in a two-level system that has been prepared initially by an impulse excitation. What is remarkable is that this is an example where the SF pulse temporal width τ_{SF} is much less than the pump width τ_P even though the two pulses temporally overlap, i.e., the SF process begins late and terminates early with respect to the pump time duration. Pulses of this type have been observed [8.15-17] in CO_2-pumped CH_3F.

Figures 8.22 and 23 are isometric representations of pump pulse and SF pulse copropagation and interaction via the nonlinear medium. These figures exhibit details of the dynamic mutual pulse reshaping (self-defocusing) during SF buildup. They show the pulse intensities as functions of the radial coordinate ρ and retarded time τ for two different penetrations ($z = 4.4$ cm and $z = 5.3$ cm, respectively) into the high-gain medium. The injected pulse is initially radially and temporally Gaussian. Both the pump pulse and the SF pulse are seen to exhibit considerable self-defocusing with ringing following the main SF peak. At the larger optical thickness, Fig.8.23, a large postpulse appears in both the pump and SF pulse propagation. This is due to energy feedback from the SF to the pump transition. The postpulses overlap, and so the two-photon RCR effects are active and quite significant in the dynamic evolution and coupling between the pump and SF pulses. This effect is due entirely to the coherence in the dynamical evolution of the system.

Different initial shape distributions for the pump pulse greatly affect the subsequent shapes of the pump and SF pulses. For example, if the initial temporal distribution of the injected pulse is Gaussian and the initial radial distribution is

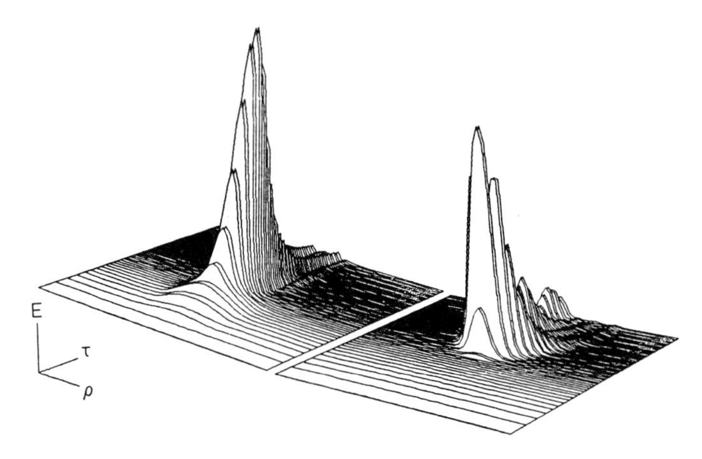

Fig.8.22. Energy E of the pump (*left*) and SF (*right*) pulses as a function of the radial coordinate ρ and retarded time τ at penetration z=4.4 cm. The parameters are as for Fig.8.5a

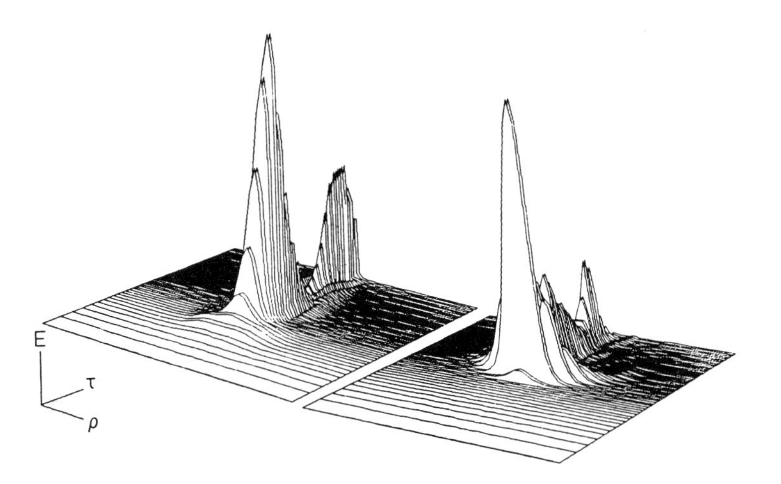

Fig.8.23. Energy E of the pump (*left*) and SF (*right*) pulses as a function of ρ and τ at penetration z=5.3 cm. The parameters are the same as for Fig.8.22

characterized by the parameter $v = 3$, (8.73), it is observed that the injected pulse undergoes considerable reshaping, due to propagation, to a more Gaussian radial distribution, and the SF pulse exhibits strong self-defocusing in the wings of the tail region. However, if the initial radial distribution of the injected pulse is Gaussian and the initial temporal distribution is half-Gaussian, with a sharp temporal cutoff as the second temporal half of the pulse, the SF pulse rises extremely sharply, in comparison to the other cases analyzed, and tapers off, with strong self-defocusing indicated in the wings of the pulse tail. Pump pulses of this type are generated using a plasma switch [8.28-30] and the corresponding SF pulses with steep rise have been observed.

8.5 Conclusions Concerning Deterministic Three-Level Superfluorescence

The effects presented here clearly demonstrate the coherence and deterministic effects on SF pulse evolution of injection pump pulse characteristics and conditions in the regime $\tau_P \geqslant \tau_R$. It is suggested that effects of the type discussed here may have in fact been operative in some of the SF experiments and their results which were published earlier [8.7-27]. The pump pulse was taken as purely coherent in these calculations. To determine whether or not effects like those reported here are indeed operative in a given experiment, it is crucial to determine the degree of coherence of the pumping process as well as its temporal duration [8.61].

Furthermore, and perhaps of greater importance, we have demonstrated the control and shaping of the SF pulse which evolves by specification of particular initial characteristics and conditions for the pumping pulse that is injected into the nonlinear medium to initiate SF emission. These manifestations and others of the same class we call the control of light by light via a nonlinear medium. This phenomenon constitutes a method for nonlinear information encoding, or information transfer from the injection pulse initial characteristics to corresponding SF pulse characteristics which evolve due to propagation and interaction in the nonlinear medium [8.69,70].

8.6 Quantum Initation: Calculational Results and Delay-Time Statistics

We now address the effects of quantum statistics of the SF spontaneous relaxation process. We adopt the computational formulation developed for the monochromatic two-level SF by *Mattar* and co-workers [8.50-52] and *Watson* et al. [8.64,65] and the two-color SF formulation developed by *Haake* et al. [8.77,81,82,86] to explain *Florian* et al.'s [8.87] first observation of polychromatic SF.

Amplified quantum initiation statistics in the highly nonlinear regime of SF pulse evolution are presented in Fig.8.24 for the separation in retarded time τ of the main peak of the SF pulse from that of the pump pulse. Here we have plotted the role of the initial temporal width of the pumping pulse τ_p versus ε, where

$$\varepsilon = \frac{\sigma(\Delta\tau_D)}{\langle\Delta\tau_D\rangle} \equiv \frac{\left[\sum_{i=1}^{NR} (\Delta\tau_d^i - \overline{\Delta\tau}_D)^2/NR\right]^{\frac{1}{2}}}{\overline{\Delta\tau}_D} [1 \pm (NR - 1)^{-\frac{1}{2}}] \quad , \qquad (8.75)$$

and where $\overline{\Delta\tau}_D$ is the mean value of the (SF) pulse peak delay with respect to the pump pulse peak, σ is the standard deviation, and NR is the number of elements in the statistical ensemble, which in this case is 10.

Since the size of the ensemble is quite small, the statistics presented here can at best cite trends, so we regard these results as quite preliminary. The case here is referred to as the highly nonlinear regime, since the pump pulse and the SF pulse

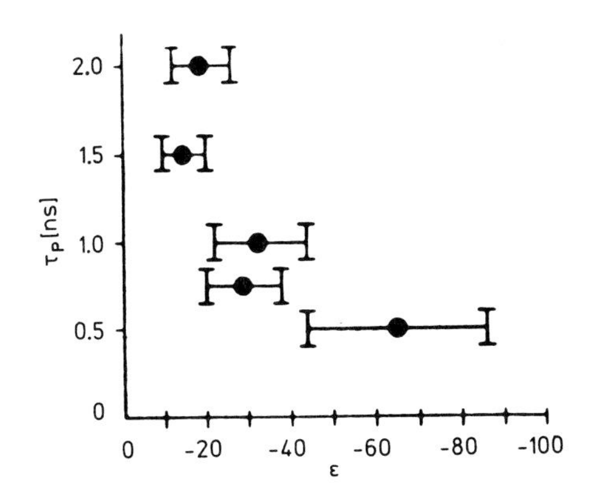

Fig.8.24. SF delay statistics for various pump pulse initial temporal widths τ_P versus ε. $F_P=0.7$; $\theta_P=2\pi$; $T_1=80$ ns; $T_2=70$ ns; $\tau_{RSF}=19.5$ ps; L=5.6 cm

completely overlap for each member of the ensemble. Thus, the two pulses are strongly interacting in this regime, via the nonlinear medium, and two-photon processes that enhance the Q matrix elements are strongly operative. The highly nonlinear regime does not correspond to SF as studied previously, but is quite interesting in itself. As preliminary results, the calculation is far less expensive than that needed to analyze the nonlinear regime of well-separated pulses. This, of course, will soon follow [8.79].

The main deduction from Fig.8.24 is that the quantum statistics of initiation are manifestly important only for the pump pulses corresponding to the shorter initial temporal duration τ_P. This supports our earlier hypothesis that amplified quantum initiation is important only for conditions such that $\tau_P/\tau_R \lesssim 1$.

The overlap of the two pulses is measured in terms of the fluctuation of the delay difference; i.e., is given with $\tau_{Da}=\tau_{DP}$ and $\tau_{Db}=\tau_{DSF}$ by

$$\sigma(\tau_{Da}-\tau_{Db}) = \sigma(\Delta\tau_D) \quad , \qquad \text{with} \tag{8.76}$$

$$\Delta\tau_D = \tau_{Da}-\tau_{Db} \quad \text{and} \quad \Delta\tau_D^i = \tau_{Da}^i - \tau_{Db}^i \quad \text{and}$$

$$\overline{\Delta\tau_D} = \sum_{i=1}^{NR} \Delta\tau_D^i/NR = \langle\Delta\tau_D\rangle \quad . \tag{8.77}$$

The normalization can be done with respect to $\langle\Delta\tau_D\rangle$, the average delay difference:

$$\frac{\sigma(\Delta\tau_D)}{\langle\Delta\tau_D\rangle} = \frac{\sum_{i=1}^{NR}(\Delta\tau_D^i - \overline{\Delta\tau_D})^2/NR}{\langle\Delta\tau_D\rangle}[1 \pm (NR-1)^{-\frac{1}{2}}] \quad ; \tag{8.78}$$

or the normalization can be constructed with regard to the arithmetic mean of the two average SF delay times:

$$\frac{\sigma(\Delta\tau_D)}{(1/2)(\overline{\tau_{Da}}+\overline{\tau_{Db}})} = \frac{\sum_{i=1}^{NR}(\Delta\tau_D^i - \overline{\Delta\tau_D})^2/NR}{(1/2)(\overline{\tau_{Da}}+\overline{\tau_{Db}})}[1 \pm (NR-1)^{-\frac{1}{2}}] \quad . \tag{8.79}$$

Equivalently, the synchronization can measured by the introduction of an auxiliary variable, similarly to *Haake* and *Reibold* [8.86],

$$T_H = (1/2)(\tau_{Da} + \tau_{Db}) - (1/2)\langle\tau_{Da} + \tau_{Db}\rangle \quad , \tag{8.80}$$

where <...> denotes the average over all the NR trajectories.

For the condition that $\tau_P/\tau_R \gg 1$, the coherence of the pump itself overwhelms the fluctuations due to quantum initiation which would otherwise become amplified in the nonlinear regime, see Table 8.4.

Table 8.4. Effects of the temporal pulse length on the relative delay statistics for a given τ_R, at a specific propagation length z_{71} a well-defined gain length Fresnel number, and thus a specific atomic density

Simulation number	TBW $=2/\tau_P$	τ_P [ns]	$\sigma(\Delta\tau_D)/\langle\Delta\tau_D\rangle$	$\sigma(\Delta\tau_D)/[(\tau_{DP} + \tau_{DSF})/2]$
747	4.0	0.50	65.47 ± 21.83	0.60 ± 0.20
746	8/3	0.75	29.04 ± 9.68	0.33 ± 0.11
745	2.0	1.0	32.59 ± 10.86	0.19 ± 0.06
744	4/3	1.50	15.32 ± 5.11	0.16 ± 0.06
743	1.0	2	19.51 ± 6.50	0.29 ± 0.10
717	0.5	4	213.57 ± 80.72	0.14 ± 0.01

$\sigma(\Delta\tau_D)$: variance of delay difference for the statistical ensemble [ns]
$\sigma(\Delta\tau_D)/\langle\Delta\tau_D\rangle$: normalized variance

An interesting effect evident in our computations is that significant statistical variation of the pump pulse peak from turn-on (i.e., $\tau=0$) was observed. Thus, the pump pulse statistics are also important, at least in the highly nonlinear regime, as well as statistics for the SF pulse.

The Fresnel variation of the standard deviation normalized to either the average delay difference $\Delta\tau_D$ or the average of the arithmetic mean of the two delays, as introduced by *Mattar* and co-workers [8.77,81,82] for two-color SF, is

$$\langle\bar{\tau}_D\rangle = \langle(1/2)(\tau_{DP} + \tau_{DSF})\rangle = (1/2)[\langle\bar{\tau}_{DP} + \bar{\tau}_{DSF}\rangle] \quad . \tag{8.81}$$

In Table 8.5, corresponding data are listed for the three different regions of propagation: (a) the SF buildup, (b) completely evolved SF partially overlapping the pump pulse, and (c) the highly nonlinear regime where the two pulses completely overlap (additional nonlinear two-photon processes also take place and compete with the SF process).

The dependence of

$$\langle difdel\rangle = \langle\tau_P - \tau_{SF}\rangle \quad , \tag{8.82}$$

on Fresnel number varies totally from one zone to the other, illustrating the occurrence of different nonlinear processes concomitantly; whereas the dependence of the variance of difdel normalized to the mean of the average delays, which dependence is an exponential decay, does not vary in shape.

Table 8.5. Gain Fresnel number dependence of relative delay statistics for a given τ_R, a well-defined N, and a specific temporal and radial pump pulse shape for three different propagation distances. A: nonlinear region and partially developed SF; B: nonlinear region with full SF evolution (partial overlapping of the SF and pump pulses; C: highly nonlinear region (strong overlapping)

Region	Simulation number	F_{gSF}	$(1/r_P)$ [cm^{-1}]	$\sigma(\Delta\tau_D)/\langle\Delta\tau_D\rangle$	$\sigma(\Delta\tau_D)/[(\tau_{DP}+\tau_{DSF})/2]$
A	749	1.62	2.35	(40.44 ±13.48)	(3.09 ± 1.03)
	748	1.07	2.9	(29.35 ± 9.78)	(3.25 ± 1.08)
	733	0.72	3.527	(35.81 ±11.94)	(3.94 ± 1.31)
	734	0.59	3.9	(28.89 ± 9.63)	(3.76 ± 1.25)
	717	0.495	4.237	(28.69 ± 9.50)	(3.92 ± 1.31)
	718	0.27	5.5771	(31.04 ±10.25)	(6.41 ± 2.14)
	719	0.23	6.25191	(34.34 ±11.44)	(8.73 ± 2.91)
	720	0.12	8.7	(40.81 ±13.60)	(13.73 ± 4.58)
B	749	1.94	2.35	142.63 ±47.54	1.881± 0.604
	748	1.28	2.9	122.57 ±40.86	2.07 ± 0.690
	733	0.683	3.527	65.57 ±21.86	2.31 ± 0.77
	734	0.704	3.9	59.82 ±19.94	2.37 ± 0.79
	717	0.593	4.237	65.23 ±21.74	3.04 ± 1.01
	718	0.322	5.5771	58.87 ±12.95	3.55 ± 1.183
	719	0.275	6.25191	36.87 ±12.29	4.28 ± 1.429
	720	0.142	8.7	49.13 ±16.38	10.67 ± 3.56
	735	0.086	11.1542	5.690± 1.896	1.37 ± 0.46
C	749	2.26	2.35	40.89 ±13.63	1.161± 0.387
	748	1.49	2.9	73.72 ±24.57	1.463± 0.488
	733	1.005	3.527	73.27 ±24.57	1.463± 0.488
	734	0.82	3.9	91.967±30.65	1.692± 0.564
	717	0.69	4.237	213.57 ±80.72	1.512± 0.571
	718	0.375	5.5771	83.47 ±27.82	2.266± 0.755
	719	0.320	6.25191	43.90 ±14.63	2.147± 0.716
	720	0.165	8.7	27.742±77.80	3.209± 1.069
	735	0.100	11.1542	11.24 ± 3.746	2.555± 0.852

We also studied the fluctuations in conjunction with the *two* Langevin forces applied simultaneously. We studied first the Fresnel number variation for a given density. The evolution statistics are summarized in Tables 8.6 and 7 for two propagation lengths. The first is where SF is still building up (weak overlap) and the second one is in the highly nonlinear region where the SF and pump pulses strongly overlap, leading to two-photon processes. Except for (i) simulation 761 where the two Langevin forces are equal and (ii) simulation 768 where no Langevin force was considered in the pump transition, the pump Langevin force is 10 times weaker than the SF Langevin force.

One can conclude that, whenever the Fresnel number is reduced, the delay of the pump peak and the uncertainties (standard deviations) of the pump delay are reduced, whereas the SF delay, its standard deviation, and uncertainties increase. Further work is continuing, for both the nonlinear regime and the highly nonlinear regime and will be reported shortly [8.79].

Table 8.6. Gain Fresnel number dependence of relative delay statistics for a given τ_R, well-defined N, and a specific radial pump pulse shape for a propagation distance z_{61}

Simulation number	$TB = r_P^{-1}$ [cm^{-1}]	$1/r_P^2$ [cm^{-2}]	$\bar{\tau}_{DP}$ [ns]	σ_{DP} [ns]	$\bar{\tau}_{DSF}$ [ns]	σ_{DSF} [ns]
761	3.527	12.44	8.46	0.736 ±0.147	8.75	2.21 ±0.44
764	3.527	12.44	8.21	1.07 ±0.21	7.50	17.2 ±3.43
762	4.237	17.95	8.20	0.864 ±0.173	7.92	1.69 ±0.34
765	5.58	31.14	8.11	0.696 ±0.14	7.99	1.68 ±0.34
766	6.25	39.06	8.02	$(0.134 \pm 0.027)10^{-4}$	8.03	1.86 ±0.373
763	8.7	75.6	7.64	zero	8.20	2.18 ±0.44
767	11.1542	124.4	7.37	$(0.264 \pm 0.053)10^{-4}$	8.39	2.04 ±0.408
768	11.1542	124.4	7.37	$(0.264 \pm 0.053)10^{-4}$	9.40	1.15 ±2.30

	$[\text{variance (difdel)}]^{1/2}$ [ns]	$(1/2)(\tau_{DP} + \tau_{DSF})$ [ns]	$\{\text{VAR(difdel)}/[(\tau_{DP} + \tau_{DSF})/2]\}^{1/2}$
761	0.153	8.61	1.78 ± 0.36
764	0.06	7.86	0.76 ± 0.15
762	0.095	8.06	1.18 ± 0.24
765	0.115	8.05	1.43 ± 0.29
766	0.150	8.025	1.87 ± 0.37
763	0.178	7.92	2.25 ± 0.45
767	0.171	7.88	2.17 ± 0.43
7.68	0.108	8.385	1.29 ± 0.26

We now examine the fluctuations associated with the SF buildup when the pump pulse length is reduced with respect to τ_R for constant Fresnel number and constant density. The resulting statistics are listed in Tables 8.8 and 9.

Unusual trajectories, where the absolute peak of the output power is not the first peak, appear whenever the two applied Langevin forces are equal and whenever the ratio τ_P/τ_R decreases. As mentioned earlier, these pulse distortions can be interpreted as phase waves, which were discussed by *Hopf* [8.76] and by *Watson* et al. [8.64,65]. One should note that the phase waves are more in evidence in the two-level case where the pump that creates the inversion is infinite and does not deplete. These unusual trajectories imply that for any temporal incoherence in the pump whatsoever, the SF structure is distorted (i.e., not shielded from the fluctuations in the pump pulse) unless strong diffraction effects are present.

8.7 Conclusions Concerning Effects of Resonance Diffraction and Copropagation on Superfluorescence Evolution

The interplay of diffraction and deterministic/quantum initiation demonstrates that dynamic pump depletion and cross coupling (energy transfer and conversion efficiency) are crucially dependent on the gain Fresnel number of the SF and on the ratio of the

Table 8.7. Gain Fresnel number dependence of relative delay statistics for a given τ_R, well-defined N, and a specific radial pump pulse shape for a longer propagation distance z_{71} such that $z_{71} > z_{61}$

Simulation number	$Tb\rho_N(r_P^{-1})$ $[cm^{-1}]$	F	$\bar{\tau}_{DP} = \langle\tau_{DP}\rangle$ [ns]	$\bar{\sigma}_{DP}/\bar{\tau}_{DP}$	$\bar{\tau}_{DSF}$ ns	$\bar{\sigma}_{DSF}/\bar{\tau}_{DSF}$
761	3.527	1.0	8.64	$(0.28 \pm 0.06)\times 10^{-4}$	8.51	1.6 ± 0.32
764	3.527	1.0	8.18	1.25 ± 0.25	7.26	20.6 ± 4.12
762	4.237	0.69	8.20	1.14 ± 0.23	7.85	1.44 ± 0.29
765	5.58	0.40	8.18	0.81 ± 0.16	7.91	1.54 ± 0.31
766	6.25	0.32	8.14	0.96 ± 0.19	7.95	1.63 ± 0.33
763	8.7	0.165	7.83	0.83 ± 0.166	8.09	1.70 ± 0.34
767	11.542	0.10	7.37	$(0.264 \pm 0.053)\times 10^{-4}$	8.28	1.71 ± 0.34
768	11.542	0.10	7.37	$(0.264 \pm 0.053)\times 10^{-4}$	9.30	2.37 ± 0.47

	$\sigma(\Delta\tau_D)/\langle\Delta\tau_D\rangle$	$\sigma(\Delta\tau_D)$ [ns]	$\langle\tau_D\rangle = (\bar{\tau}_{DP} + \bar{\tau}_{DSF})/2$ [ns]	$\sigma(\Delta\tau_D)/\langle\bar{\tau}_D\rangle$
761	106.9 ± 21.4	0.135	8.57	1.58 ± 0.35
764	163.27 ± 32.6	1.50	7.72	19.43 ± 3.88
762	15.74 ± 3.1	0.055	8.025	0.78 ± 0.14
765	28.3 ± 5.6	0.075	8.045	0.93 ± 0.19
766	39.2 ± 7.8	0.074	8.045	0.92 ± 0.18
763	$-(72.36 \pm 14.6)$	0.191	7.96	2.4 ± 0.48
767	$-(15.7 \pm 3.1)$	0.141	7.815	1.8 ± 0.36
768	$-(11.45 \pm 2.2)$	0.220	8.335	2.63 ± 0.53

$$\sigma(x) = \mathrm{Var}(x) = E(x^2) - |E(x)|^2$$
$$E(x) = \int x f(x) dx$$
$$E(x^2) = \int x^2 f(x) dx$$

Table 8.8[a]. Effect of the pump pulse length on the delay statistics for a given Fresnel number, a given τ_R, and specific temporal and radial pump pulse shape for a given propagation distance z_{61}

Simulation number	TBW $[ns^{-1}]$	$\bar{\tau}_{DP}$ [ns]	σ_{DP} [ns]	$\bar{\tau}_{DSF}$ [ns]	σ_{DSF} [ns]	$\langle\tau_{DP}\rangle$ [ns]
762	0.5	8.20	0.64 ± 0.173	7.92	16.9 ± 0.34	
771	1.0	5.18	43.88 ± 8.78	5.41	36.31 ± 7.26	7.28
772	5/3	3.24	16.7 ± 3.34	4.94	30.30 ± 6.06	7.10
770	2.00	5.04	25.67 ± 5.135	5.40	27.35 ± 5.47	14.1

$\sigma_{DP}/\langle\tau_{DP}\rangle$	τ_{DSF} [ns]	σ_{DSF} [ns]	$[\mathrm{VAR}\ (\mathrm{difdel})]^{1/2}$	$(\tau_P+\tau_{SF})/2$ [ns]	$\{\mathrm{VAR}(\mathrm{Difdel})/[(\tau_P+\tau_{SF})/2]\}^{1/2}$
			0.095	8.06	1.18 ± 0.24
$(0.107 \pm 0.021)\times 10^{-4}$	9.8	7.7 ± 1.54	1.631	4.09	0.40 ± 0.09
$(0.11 \pm 0.021)\times 10^{-4}$	18.06	8.37 ± 1.67	3.26	7.97	0.41 ± 0.08

[a]Note that there are more columns in Table 8.7 than in Table 8.5.

Table 8.9[a]. Effect of the pump pulse length on the delay statistics for a given Fresnel number, a given τ_R, and specific temporal and radial pump pulse shape for a longer propagation distance z_{71} such that $z_{71} > z_{61}$

Simulation number	$TBW=2/\tau_P$ [ns^{-1}]	$\bar{\tau}_{DP}=\langle\tau_{DP}\rangle$ [ns]	$\bar{\sigma}_{DP}/\bar{\tau}_{DP}$	$\bar{\tau}_{DSF}$ [ns]	$\bar{\sigma}_{DSF}/\bar{\tau}_{DSF}$	$\sigma(\Delta\tau_D)/\langle\Delta\tau_D\rangle$
762	0.5	8.20	1.14 ± 0.23	7.85	1.44 ±0.29	15.74 ± 3.15
771	1.0	4.23	50.85 ±10.17	5.42	36.31 ±7.26	251.1 ±50.21
772	1.67	3.07	15.90 ± 3.18	5.04	33.44 ±6.7	79.1 ±16.49
770	2.00	2.55	28.37 ± 5.67	5.31	26.7 ±5.33	49.42 ±10.31

	$\lvert\tau^a_{DP}\rvert$ [ns]	$\lvert\sigma^a_{DP}\rvert$ [ns]	$\lvert\tau^a_{DSF}\rvert$ [ns]	$(\sigma_{DSF}/\tau_{DSF})$	$\langle\tau_D\rangle = (\bar{\tau}_{DP}+\bar{\tau}_{DSF})/2$ [ns]	$\sigma(\Delta\tau_D)/\langle\Delta\tau_D\rangle$
762	7.85	1.14 ± 0.23	7.85	1.44 ±0.29	10.055	0.68 ±0.14
771	7.37	$(0.26 \pm 0.05)\times10^{-4}$	8.25	4.36 ±0.87	2.98	0.68 ±0.14
772	7.1	$(01.11 \pm 0.021)\times10^{-4}$	9.38	8.27 ±1.65	1.73	42.6 ±8.51
770	7.1	$(01.11 \pm 0.02)\times10^{-4}$	8.82	10.31 ±2.06	1.50	38.2 ±7.6

[a]Note that Table 8.8 has more variables than Table 8.6.

pump temporal duration to the SF cooperation time. We are now analyzing further the nonlinear interaction as affected by the SF and/or the pump detuning, and thus examining the enhanced role of dispersion.

8.8 The Role of Dispersion

As motivation for this calculation, let us recall *LeBerre* and co-workers' [8.88-94] and *Teichmann* and co-workers' [8.95-97] analysis of CW (single) laser radiation propagating in a two-level saturable absorber. The relevant gain parameter was found to be $\tilde{\alpha} = \alpha[1 + (\Delta\omega T_2)^2 + I/I_s]^{-1}$. One can translate this back to the double (pulse) SIT regime as a universal parameter, gain Fresnel number $F_{ga,b}$, where the concept of a normalized Beer length $\alpha_{a,b}$ is given in terms of detuning $\Delta\omega_{a,b}$ and the effective value of the on-axis (peak) Rabi frequency $\nu^{a,b}_{Rabi}$ (equivalently, the on-axis area which equals the product of the Rabi frequency and the pulse length) divided by the equivalent of a CW saturation intensity $I_s = (|\mu|^2/\hbar^2) \times T_1T_2$. Thus, the new gain Fresnel number $F_{ga,b}$ is defined in terms of

$$\tilde{\alpha}_{a,b} = \alpha_{a,b}\left[1 + (\Delta\omega_{a,b}T_2)^2 + \nu^{a,b}_{Rabi}\,\tau_P\left(\frac{\mu_{a,b}}{\hbar}\right)^2 (T_1^{a,b}T_2^{a,b})\right]^{-1} . \tag{8.83}$$

Note that the value of $\tilde{\alpha}_{a,b}$ is independent of the sign of the detuning.

The dispersion problem is dealt with by carrying out the calculation first with only the pump detuned, then with the pump resonant and the SF off-resonant. Calculations being carried out at present with equal detuning in both transitions (i.e., à la Raman) but for a nonvirtual upper level will be reported shortly.

Feld's semiclassical (propagational) model [8.46] for SF initiated with a deterministic tipping angle can be considered as equivalent to the propagation of a very small optical area in an inverted medium. The initial tipping angle in the SF calculation gives rise to a small SF field after a brief propagation in the inverted atomic system. Correspondingly, in the three-level problem one can effectively study the copropagation of two pulses of quite different input areas: the first one (the pump) is much larger than the second one (the SF). This is the optically pumped three-level SF's equivalent to a double SIT of two very different pulses, a strong pump E_a and a weak probe E_b, such that $E_b/E_a = 1/100$, initially.

As realized by *Mattar* et al. and *Mattar* [8.98,99], the SF plays an identical role to the weak probe's role in spectroscopy. To insure that the two pulses overlap while propagating, that is, that they do not get out of synchronization (while coherently exchanging energy back and forth with the resonant or off-resonant transitions), one must select a Beer length ratio that is much larger than unity. This condition $\beta = \alpha_b/\alpha_b > 1$ translates into $\alpha_b > \alpha_a$, which means (a) $\mu_b^2\omega_b > \mu_a^2\omega_a$ and (b) $g_b > g_a$, since $g_{a,b} = (c\tau_p)\alpha_{a,b}$. This condition must also be satisfied for off-resonant SF as well as $\tilde{\alpha}_b > \alpha_a$ with $\tilde{\alpha}_b = \alpha_b\{1 + [(\Delta\omega_b)\tau_p]^2\}^{-1} > \alpha_a$. That is, the SF pulse sees a larger optical thickness than the pump does. To stress the importance of β in the growth of the SF pulse (equivalently, the growth of the weak probe), a family of calculations in both uniform plane-wave and nonplanar propagations have been carried out. The larger the value of β, the greater the overlap; thus the strength (peak value) of the SF is enhanced and the temporal delay relative to the pump pulse is reduced.

The weak probe or SF detuning calculations have been made for an on-axis 2π pump pulse and an SF-equivalent weak probe of 0.02π with $\beta = \alpha_b/\alpha_a = 30$. Otherwise, the parameters are identical to those of Fig.8.4 as, for example, the input pulses had Gaussian profiles in both ρ and τ. The FWHM were $r_p = 0.24$ cm and $\tau_p = 4$ ns, respectively, etc. The detuning $(\Delta\omega_b)\tau_p$ varied from zero (exact resonance) to 0.45 with the pulse reciprocal length TBWA = TBWB = 0.05 ns^{-1}. Since the ratio of the maximum detuning to the initial pulse bandwidth is 9, almost an order of magnitude, special care, as recognized by *Berman* et al. and *Mattar* [8.98,99], must be taken to insure adequate temporal resolution.

To illustrate the physics, we review the on-resonance coherent interaction in the regime where diffraction is expected to play a role. The pump is clearly seen to deplete while the SF grows/builds up. The pump can experience an on-axis energy enhancement due to a redistribution of flux from the peripheral shells towards the beam center. This trend of self-focusing can slow down the rate of pump depletion. The energy, i.e., the output power OPOWRA,B is displayed in Fig.8.25 as a function of time for various propagation distance η in the simulation LR2CFS 1140. The peak

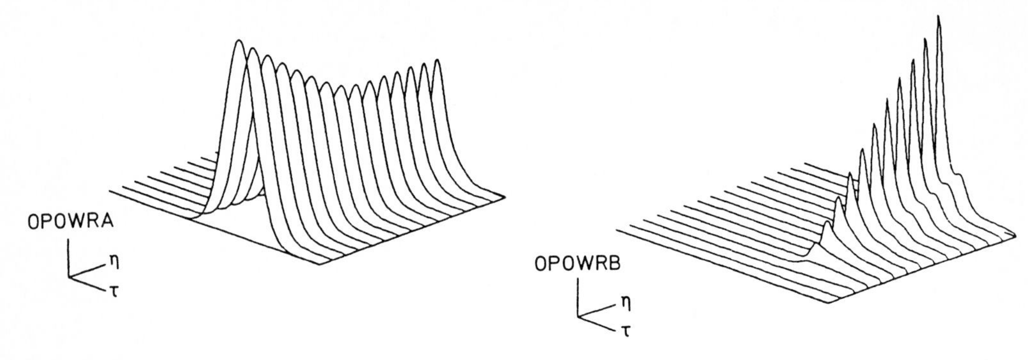

Fig.8.25. The coherent exchange of energy from the depleting pump to the SF is evident in this isometric representation of the output power of each beam as a function of the propagation distance (optical thickness)

of OPOWRA is clearly seen to decrease whereas the SF peak OPOWRB is seen to increase nonlinearly . For sufficiently large $\alpha_a l$, OPOWRB buildup may slow down, taper, or even change sign.

In Figs.8.26a,b partial output powers OPWAPA, OPWBPB are displayed to illustrate the effect of a varying detector pinhole radius on the output by including a larger number of the beam shells. This effect is predominant for large Fresnel numbers where the diffraction coupling is relatively weak. In Figs.8.26c,d the energy of

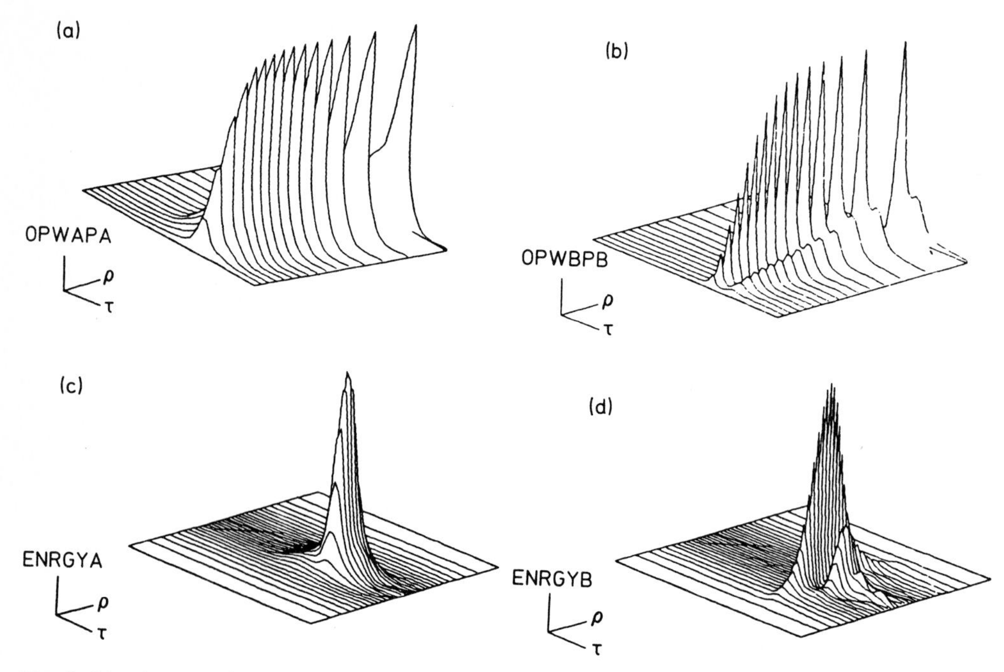

Fig.8.26a-d. The partial output powers OPWAPA and OPWBPB emphasize the contribution of the nonuniform atomic density. The diameter of the centered aperture is variable

each laser pulse (pump and SF) is presented as an isometric plot versus ρ and τ at the exit plane η of the cell to illustrate the nonlinear buildup of the SF at the expense of the pump; the initial ratio of the energies has changed completely.

The time integrated pulse area (THETAA, THETAB) and the time integrated pulse energy (ENERA, ENERB) are isometrically plotted in Fig.8.27 versus ρ and η to illustrate (i) the (longitudinal and transverse) reshaping, (ii) the cooperative transfer of energy from one beam to the other as well as from the outer rim shells of one beam to its inward core shell causing an energy redistribution at the tail of the pulse and an on-axis enhancement for the SF. One should also note that the on-axis pump depletion slows down beyond a certain optical thickness at which the SF has built up considerably and has reached a certain saturation plateau, and the pump experiences beam narrowing and inward radial flux (transverse energy current, $|E|^2\partial\phi/\partial\rho$) flow.

In Fig.8.28, the two output powers are contrasted for three different penetration depths, namely $\alpha l_{51} < \alpha l_{61} < \alpha l_{71}$. The buildup SF Rabi frequency is seen to have become larger than that of the pump and the SF pulse length to have been reduced. Both the pump and the SF pulses have strongly departed from the input Gaussian shape: severe pulse distortion and pulse compression is clearly noticeable. The peak temporal delay between the pump and the SF is seen to decrease as a function of l displaying the different oscillator strengths $\beta = \alpha_{23}/\alpha_{12} = \alpha_b/\alpha_a = \mu_{23}^2\omega_{23}/\mu_{12}^2\omega_{12} \neq 1$.

The effect of the diffraction coupling can be seen by comparing the output power evolution at the end of the nonlinear resonant cell for different Fresnel numbers. The beams reshape differently depending on the amount of communication between the various concentric shells. The associated self-action phenomena such as self-phase modulation (phase encoding due to induced differential polarization) and self-focusing or self-defocusing are strongly modified by the choice of the Fresnel number for a given atomic number density, that is, a given nonlinear Beer length.

The additional effect of dispersion through the SF detuning, which reduces the SF gain since $\tilde{\alpha}_b(\Delta\omega_b) < \alpha_b(\Delta\omega_b = 0)$, has been systematically studied in both plane-wave and nonplanar configurations. The SF detuning reduces β, the SF to pump effective Beer length ratio. That is, $\tilde{\beta} = \tilde{\alpha}_b/\alpha_a$ is smaller than the on-resonance β. The SF grows at a smaller rate causing the pump to deplete less. The relative cross-amplification E_b/E_a is modified with the detuning; as the detuning increases, the cross-amplification, which measures the coherent energy exchange, changes from its maximum on-resonance value to a smaller value.

Figure 8.29 illustrates the symmetrical role of the SF detuning on the SF output power OPOWRB at z_{61} in the uniform-plane-wave regime. The SF detunings studied are $\Delta\omega_b = 0.0$, ± 1.43, ± 3.43, and ± 6.86 rad/ns from the simulations LR2CFS 884; 921, 927; 923, 929; and 926, 931.

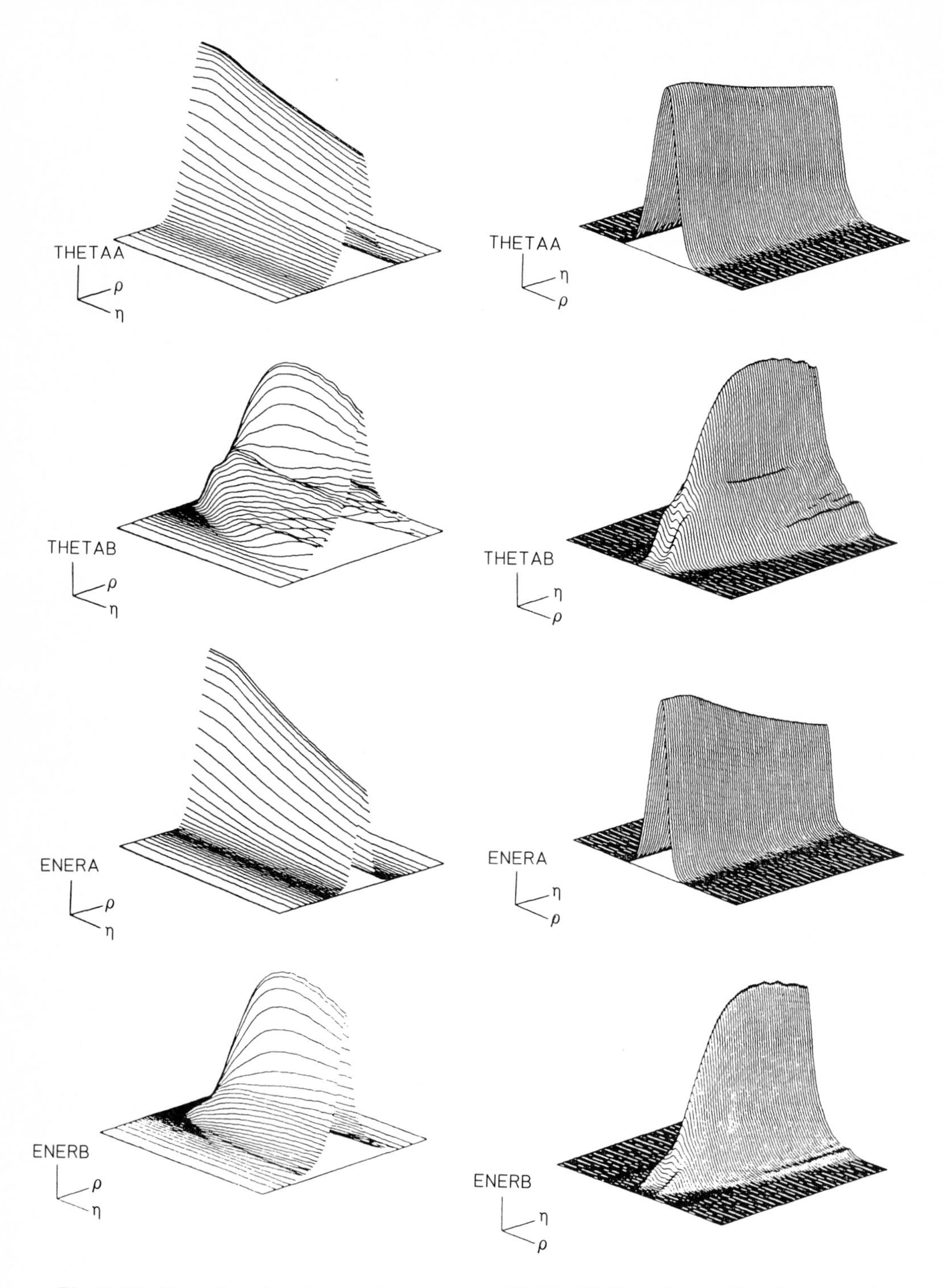

Fig.8.27. From the time integrated energies ENERA, ENERB and the time integrated field areas, THETAA, THETAB, one can appreciate the role of (nonuniform) transverse variation as well as the cross-depletion of the pump to the growing SF

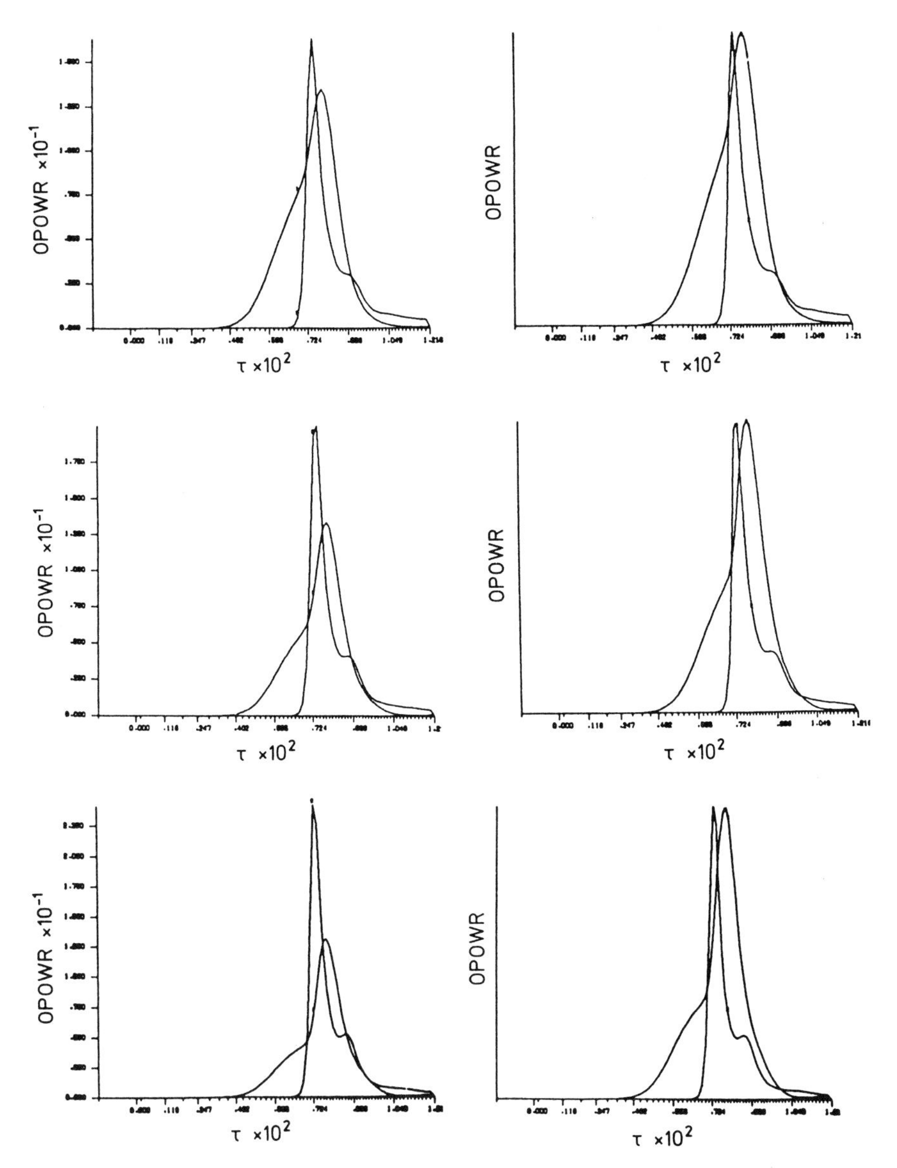

Fig.8.28. Comparison of the radially integrated energies of the pump and SF pulses. The output powers are plotted against τ for three different propagation distances z_{51}, z_{61}, and z_{71}. In the left-hand graphs, the scale is local and is in normalized Rabi units, while in the right-hand graphs the peaks have been normalized to clearly display the overlap and the relative peak delay

Figure 8.30 summarizes the interplay of propagational effects, $\alpha_b l$, diffraction, and dispersion in the SF buildup. Figure 8.30 is the off-resonant equivalent of Fig.8.28: the SF magnification is more subdued (less pronounced) here.

It is worth noting that the symmetrical variation of the spectrum found in the uniform-plane-wave regime does *not* remain when diffraction is included. This asym-

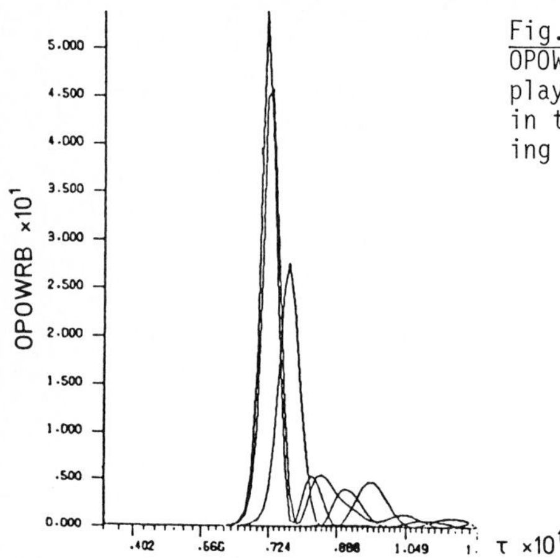

Fig.8.29. The variation of the SF output power OPOWRB (defined in Rabi frequency units) displays a perfect symmetry around the resonance in the uniform-plane-wave regime. As the detuning increases, the SF peak value decreases

metry can be readily appreciated by examining the single-field two-level SIT interaction. For on-resonance SIT the field cannot develop a phase ϕ unless diffraction is present. The diffraction initiates the phase variations essential to transverse flow. A minimal diffraction coupling is necessary to initiate resonant phase.[1] Once phase derivatives develop, an instantaneous detuning $(\partial\phi/\partial\tau)$ arises, driving a complex polarization. Subsequently, the complex polarization enhances the evolution of the phase. Off-resonance SIT self-focusing and defocusing would occur for a nonuniform input through the complex polarization term in the two-level Bloch equation even if the diffraction coupling were absent (i.e., at very large or infinite Fresnel numbers where the shell model is predominant). The resulting variation of either the output power or of the SIT time-integrated on-axis energy density in terms of the detuning is asymmetrical.[2] This competition between diffraction and dispersion

1 For short optical thicknesses, the diffraction-induced phase can be evaluated using a perturbational treatment. For a one-field two-level system, the on-axis phase has been calculated in terms of the accumulated diffraction from the adjacent shells. Details of the treatment can be found in [8.53]. In summary, the phase is given by

$$\phi = \arctan\left[\int_0^{\eta} \nabla_T^2(\mathrm{Re}\{E\})\, d\eta'/F_g(\mathrm{Re}\{E\})\right]$$

$$\simeq \arctan\left[\nabla_T^2(E_{input})\eta/(F_g\, E_{input})\right] .$$

Even though the incoming nonuniform (e.g., Gaussian) field E_{input} can be collimated, after a short propagation distance the field acquires a phase variation of a definite sign. To alter this sign one would need to consider an input beam through a (converging or diverging) lens with a radial (focusing or self-defocusing) phase.

2 Equivalently, one can elaborate by relating two different aspects of the medium polarization to the observed asymmetric behavior. The absorptive properties of the medium are a resonant function of the frequency offset: a maximum rate of energy exchange between field and matter occurs at resonance and the dispersive properties of the medium increase with the detuning. The development of a dispersive part of the polarization leads to self-refraction directly while a secondary effect is the evolution of phase modulation, which in turn leads indirectly to self-refraction. Note that the absorption and dispersion are related by the usual Kramers-Kronig relations.

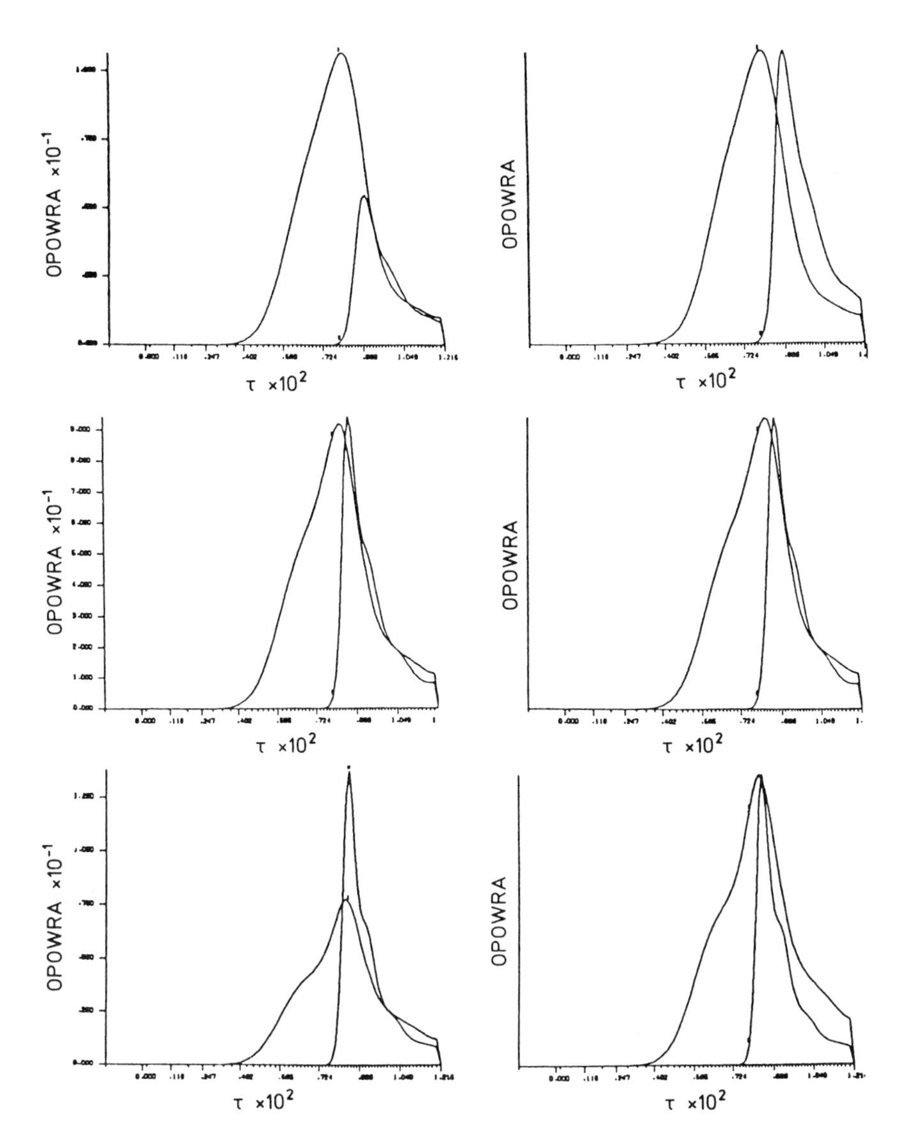

Fig.8.30. The pump and SF output powers OPOWRA and OPOWRB are contrasted for an SF detuning of $\Delta\omega_b=7\pi$ for the propagation distances z_{51}, z_{61}, and z_{71}. As z increases, the SF peak increases, but its relative peak delay and pulse length decrease. Also, as z increases, the temporal OPOWR shape becomes distorted. In the left column, Rabi frequency units are used, while in the right column a maximum renormalization scale helps to clearly exhibit the increase of overlap of the two pulses as z increases

was rigorously calculated by *Mattar* and *Newstein* [Ref.8.53a, Figs.8,9,13], and was observed in Na by *Gibbs* and *Bölger* [8.55]. Such asymmetry is displayed in [Ref.8.56, Fig.2]. Note that the phase variation due to the off-resonance effect, which is independent of the sign of detuning, can either interfere constructively with or be canceled by the diffraction-induced phase. In the more general two-field three-level problem, the same interplay of the physical processes of dispersion and diffraction takes place. Intuitively, one would expect the same asymmetric dependence for the output when both diffraction and dispersion are present.

8.9 Conclusions Concerning the Mutual Influence of Dispersion and Diffraction on the Superfluorescence Buildup

The spectral dependence of the SF has been demonstrated. Further calculations for stronger pumps and variable SF detunings are being planned to assess the role of the dynamic Stark effect on the optimum SF buildup. A stronger pump might experience coherent pulse breakup, which would modify the peak Rabi frequency associated with each subpulse.

The fundamental problem of coherent transfer of energy from the pump pulse to the SF pulse has been rigorously addressed for the first time. These preliminary results reveal the dynamic influence of the SF detuning. This study elucidates why the SF detuning has become important as a means of tuning to the frequency at which the SF gain becomes important for a given pump strength. One finds a range of SF detuning for given nonlinear propagation distances for which gain occurs and other detunings for which absorption occurs.

8.10 Final Conclusion

The effects of either transverse or pumping effects in the propagational model have not yet been studied experimentally and only a few theoretical works treat the subject. In this chapter, we have reported the first computational results for transverse spatially resolved SF generation from optically pumped three-level systems.

Two immediate applications of the above analysis would be the description of probe evolution for nonlinear spectroscopy and the buildup of Stokes emission in a unidirectional Raman forward amplifier. The Stokes initation could be either a weak deterministic signal or quantum fluctuations. The methodology applied herein would enable one to study, with little modification, transient and transverse variations of pump depletion in Raman amplifiers. The Stokes buildup is analogous to the SF buildup. There is a difference between the two signals (SF and Stokes) in the role of dispersion. In the Raman amplifier both the pump and Stokes lines are equally detuned from a virtual level, whereas the pump and the SF can experience different asymmetrical frequency offsets from a real upper level. Thus, the current description can unify different phenomena which stem from entirely different physical applications even though they were studied and reported as independent processes. Through the realistic modeling of coherent energy transfer between two different wavelength fields interacting cooperatively with a nonlinear media (e.g., the pump signal and the SF pulse) one may also understand the physics of information encoding.

Acknowledgments. FPM gratefully acknowledges extensive discussions with Prof. M.S. Feld, Prof. H.M. Gibbs, and Dr. S.L. McCall on the physics of two-level superfluorescence. He also greatly appreciates Prof. H.A. Haus's continuous interest on his behalf and sustained guidance on the subject of single and double beam propagation in nonlinear media. FPM thanks Prof. J.H. Eberly for having introduced him to the two-laser three-level SIT interaction problem and is indebted to Prof. F. Haake for introducing him to the two-color superfluorescence calculation; the currently reported work has benefited from a collaboration which was sponsored by DAAD (Deutscher Akademischer Austausch Dienst). FPM also acknowledges numerous enlightening discussions with Prof. P.R. Berman on the strong pump weak probe interaction and the role of propagation in Rabi-side bands generation. Professor M.S. Feld's hospitality at the Spectroscopy Lab has been a great source of motivation and encouragement. It has led to an appreciation of the experimental subtleties which help to refine the computer modeling. Fruitful discussions with Prof. M. Lax at CCNY on the physics and numerics of fluctuation processes are also appreciated. The skillful and laborious word processing effort of C. Delvin is joyfully acknowledged. The authors are grateful for the careful reading and editing of Dr. H. Lotsch, Mr. C.-D. Bachem, and Ms. D. Hollis at Springer-Verlag.

Part of this work was performed at the MIT Regional Laser Center which is a National Science Foundation Regional Instrumentation facility.

The numerical algorithm was initiated under the auspices of the Office of Naval Research and the Department of Energy at the Physics Department of the University of Rochester, Rochester, NY. The development of the code applied to this study was sponsored by the Army Research Office DAAG23-79-C-0148 and the Office of Naval Research N000-14-80-C0174 at the Polytechnic Institute of New York, Brooklyn NY; present stipend supported by the Office of Naval Research N00014-77-C-0553 at NYU, and jointly by the Air Force Office of Scientific Research and the Army Research Office under ARO administered grant DAAG-29-84-I-0137 as well as by the National Science Foundation (NSF-PHY-84-06107 and NSF-PHY-85-12051). Partial support in 1984 was extended to City College of New York by the Army Research Office Contract DAAG-29-85-K-0112 and the Department of Energy Grant DE-FG02-84-ER45058. The data reduction software management has been carried out under the auspices of Battelle Colombus Labs for MICOM, NV&EOL, and CRDC. The fluctuation calculations were supported by the Directed Energy Directorate at Redstone Arsenal, Huntsville, Alabama and were carried out at the BMD-ARC computing center.

References

8.1 R.H. Dicke: Phys. Rev. **93**, 99 (1954); In Proc. 3rd Int. Conf. on Quantum Electronics, Paris, 1963, ed. by P. Grivet, N. Bloembergen (Columbia University Press, New York 1964)
8.2 S.L. McCall: Ph.D. Thesis in Physics, University of California, Berkeley (1968)
8.3 D.C. Burnham, R.Y. Chiao: Phys. Rev. **188**, 667 (1969)
8.4 E.L. Hahn: Phys. Rev. **80**, 580 (1950)
8.5 R. Friedberg, S.R. Hartmann: Phys. Lett. A**37**, 285 (1971); A**38**, 227 (1972)
8.6 A. Flusberg, F. Mossberg, S.R. Hartmann: In *Cooperative Effects in Matter and Radiation*, ed. by C.M. Bowden, D.W. Howgate, H.R. Robl (Plenum, New York 1977) p.37
8.7 N. Skribanowitz, I.P. Herman, J.C. MacGillivray, M.S. Feld: Phys. Rev. Lett. **30**, 309 (1973)
8.8 I.P. Herman, J.C. MacGillivray, N. Skribanowitz, M.S. Feld: In *Laser Spectroscopy*, ed. by R.G. Brewer, A. Mooradian (Plenum, New York 1975) p.379
8.9 M. Gross, C. Fabre, P. Pillet, S. Haroche: Phys. Rev. Lett. **36**, 1035 (1976)
8.10 A. Flusberg, T. Mossberg, S.R. Hartmann: Phys. Lett. **58**A, 373 (1976); Phys. Rev. Lett. **38**, 59 (1977)
8.11 Q.H.F. Vrehen, H.M.J. Hikspoors, H.M. Gibbs: Phys. Rev. Lett. **38**, 764 (1977)

8.12 M. Gross, J.M. Raimond, S. Haroche: Phys. Rev. Lett. **40**, 1711 (1978)
S. Haroche: In *Coherence and Quantum Optics IV*, ed. by L. Mandel, E. Wolf (Plenum, New York 1978) p.539
8.13 J. Okada, K. Ikeda, M. Matsuoka: Opt. Commun. **27**, 321 (1978)
8.14 A. Crubellier, S. Liberman, P. Pillet: Phys. Rev. Lett. **41**, 1237 (1978)
8.15 A.T. Rosenberger, S.J. Petuchowski, T.A. DeTemple: In *Cooperative Effects in Matter and Radiation*, ed. by C.M. Bowden, D.W. Howgate, H.R. Robl (Plenum, New York 1977) p.15
8.16 J.J. Ehrlich, C.M. Bowden, D.W. Howgate, S.H. Lehnigk, A.T. Rosenberg, T.A. DeTemple: In *Coherence and Quantum Optics IV*, ed. by L. Mandel, E. Wolf (Plenum, New York 1978) p.923
8.17 A.T. Rosenberg, T.A. DeTemple: Phys. Rev. A**24**, 868 (1981)
8.18 H.M. Gibbs, Q.H.F. Vrehen, H.M.J. Hikspoors: Phys. Rev. Lett. **39**, 547 (1977)
8.19 H.M. Gibbs, Q.H.F. Vrehen, H.M.J. Hikspoors: In *Laser Spectroscopy III*, ed. by J.L. Hall, J.L. Carlsten, Springer Ser. Opt. Sci., Vol.7 (Springer, Berlin, Heidelberg 1977) p.213
8.20 Q.H.F. Vrehen, H.M.J. Hikspoors, H.M. Gibbs: In *Coherence and Quantum Optics IV*, ed. by L. Mandel, W. Wolf (Plenum, New York 1978) p.543
8.21 H.M. Gibbs: "Superfluorescence Experiments", in *Coherence in Spectroscopy and Modern Physics*, ed. by F.T. Arecchi, R. Bonifacio, M.D. Scully (Plenum, New York 1978) p.121
8.22 Q.H.F. Vrehen, M.F.H. Schuurmans: Phys. Rev. Lett. **42**, 224 (1979)
8.23 Q.H.F. Vrehen: In *Laser Spectroscopy IV*, ed. by H. Walther, K.W. Rothe, Springer Ser. Opt. Sci., Vol.21 (Springer, Berlin, Heidelberg 1979) p.471
8.24 M. Gross, P. Goy, C. Fabre, S. Haroche, J.M. Raimond: Phys. Rev. Lett. **43**, 343 (1979)
8.25 C. Brehignac, Ph. Cahuzac: J. Phys. (Paris) Lett. **40**, L-123 (1979)
8.26 Ph. Cahuzac, H. Sontag, P.E. Toschek: Opt. Commun. **31**, 37 (1979)
8.27 A. Crubellier, M.S. Schweighofer: In *Coherence and Quantum Optics IV*, ed. by L. Mandel, E. Wolf (Plenum, New York 1978) p.567
A. Crubellier, S. Liberman, D. Mayou, P. Pillet, M.G. Schweighofer: Opt. Lett. **7**, 16 (1982)
8.28 F.T. Arecchi, E. Courtens: Phys. Rev. A**2**, 1730 (1970)
8.29 N.E. Rehler, J.H. Eberly: Phys. Rev. A**3**, 1735 (1971)
8.30 J.H. Eberly: Am. J. Phys. **40**, 1374 (1972). This is one of the earliest indications of the physical volume (thus transverse) and angular distribution effects in coherence brightening. This can be found as Eqs. 14-17
8.31 J.C. MacGillivray, M.S. Feld: Phys. Rev. A**14**, 1169 (1976)
8.32 R. Bonifacio, P. Schwendimann, F. Haake: Phys. Rev. A**4**, 302 (1971); **4**, 858 (1971)
R. Bonifacio, L.A. Lugiato: Phys. Rev. A**11**, 1507 (1975); **12**, 587 (1975)
See also [8.15-17]
8.33 J.C. MacGillivray, M.S. Feld: In *Cooperative Effects in Matter and Radiation*, ed. by C.M. Bowden, D.W. Howgate, H.R. Robl (Plenum, New York 1977) p.1:
In 1977, J.C. MacGillivray and M.S. Feld showed that a Gaussian average of plane wave solutions removes most of the ringing but results in a much longer tail than observed (private communication to H.M. Gibbs, Spring 1977)
8.34 J.C. MacGillivray, M.S. Feld: In *Applications in Lasers to Atomic, Molecular and Nuclear Physics*, ed. by Institute of Spectroscopy, USSR Academy of Science (Nauka, Moscow 1979) p.117
8.35 J.C. MacGillivray: D.Sc. Thesis, M.I.T. (1980)
8.36 E. Ressayre, A. Tallet: Phys. Rev. Lett. **30**, 1239 (1973); **37**, 424 (1976); Phys. Rev. A**15**, 2410 (1977); **18**, 2196 (1978); In *Coherence and Quantum Optics IV*, ed. by L. Mandel, E. Wolf (Plenum, New York 1978) p.799
Independently, Ressayre and Tallet concluded that superfluorescence cannot be described accurately by a one- or two-mode model, i.e., geometrical and propagational effects are essential to a quantitative description
8.37 R. Glauber, F. Haake: Phys. Lett. **68**A, 29 (1978)
8.38 R. Haake, H. King, G. Schroder, J. Haus, R. Glauber, F. Hopf: Phys. Rev. Lett. **42**, 1740 (1979)
8.39 F. Haake, J. Haus, H. King, G. Schroder, R. Glauber: Phys. Rev. Lett. **45**, 558 (1980)

8.40 F. Haake: In *Laser Spectroscopy IV*, ed. by H. Walther, K.W. Rothe, Springer Ser. Opt. Sci., Vol.21 (Springer, Berlin, Heidelberg 1979) p.451
8.41 M.F.H. Schuurmans, D. Polder, Q.H.F. Vrehen: J. Opt. Soc. Am. **68**, 699 (1978)
8.42 D. Polder, M.F.H. Schuurmans, Q.H.F. Vrehen: Phys. Rev. A**19**, 1192 (1979)
8.43 Q.H.F. Vrehen, J.J. der Weduwe: Phys. Rev. A**24**, 2857 (1981)
8.44 F.A. Hopf: Phys. Rev. A**20**, 2054 (1979)
8.45 R. Bonifacio, J.D. Farina, L.M. Narducci: Opt. Commun. **31**, 377 (1979). This paper uses mean field theory (MFT), which neglects propagation and predicts *no* ringing for the low-density Cs regime. Therefore the mode-competition transverse effects that they treat apply only to high-density MFT oscillatory SF and are *not* directly relevant to the Cs data.
8.46 M.S. Feld, J.C. MacGillivray: In *Coherent Nonlinear Optics*, ed. by M.S. Feld, V.S. Letokhov, Topics Curr. Phys., Vol.21 (Springer, Berlin, Heidelberg 1980) p.7
8.47 R.K. Bullough, R. Saunders, C. Feuillade: In *Coherence and Quantum Optics IV*, ed. by L. Mandel, E. Wolf (Plenum, New York 1978) p.263
8.48 F.P. Mattar, H.M. Gibbs: In Proc. Int. Conf. on Lasers 1980, New Orleans, ed. by V.J. Corcoran (STS, McLean, VA 1981) p.777
8.49 F.P. Mattar, H.M. Gibbs, S.L. MacCall, M.S. Feld: Phys. Rev. Lett. **46**, 1121 (1981)
8.50 F.P. Mattar: In Abstract Digest of European Conference on Atomic Physics, Heidelberg 1981, ed. by J. Kowolski, G. Zuputlitz, H.G. Weber (European Physical Society, Geneva 1981)
8.51 F.P. Mattar: In Proc. Los Alamos Conference on Optics, SPIE **288**, 353 (1981)
8.52 F.P. Mattar, M. Cormier, H.M. Gibbs, E. Watson, S.L. McCall, M.S. Feld: 5th Int. Conf. on Laser Spectroscopy, Japser, Canada 1981 (post deadline paper); ARO Workshop on Nonlinear Oscillators, Los Alamos 1981 (unpublished); In *Coherence and Quantum Optics V*, ed. by L. Mandel, E. Wolf (Plenum, New York 1984) p.487. Note that in Fig.3, the energy, which is obtained from rigorous simulations, displays temporal ringing even in the central shell.
8.53 F.P. Mattar, M.C. Newstein: IEEE J. QE-**13**, 507 (1977); *Cooperative Effects in Matter and Radiation*, ed. by C.M. Bowden, D.W. Howgate, H.R. Robl (Plenum, New York 1977) p.139
8.54 S.L. McCall, E.L. Hahn: Bull. Am. Phys. Soc. **10**, 1189 (1965); Phys. Rev. Lett. **28**, 308 (1967); Phys. Rev. **183**, 457 (1969); Phys. Rev. A2, 861 (1970)
8.55 B. Bölger, L. Baede, H.M. Gibbs: Opt. Commun. **18**, 199 (1976)
8.56 H.M. Gibbs, B. Bölger, F.P. Mattar, M.C. Newstein, G. Forster, P.E. Toschek: Phys. Rev. Lett. **37**, 1743 (1976)
8.57 F.M. Mattar, M.C. Newstein, P.E. Serafim, H.M. Gibbs, B. Bölger, G. Forster, P.E. Toschek: In *Coherence and Quantum Optics IV*, ed. by L. Mandel, E. Wolf (Plenum, New York 1978) p.143
8.58 J.J. Bannister, H.J. Baker, T.A. King, W.G. McNaught: Phys. Rev. Lett. **44**, 1062 (1980)
8.59 D.J. Heinzen, J.E. Thomas, M.S. Feld: In *Coherence and Quantum Optics V*, ed. by L. Mandel, E. Wolf (Plenum, New York 1984) p.81; Phys. Rev. Lett. **54**, 677 (1985); in *Laser Spectroscopy VII*, ed. by T.W. Hänsch, Y.R. Shen, Springer Ser. Opt. Sci., Vol.49 (Springer, Berlin, Heidelberg 1985) p.290
8.60 Panel discussion in *Cooperative Effects in Matter and Radiation*, ed. by C.M. Bowden, D.W. Howgate, H.R. Robl (Plenum, New York 1977) p.357
8.61 C.M. Bowden, C.C. Sung: Phys. Rev. A**18**, 1558 (1978); Phys. Rev. A**20**, 2033 (1979)
8.62 C.M. Bowden, C.C. Sung: Phys. Rev. Lett. **50**, 156 (1983); Opt. Commun. **45**, 273 (1983)
8.63 C.M. Bowden: In *Laser Physics*, ed. by J.D. Harvey, D.F. Walls, Lecture Notes in Physics, Vol.182 (Springer, Berlin, Heidelberg 1983)
8.64 E.A. Watson, H.M. Gibbs, F.P. Mattar, M. Cormier, S.L. McCall, M.S. Feld: J. Opt. Soc. Am. **71**, 1589 (1981)
8.65 E.A. Watson, H.M. Gibbs, F.P. Mattar, M. Cormier, Y. Claude, S.L. McCall, M.S. Feld: Phys. Rev. A**27**, 1427 (1983)
Figure 4, which corresponds to [Ref.8.63, Fig.3], has been greatly reduced, which unfortunately prevents one from seeing clearly that on-axis ringing is still present. By aperturization (i.e., by changing the radial width of the

detector) one would be able to observe the ringing and how the integration over the various shells leads to washing out of the ringing. The implication in the conclusion that no ringing is expected in SF is not consistent with the rigorous calculations and is incorrect. This exhibition of ringing from small regions of the SF output has recently been confirmed experimentally in Rb by Heinzen et al. [8.56]. The absence of ringing was unambiguously demonstrated as primarily due to a spatial averaging (shell) effect
8.66 C.M. Bowden, F.P. Mattar: In Proc. Los Alamos Conference on Optics, SPIE **288**, 364 (1981)
8.67 F.P. Mattar, C.M. Bowden: In Summary Digest Int. Conf. Excited States and Multi-Resonant Nonlinear Optical Processes in Solids, Aussois, France 1981 (Les Ed. de Physique, Les Ulis, France 1981) p.59
8.68 F.P. Mattar, C.M. Bowden, Y. Claude, M. Cormier: In Proc. 1981 Int. Conf. on Lasers, ed. by C.B. Collins (STS, Mclean, VA 1982) p.331
8.69 F.P. Mattar, C.M. Bowden: Appl. Phys. B**29**, 149 (1982); Phys. Rev. A**27**, 345 (1983)
8.70 C.M. Bowden, F.P. Mattar: In *Coherence and Quantum Optics V*, ed. by L. Mandel, E. Wolf (Plenum, New York 1984) p.87
8.71 A.T. Rosenberger, H.K. Chung, T.A. DeTemple: IEEE J. QE-**20**, 523 (1984)
8.72 R. Bonifacio, F.A. Hopf, P. Meystre, M.O. Scully: Phys. Rev. A**12**, 2568 (1975)
8.73 J.R.R. Leite, R.S. Sheffield, M. Ducloy, R.D. Sharma, M.S. Feld: Phys. Rev. A**14**, 1151 (1976)
8.74 F.P. Mattar, C.M. Bowden, C.C. Sung: In *Coherence and Quantum Optics V*, ed. by L. Mandel, E. Wolf (Plenum, New York 1984) p.507
8.75 F.P. Mattar: In Proc. 1983 Los Alamos Optics Conf., ed. by R.S. McDowell, S.C. Stotlar, SPIE **380**, 508 (1983)
8.76 F. Hopf: Phys. Rev. A**20**, 2064 (1979)
8.77 F. Haake, R. Reibold, F.P. Mattar: J. Opt. Soc. Am. B**1**, 547 (1984)
8.78 R.J. Glauber: Phys. Rev. **130**, 2529 (1963); **131**, 2766 (1963)
8.79 F.P. Mattar, C.M. Bowden, C.C. Sung: To be published
8.80 F.P. Mattar: Appl. Phys. **17**, 53 (1978)
F.P. Mattar, M.C. Newstein: Comput. Phys. Commun. **20**, 139 (1980); **32**, 225 (1984)
F.P. Mattar, J.H. Eberly: In *Laser-Induced Processes in Molecules*, ed. by K.L. Kompa, S.D. Smith, Springer Ser. Chem. Phys., Vol.6 (Springer, Berlin, Heidelberg 1979) p.61
F.P. Mattar: In *Optical Bistability*, ed. by C.M. Bowden, M. Ciftan, H.R. Robl (Plenum, New York 1981) p.503; in Proc. 10th Modeling and Simulation Conf., Pittsburgh 1978, ed. by W.G. Vogt, M.H. Mickle (Instrument Society of America, Pittsburgh 1979)
B.R. Suydam, F.P. Mattar: In Proc. 1983 Los Alamos Conf. on Optics, ed. by R.S. McDowell, S.C. Stotlar, SPIE **380**, 439 (1983)
8.81 F.P. Mattar, F. Haake: "Two-Color Superfluorescence" in Abstract Digest of 1982 Int. Conf. on Lasers, New Orleans 1982, ed. by R.E. Powell, dist. by V.J. Corcoran, ψ (Potomac Synergetic Inc.) McLean, VA
8.82 F.P. Mattar: "Fresnel Dependence of Quantum Fluctuations in Two-Color Superfluorescence from Three-Level Systems" in Abstract Digest of the Southwest Conference on Optics, Albuquerque 1985; also submitted to Phys. Rev. A
8.83 H.C. Van de Hulst: *Light Scattering by Small Particles* (Wiley, New York 1957) pp.179-183
8.84 H.M. Gibbs, R.E. Slusher: Phys. Rev. A**5**, 1634 (1972); A**6**, 2326 (1972)
8.85 P.D. Drummond, J.H. Eberly: Phys. Rev. A**25**, 3446 (1982)
8.86 F. Haake, R. Reibold: Phys. Lett. **92**A, 29 (1982); Phys. Rev. A**29**, 3208 (1984)
8.87 R. Florian, L.O. Schwan, D. Schmid: Solid State Commun. **42**, 55 (1982); Phys. Rev. A**23**, 2709 (1984)
8.88 M. LeBerre, E. Ressayre, A. Tallet: Phys. Rev. A**25**, 1604 (1982)
8.89 M. LeBerre, F.P. Mattar, F. Ressayre, A. Tallet: Proc. of the Max Born Centenary Conf., ed. by R.E. Powell, SPIE **369**, 269 (1983); In Abstracts Digest of the 1982 Int. Conf. on Lasers, New Orleans 1982, dist. by V.J. Corcoran, ψ (Potomac Synergetic Inc.) McLean, VA
8.90 M. LeBerre, E. Ressayre, A. Tallet: In *Coherence and Quantum Optics V*, ed. by L. Mandel, E. Wolf (Plenum, New York 1984) p.331

8.91 M. LeBerre, F. Ressayre, A. Tallet, H.M. Gibbs, M.C. Rushford, F.P. Mattar: In *Coherence and Quantum Optics V*, ed. by L. Mandel, E. Wolf (Plenum, New York 1984) p.347
8.92 M. LeBerre, E. Ressayre, A. Tallet: Phys. Rev. A**29**, 2669 (1984)
8.93 M. LeBerre, E. Ressayre, A. Tallet, K. Tai, H.M. Gibbs, M.C. Rushford, N. Peyghanbarian: J. Opt. Soc. Am. B**1**, 591 (1984)
8.94 M. LeBerre, E. Ressayre, A. Tallet, F.P. Mattar: J. Opt. Soc. Am. B2, 956 (1985)
8.95 J. Teichmann, F.P. Mattar: Abstract Digest Annual Meeting of the Canadian Association of Physicists and Canadian Astronomical Society, Victoria, B.C., Canada 1983
8.96 J. Teichmann, Y. Claude, F.P. Mattar: Opt. Commun. **54**, 33 (1985)
8.97 J. Teichmann, Y. Claude, F.P. Mattar, C. Bardin, J.P. Babuel-Peyrissac, J.P. Marinier: In Abstracts Digest 1984 DEAP Meeting of the American Physical Society, University of Connecticut, Storrs, CT 1984
J.P. Babuel-Peyrissac, J.P. Marinier, C. Bardin, F.P. Mattar, J. Teichmann, Y. Claude, B.R. Suydam: In Refereed Proc. Southwest Conf. on Optics, Albuquerque 1985, ed. by R.S. McDowell, S.C. Stotler, SPIE **540**, 569 (1985)
8.98 F.P. Mattar, E.J. Robinson, P.R. Berman: "Transverse and Dispersion Effects in the Strong Pump Weak Probe Interaction", in Abstract Digest of the Southwest Conference on Optics, Albuquerque 1985; also submitted to Phys. Rev. A
8.99 F.P. Mattar: "Propagation Effects in Strong Pump-Weak Probe Interactions", in *Laser Spectroscopy VII*, ed. by T.W. Hänsch, Y.R. Shen, Springer Ser. Opt. Sci., Vol.49 (Springer, Berlin, Heidelberg 1985) p.218

Additional References

Reviews of superfluorescence:

Ernst, V, Stehle, P.· Emission of radiation from a system of many excited atoms. Phys. Rev. **176**, 1456 (1968). The dependence of superadiance on the shape and size of the sample was indicated
Gross, M., Haroche, S.: Superfluorescence, an essay on the theory of collective spontaneous emission. Phys. Rep. **93**, 301-396 (1982)
Schuurmans, M.F.H., Vrehen, Q.H.F., Polder, D., Gibbs, H.M.: Superfluorescence, in *Advances in Atomic and Molecular Physics*, Vol.17, ed. by D.R. Bates, B. Bederson (Academic, New York 1981) pp.168-228
Vrehen, Q.H.F., Gibbs, H.M.: "Superfluorescence experiments", in *Dissipative Systems in Quatnum Optics*, ed. by R. Bonifacio, Topics Curr. Phys., Vol.27 (Springer, Berlin, Heidelberg 1981) pp.111-147

Analytical treatments of superfluorescence:

Andreev, A.V., Emel'yanov, V.I., Il'inskii, Yu.A.: Collective spontaneous emission (Dicke superradiance). Sov. Phys.-Usp. **23**, 493-514 (1980)
Andreev, A.V., Il'inskii, Yu.A.: Superradiance of extended systems. Izv. Aka. Nauk SSSR, Ser. Fiz. **46**, 985-989 (1982) [English Transl.: Akademiia Nauk SSSR, Bull., Phys. Ser. **46**, 150-154 (1982)]
Crubellier, A., Liberman, S., Pavolini, D., Pillet, P.: Superradiance and subradiance: I – Interatomic interference and symmetry properties in three-level atoms. J. Phys. **18**, 3811-3833 (1985)
Crubellier, A., Liberman, S., Pillet, P.: Superradiance theory and random polarization. J. Phys. **17**, 2771 (1984)
Crubellier, A., Liberman, S., Pillet, P., Schweighofer, M.G.: Experimental study of quantum fluctuations of polarization in superradiance. J. Phys. B**14**, L177-182 (1981)
Crubellier, A., Pavolini, D.: Superradiance and subradiance: II – Atomic systems with degenerate systems. J. Phys. B, to be published

Emel'yanov, V.I., Seminogov, V.N.: Effect of pump depletion on Superradiance in Raman Scattering of Light. Sov. J. Quantum Electron. **9**, 383-385 (1979)
Foerster von, T., Glauber, R.J.: Quantum theory of light propagation in amplifying media. Phys. Rev. A3, 1484-1511 (1971)
Gabitov, I.P., Zakharov, V.E., Mikhailov, A.V.: Non linear theory of superfluorescence. Sov. Phys.-JETP **59**, 703-709 (1984)
Hauko, L., Benard, D.J., Davis, S.J.: Observation of superfluorescent emission of the B-X system in I_2. Opt. Commun. **30**, 63-65 (1979)
Herman, B.J., Drummon, P.D., Eberly, J.H., Sobolewska, B.: Coherent propagation and optical pumping in three-level systems. Phys. Rev. A**30**, 1910-1924 (1984)
Karnyukhin, A.V., Kuz'min, R.N., Namiot, V.A.: Superradiance in a two-dimensional model. Sov. Phys.-JETP **57**, 509-517 (1983)
Kocharovsky, V.V., Kocharovsky, V.V.: Superradiance statistics for three-dimensional samples. Opt. Commun. **53**, 345-348 (1985)
Laptev, V.D.: Effect of spatial inhomogeneity on the kinetics and spectrum of a superradiation pulse of an extended system. Opt. Spectrosc. (USSR) **55**, 449-451 (1983)
Laptev, V.D., Reutova, N.M., Sokolov, I.V.: Influence of transverse inhomogeneity of the radiation field and of the active medium on the dynamics of superradiance from an extended system. Sov. J. Quantum Electron. **13**, 1372-1375 (1983)
Laptev, V.D., Reutova, N.M., Sokolov, I.V.: Initial state of a superradiance pulse in the case of delayed or prolonged excitation of matter. Kvantovaya Elektron. (Moscow) **11**, 1646-1650 (1984)
Laptev, V.D., Sokolov, I.V.: Dynamics of superradiance in a thin dielectric waveguide. Kvantovaya Elektron. (Moscow) **11**, 1881-1882 (1984)
Leonardi, C., Peng, J.S., Vaglica, A.: Beats in Dicke superradiant emission. J. Phys. B**15**, 4017 (1982)
Leonardi, C., Vaglica, A.: Coherent trapping and beats in superfluorescence. J. Phys. B**14**, L307 (1981)
Leonardi, C., Vaglica, A.: Fluctuations in superfluorescent beating. Opt. Commun. **53**, 340 (1985)
Lizin, I.M., Makhviladze, T.M., Shelepin, L.A.: Superradiance effects in molecular systems, in coherent cooperative processes. Tr. Fiz. Inst., Akad. Nauk SSSR **87**, 21-36 (1976) [English Transl.: Proc. Lebedev Inst. Phys. **87**, 19-34 (1978)]
Mantsyzov, B.I., Bushuev, V.A. Kuz'min, R.N., Serebryakov, S.L.: Superradiance in extended media. Sov. Phys.-JETP **58**, 498-502 (1983)
Marek, J.: Observation of superradiance in Rb vapor. J. Phys. B**12**, L299 (1979)
Molander, W.A., Stroud, Jr., C.R.: J. Phys. B**15**, 2109 (1982)
Mostowski, J., Sobolewska, B.: Delay-time statistics in superfluorescence initiation for large Fresnel number. Phys. Rev. A**28**, 2573-2575 (1983) and in *Coherence and Quantum Optics V*, ed. by L. Mandel, E. Wolf (Plenum, New York 1984) pp.497-502
Mostowski, J., Sobolewska, B.: Initiation of superfluorescence from a sphere. Phys. Rev. A**28**, 2943-2952 (1983)
Mostowski, J., Sobolewska, B.: Transverse effects in stimulated Raman scattering. Phys. Rev. A**30**, 610-612 (1984)
Pavolini, D., Crubellier, A., Pillet, P., Cabaret, L., Liberman, S.: Experimental evidence of subradiance. Phys. Rev. Lett. **54**, 1917 (1985)
Ponte Goncalves, A.M., Tallet, A., Lefebvre, R.: Superradiant effects on pulse propagation in resonant media. Phys. Rev. **188**, 576 (1969)
One should note that this relatively unknown paper investigates the enhancement role of cooperative radiative interactions, such as superradiant spontaneous emission effects, in the propagation of light pulses in resonant media (Self Induced Transparency, SIT) and in the evolution of photon echoes. It displays the authors' awareness of the propagational aspect of superfluorescence on one hand, but it also shows their lack of full understanding of the SF by the fact that they had not realized that the SF and SIT master Maxwell-Bloch equations are identical with, however, different initial conditions. These conditions can be summarized for SIT as $\theta \neq 0$, $P = 0$, while for SF, θ is the time-integrated pulse area and P is the complex slowly varying envelope
Prasad, S., Glauber, R.J.: Diffractive effects in pulse propagation through a resonant medium. Phys. Rev. A**31**, 1575-1582 (1985)

Analytical treatment of SF and SIT by linearizing the problem and treating diffraction à la A. Papoulis's *Systems and Transforms with Applications in Optics* (McGraw-Hill, New York 1968) pp.140-175 using Hankel Transforms

Prasad, S., Glauber, R.J.: Initiation of superfluorescence in a larger sphere. Phys. Rev. A**31**, 1583-1597 (1985)

Raymer, M.G., Walmsley, I.A., Mostowski, J., Sobolewska, B.: Quantum theory of spatial and temporal coherence properties of stimulated Raman scattering. Phys. Rev. A**32**, 332 (1985). This is an analytical treatment of Stokes buildup in a forward Raman amplifier

Ryschka, M., Marek, J.: Quantum beats in superradiance in sodium vapor. J. Phys. B**13**, L491 (1980)

Ryschka, M., Marek, J.: Observation of quantum beats in superradiance on the $5^2D_{3/2}$ $-6^2P_{1/2}$ transition in cesium vapours. Phys. Lett. **86**A, 98 (1981)

Sobolewska, B.: Initiation of superfluorescence in a three-level swept-gain amplifier. Opt. Commun. **46**, 170-174 (1983)

This paper treats SF initiation in a three-level system. This was derived subsequent to the work of *Bowden* and *Sung* [8.62].

Trifonov, E.D., Zoitsev, A.I., Malikov, R.F.: Superradiance of an extended system. Sov. Phys.-JETP **49**, 33-38 (1979)

Verkhoglyad, A.G., Krivoshchekov, G.V., Kurbatov, P.F.: Nature of Xe 1 laser line superradiance in the presence of a buffer gas. Kvantovaya Elektron. (Moscow) **11**, 291-298 (1984)

Walmsley, I.A., Raymer, M.G.: Experimental study of the macroscopic quantum fluctuations of partially coherent stimulated Raman scattering. Phys. Rev. A**33**, 382-390 (1986)

Zaitsev, A.I., Malyshev, V.A., Trifonov, E.D.: Superradiance of a multiatomic system with allowance for the Coulomb interaction. Sov. Phys.-JETP **57**, 275-281 (1983)

Zinov'ev, P.V., Lopina, S.V., Naboikin, Yu.V., Silaeva, M.B., Samartsev, V.V., Sheibut, Yu.E.: Superradiance in a diphenyl crystal containing pyrene. Sov. Phys.-JETP **58**, 1129-1133 (1983)

Zverev, V.V.: Calculating the statistical characteristics of the superradiance from an ensemble of multilevel molecules. Opt. Spectrosc. (USSR) **54**, 432-434 (1983)

Subject Index

Absorption and dissociation data 60
Absorption of energy 9,20,29,38,39,174,
183,184
Activated complex 98,987
Activation energy 49,50,54,72
Adiabatic inversion 5,215,216,217,220
Adiabatic variations of the laser pulse
amplitude 6
Allylmethylether 70
Angular and velocity distribution for
SF_5 108
Anharmonic interactions 167
Anharmonic rotation-vibrational cou-
plings 210
Anharmonic shift 180
Anharmonic splitting 180
Average dissociation lifetime 99
Average translational energy 112

Balance of up-excitation and dissocia-
tion 105
Bandwidths 183
Bixon-Jortner model 168
Bloch equations 163,198
Boltzmann equation 159,163,172,174,198,
203
Bond-selective, or mode-selective laser
chemistry 3,52
Bottleneck 56,104,172,176,182,186

CF_2CFCl high-pressure falloff 209
CF_2HCl decomposition 56,169,178
CF_3D 30,56
CF_3I 112
CH_3COF_3 30
CH_3NC to CH_3CN 52
Changeover 207
Chemical activation 100
Chemionization 47
Chloroethane 65
CO_2 laser 151
Coarse-grained time average 203
Coherent vs. incoherent excitation 2,5,
204,209
Collimated beams 15
Collisional effects 6,10,61,63,69,177
Collisional redistribution of rotational
and vibrational energy 65
Collisionless dissociation 2
Collisionless unimolecular process 95
Collisions 68,159
Comparison of theory with experiment 169,
173
Competing dissociation channels 117
Competition between dissociation and
up-excitation 114
Competitive dissociation channels 108
Continuously tunable CO_2 laser 4,151
Coupling 168
Critical configuration 98
CrO_2Cl_2 56
Cross section 13,102,174
Cyclopropane 184,210

D_2CO 31,65
Data-analysis techniques for reaction 16
Data-reduction techniques 11,71
Decomposition yield 184
Density of vibrational states 48,49,50,
55,72,102
Density operator 199
Dependence of CF_2 170
Dephasing rate 170
Dephasing relaxation rate 202
Dephasing width 166
Deviations from thermal behavior 189
Dipole coupling correlation functions
167,201
Discrete levels 172
Discriminating reaction mechanism 197
Dissociation channel 96
Dissociation continuum 96
Dissociation lifetimes 115
Dissociation of SF_6 151
Dissociation probability 31,72
Dissociation products 112
Dissociation yield 105
Distribution of energy 100,105,112,174
over vibrational states 45,48,57,60,
63,68
Dynamics of dissociation 96

Electric and magnetic fields 56
Elimination reactions 70
3-center elimination 112,119
4-center elimination 113
Emission and absorption coefficients 165
Energy randomization 210

Equilibration of energy 171
Equivalent state 166
Ergodic assumption 194
Ethyl acetate 70
Ethyl isocyanide 69
Ethylvinylether 52,70
Excitation mechanism 96

Fermi Golden Rule 165,200,205
Fine structure of the spectrum of excitation 149
Flow of energy 159
Fluence 2,13,14,105,164,176
Fluence dependence of absorption 25,28, 32,35,40,165
Fluence distribution function 17
Focusing 14,15,17,18,19
Fraction of molecules pumped 23
Fragmentation 184
Frequencies 184

Gas cell experiments 95
Gas sample properties 16
Gas-dynamic cooling 149,151,152
Gaussian beams 12,14,17
Generalized master equation 201

Halogenated methanes 115
Heat bath feedback 163,200
Heating effects 160,185
HF luminescence 152,153
High-pressure falloff 177,178
Hot band shifts 162

Incoherent evolution 207
Incoherent stepwise multiphoton process 113
Inelastic energy loss 204
Infrared chemiluminescence 47
Intensity dependence 162,167,170,176
Intensity-dependent "bottleneck" 103
Interface 170,171
Intermolecular V-V energy exchange 70
Intramolecular energy relaxation 65,162, 163,164,192
Intramolecular relaxation constants 186
Intramolecular relaxation matrix 164
Intramolecular relaxation time scale 198
Ionization 47
Isomerization and dissociation of cyclopropane 185
Isomerization of cyclopropane 183
Isopropyl bromide 70
Isotopic selectivity 17,62,65,69,97,150

Kassel model 193,194

Landau-Teller collision mechanism 177, 188
Large polyatomic molecules 26,30,55,56
Laser chemistry 4,5
Laser isotope separation 1,2,3,4,5,9,160, 179,215,219,220
Laser properties 16
Laser-beam properties 71
Laser-excited population distribution 102
Laser-induced fluorescence (LIF) 46,47, 71,96
Laser-induced isomerization 69
Laser-specific effects 192
Level of excitation 106
Level shift 183
Level system 215,216,218
Limits to the applicability of rate equations 205
Line broadening factors 163
Low-fluence spectral absorption in SF_6 33
Low-intensity absorption cross section 48
Luminescence signal 153,154,155

Major MPD products 108
Messiah's adiabatic condition 218,219, 220
Messiah's sudden condition 216
Microscopic reversibility 164,165,192
Mode-selective 97
Mode-selective chemistry 159
Molecular and laser properties 19,45
Molecular beam 47
Molecular beam method 95
Molecular collisions 118
Molecular elimination 69
Molecular properties 71,72
Mori projection operator formalism 201
Morse oscillator potential 160
Multiphoton absorption (MPA) 10,20,26,31, 32,62,71
Multiphoton dissociation (MPD) 9,10,20, 44,45,48,51,52,54,61,62,68,71,95,159
Multiphoton excitation (MPE) 9,20,21,24, 26,32,44,45,51,52,54,62
Multiphoton resonance structure 4,150

Non-RRKM behavior 159
Nonadjacent radiative transitions 204
Nonlinear absorption 13
Normalized absorption 21,26,29,54
Number of photons absorbed per molecule 6,14,22,23,63

O_3 and OCS 52
Off-diagonal density matrix elements 200
Optical pulse duration and shape in MPA processes 41
Optimum frequency 5
Oscillator strength 166

Periodic behavior 168
Perturbation theory 206
Phase information 200

Phase-coherence effects 199
Phenomenological rate equations 101
Polyatomic molecules 31
Population dynamics 168,206,207
Power broadening 65,176,208,210
Pressure and pressure effects 61,62,65, 67,71,72,73
Pressure broadening 65
Pulse-width effects 43
Pumped mode 161

Quasi-continuum 48,68,96,149,161,171
  coherent excitation of 2,5,164
Quasi-continuum eigenstates 209

Rabi cycling 31
Rabi frequency 203
Rabi frequency degeneracy 208
Radiation pumping of lower vibrational levels of polyatomics 149
Radiative contributions of Rabi frequency 199
Randomization of energy 106,168,194
Rate constant falloff curve 195
Rate equation 162,176,190
Rate of approach to the steady state 203
Reaction probability 16,17,18,19,51,58, 59,60,69
Real-time monitoring technique 46,47
Recommended values of SF_6 absorption 44
Redistribution of energy 166
Relative dissociation yield 108
Relaxation time 208
Restricted intramolecular relaxation 194
Restricted quantum exchange 159,209
Rotational compensation 180
Rotational hole filling 63,70,172
Rotational relaxation 62,65,66,67
Rotational structure 180
RRK model 194
RRKM (Rice,Ramsperger,Kassel,and Marcus) model of unimolecular reactions 2,3 48,98,113,114,166,171,183,185

S_2F_{10} 30,64
Saturation population for a pumped N-level system 202
sec-Butyl acetate 30,65,68
Secondary dissociation 116
Selective chemistry 186
Selective decay 197
Selectivity 210
Self-actions of laser light 6
Self-defocusing 6
Self-focusing 6,56
  cylindrically symmetric model 227
  diffraction 228
  dispersion on 231
  non-plane-wave SF analysis 227
  nonlinear channeling 231
  self-induced transparency 227
  spatial averaging 227

Semiclassical quantization of quasi-periodic trajectories 210
SF_5-F bond strength 51
SF_5 48
SF_5Cl 46
SF_5NF_2 30
SF_6 24,25,38,108
SF_6 absorption 40,43
SF_6 and $CFCl_3$ 116
SF_6 decomposition 178
Sidebands in scattered light 202
Slope of <n> versus fluence
  2/3 power 30,32,72
Small molecules 31
Small-signal absorption cross section 51
Small-signal cross section 56,61,73
Smaller species 55
SO_2 52
Spectral population distribution function $f(\nu)$ 24
Spherical-top Hamiltonian 179
Spread in molecular frequencies 189

Stable-species analysis techniques 45,46
Stark broadening 31
Statistical theories, *see* RRKM and RRK
Steady state approaches 202
Strong collision assumption 98
Strong intramolecular relaxation 193
Superfluorescence
  adiabatic following 223
  advance in time 244
  Beers absorption length F 223,271
  bivariate distribution 229
  coherent pumping 232
  copropagation of pump and SF pulses 260
  diffraction 235,242,245,249,250,266, 272,274
  dispersion effects 224,271,274
  dynamic pump depletion 228,229
  effective absorption length 247
  effective Beer length 247
  Hamiltonian 232
  Heisenberg equations of motion 232, 237,240
  high gain requirement 224
  Langevin force 229,233,239,248,265, 266
  light controlled by light 230,241
  Maxwell-Bloch equations 223
  mean field theory 223,226,227,248
  off-resonant SF 271
  nonlinear buildup 268
  output power 244,268
  parametric coupling 239
  phase waves 266
  propagational effects 226
  pulse compression 272
  pulse distortion 272
  pulse overlap 229
  pump depletion 241,269

quantum fluctuations 227,228
quantum initiation process 226
quantum statistics 263
Raman process 229
rate equation 223
ringing emission 226,227,242,244
self-defocusing 231
self-focusing 231,255,266
shape invariance 248
shell model 242,244
slowly-varying-amplitude approximation 233
spatial averaging 235
synchronization 264,271
time retardation 244
tipping angle 248
transverse effects 249
transverse energy flow 237,269
transverse reshaping 223,269
two-photon processes 263,265
weak overlap 223
weak probe 271,272
wide aperture detector 245
Systems with an additional potential energy barrier 112

T_1 208
T_2 159,164,167,207
Temperature effect 61
Tetramethyldioxetane 70
Theoretical energy distributions 175
Theory of incoherent multiphoton excitation 97
Thermal distribution 102,162,186,187
Thermal heating 101
Thermal solution 188
Thermal unimolecular reactions 98,99
Time domain 207
Time-averaged density matrix 210
Time-dependent Schrödinger equation 205
trans- to *cis*-Dichloroethylene 69
trans-Dichlorethylene 74
Transition into dense energy levels 205
Translational energy distribution 99,112 114,210
Two-frequency method, 4,149,153
Two-level model 21,22,23,31
Two-photon resonance 4,156

Ultraviolet fluorescence 46
Unfocused pulses 19
Unimolecular decay 159,160,192
Unimolecular dissociation 97

V-T collisional deactivation 64,69,70, 178
Vacuum ultraviolet fluorescence 46
Vibration-state density 53,61
Vibrational relaxation 199
Vibrational state of SF_6 molecules 149
Vinyl chloride 65
Vinyl fluoride 52

Whitten-Rabinovitch approximation 102